W9-BYB-698

INTERACTIONS

INTERACTIONS

A Journey Through the Mind of a Particle Physicist and the Matter of This World

SHELDON L. GLASHOW

WITH BEN BOVA

A Warner Communications Company

Warner Books, Inc., 666 Fifth Avenue, New York, NY 10103

A Warner Communications Company

Printed in the United States of America
This book was originally published in hardcover by Warner Books
First Trade Paperback Printing: June 1989

10 9 8 7 6 5 4 3 2 1

Library of Congress Cataloging-in-Publication Data

Glashow, Sheldon L.
Interactions : a journey through the mind of a particle physicist and the matter of this world.

Includes index.
1. Nuclear physics—Popular works. 2. Matter, Nuclear—Popular works. 3. Glashow, Sheldon L. 4. Physicists—United States—Biography. I. Bova, Ben, 1932– . II. Title.
QC778.G53 1988 539.7 87-40413
0-446-38946-3 (pbk.) (U.S.A.)
0-446-38947-1 (pbk.) (Can.)

Book design: H. Roberts
Cover photo by Al Frankevich
Cover design by Neal Pozner

Illustrations by Howard Roberts
pages: 7, 9, 13, 16, 18, 24, 25, 28, 36, 37, 44, 53, 55, 63, 64, 66, 76, 78, 79, 81, 82, 85, 88, 98, 103, 121, 157, 161, 174, 181, 183, 185, 190, 214, 224, 225, 230, 247, 292, 294, 302, 311, 318, 319, 324, 331

To Joan Glashow

To Barbara Bova

Contents

Introduction

IN THIS BOOK, I TRY TO EXPLAIN MY fascination with subatomic physics and how I became a member of the elementary-particle elite.

Science has been my life because it is the systemization of curiosity. When I was a child growing up in Manhattan, I wanted to know how a car or a clock worked, what a rainbow was, and why an uncooked egg could not be made to spin. My greatest discovery was that science can be more than a mere hobby, It could be my profession. People would actually pay me to do what I most wanted to do: to satisfy my own curiosity.

I share with most physicists a deep faith that nature is basically simple, and if we just knew nature well enough, we could describe the entire universe very simply. For centuries scientists have been working toward that goal. Some sought the smallest pieces of matter, the basic stuff of *being*. They thought they had found their goal when they discovered the atom. They were wrong. They learned of even smaller bits of matter within the atom. Even the parts of atoms had parts, and these had parts themselves!

But it is not enough to identify the basic building blocks of matter. Other scientists asked how these tiny things could organize themselves to produce the wonders of nature. They sought the basic laws of *becoming*—the forces of change such as heat, gravity, electricity, magnetism and even life itself. Could all these things be explained in terms of the ultimate particles of nature and the rules by which they combine?

In my lifetime we have seen these two approaches come together, like the two wings of an army's pincer movement closing in on the ultimate objective. We have learned that matter and energy are inextricably linked. You cannot have one without the other. To understand the basic building blocks of matter, you must understand the fundamental forces that hold them together.

That is what I have spent my life doing. I have been lucky and persistent enough to make some major contributions to this magnificent quest. That is what this book is all about: the search for the ultimate portrait of the universe, as seen through the eyes of one of the searchers.

INTERACTIONS

From Bobruisk with Love

"THE YEAR 1905 WAS A MIRACULOUS one for science."

So says my Harvard University mentor, the Nobel Prize laureate Julian Schwinger. He calls 1905 *annus mirabilis*.

In that year Albert Einstein astounded the world by demonstrating the interrelated nature of matter, energy, space and time. The bizarre implications of his theory of relativity have bemused and befuddled us ever since. Science fiction writers were surprised to learn that traveling faster than the speed of light is *verboten* (although, since light travels 186,000 miles per second, you would think they'd be happy enough).

Forty years later the citizens of Hiroshima and Nagasaki were very painfully surprised to learn how effectively matter can be converted into energy in accordance with Einstein's formula $E = mc^2$. This simple formula showed, precisely and beautifully, not only that matter and energy *are* interrelated, but also exactly how much energy is invested in any chunk of matter anywhere in the universe.

Many of Einstein's greatest ideas emerged in that miracle year of 1905. His explanation of Brownian motion convinced even the most diehard skeptics that atoms were real, and their direct influences could be seen. Of course, we have already mentioned his special theory of relativity.

Also, in that year, Einstein demonstrated the schizoid nature of light. To everyone's surprise, it turns out that light is neither a beam of particles nor a wave motion. It is a little bit of each. This seeming paradox led to the development of quantum mechanics in the 1920s, which, together with relativity, forms the two sturdy pillars upon which modern physics depends.

It is a curious footnote to this story that the Swedish Academy of Sciences never recognized Einstein's greatest accomplishment, his theory of relativity. His Nobel Prize in 1921 was awarded for the discovery that light sometimes behaves like a beam of particles. Relativity was simply too controversial and too revolutionary a discovery to be honored.

The year 1905 was an important one both for Einstein and me. My father, then sixteen, who was totally unaware of these momentous discoveries, had arrived at last, after a long voyage in steerage class, at Ellis Island, a refugee from the pogrom-ridden town of Bobruisk in czarist Russia to the land of opportunity. Lev Glukhovsky, through the inscrutable wisdom of some anonymous immigration official, became Lewis Glashow.

He was to have a rewarding but surprising life in the New World. Never completing the college education he longed for, he worked as a construction laborer, clerk and master plumber. A few years after his arrival, during which he survived a fall into a vat of molten lead, he sent for his family: his father, his sister, five of his seven brothers, his kissing cousin Bella (my mother-to-be) and her family. Little did my parents suspect that they would raise three sons, a doctor and a dentist who would fight the Germans in a *second* World War, and a professor of physics who would write this book. Life is chock-full of surprises.

In science, too, surprises come very often. In each decade over the past few centuries there have been several unexpected discoveries that forced everyone to change their ideas about the way nature behaves. It might seem as if physical science is being driven hither and yon by the exotic discoveries of its erratic practitioners, resulting in a kind of crazy patchwork quilt, with unrelated bits and pieces of theory stuck together in a desperate attempt to make it all look like a continuous sheet of knowledge.

Nothing could be further from the truth. Despite the steady peppering of surprises—in fact, because of them—our understanding of nature has continuously evolved toward greater elegance, simplicity, precision and completeness. The patchwork quilt has become a tapestry.

Surprises nourish us and uncover new knowledge. If Newton were alive today he would gracefully, although perhaps grudgingly, acknowledge that Einstein's theory of gravitation is a great improvement on his own. Our understanding of elementary-particle physics today is far, far deeper than it was when I was a lad just learning the trade. In physics, at least, things are getting clearer and clearer, although the end (if there is one) is not yet in sight.

Among all the surprises that nature presents, several threads of continuity link scientific thought through the centuries. Principal among these, as I mentioned, is the rock-bottom faith of the physicist in the underlying simplicity of nature's laws. There is really no way to prove this faith: it is like the faith at the core of every religion, one either believes it or not. Physicists believe that nature is basically simple and understandable. It was this faith that led the ancient Greeks to speculate that all matter is composed of combinations of a few fundamental elements—a notion that has survived and flourished through the ages. My colleague at Harvard, Percy Bridgman, had this to say of our faith more than half a century ago:

> Whatever may be one's opinion as to the simplicity of either the laws or the material structure of nature, there can be no question that the possessors of some such conviction have a real advantage in the race for physical discovery. Doubtless, there are many simple connections still to be discovered, and he who has a strong conviction of the existence of these connections is much more likely to find them than he who is not at all sure they are there. . . .

Albert Einstein put it more succinctly: "The eternal mystery of the universe is its understandability."

Those who believe that the universe is basically simple and elegant also tend to believe that the universe is composed of a fundamental set of building blocks from which all the things of the material world are made. The Greeks put forward the notion of the atom, from the word ατωμωσ, meaning "uncuttable."

By the late nineteenth century it had become clear to many scientists that matter was indeed composed of microscopic atoms coming in several dozen species (carbon, oxygen, gold, arsenic, etc.). Atoms were believed to be immortal, eternal, immutable and unchangeable: the ultimate and irreducible forms of matter. Atoms would react with one another according to known laws of chemical behavior to form simple compounds like table salt (sodium chloride) or even the complex organic molecules of living creatures. But a single atom could be subdivided no further—or so it was thought.

The last decade of the nineteenth century produced at least three dramatic surprises in the physical sciences which set the stage for Einstein, provided the driving forces for the revolutionary developments of our own century and showed decisively that the atom was *not* the fundamental unit of matter.

1. X-rays were discovered by Wilhelm Roentgen in 1895 and were immediately put to use for medical purposes. Soon they would be employed to study the atom itself and reveal, surprisingly, that atoms have a definite inner structure. This discovery threw considerable doubt on the notion that the atom was the fundamental building block of all matter. It's tough to be fundamental and have something smaller inside you at the same time.

2. Radioactivity was first observed in France in 1896. It took some years to realize that radioactivity was the result of the spontaneous breakdown of certain atoms, in which one kind of atom transmuted itself into another. However, the alchemist's dream of converting lead into shining gold was not quite realized. Radium, an element far more rare and expensive than gold, turns itself into cheap lead in the course of a few thousand years, but not the other way around.

3. In 1897 the English physicist J. J. Thomson discovered the first of what truly may be elementary particles: the *electron*. The electron is a ubiquitous constituent of all earthly matter. Like the shoes on God's chillun, electrons are on every atom. It was quickly recognized to be the fundamental unit of electricity. An electrical current is simply a flow of electrons through a wire.

The discovery of the electron proved that atoms could not be the fundamental building blocks of the universe. Thomson believed, correctly, that electrons form a part of the atom, which he thought is held together by electrical forces. There had to be something else in the atom as well, since electrons are thousands of times lighter than atoms. This something else had to have the opposite electrical charge to the electron, so that the atom as a whole could be electrically neutral, as it was seen to be.

The race was on to probe inside the atom and see what it was made of.

Thomson envisaged the "plum pudding" model of the atom, in which small negatively charged electrons were the plums that were somehow embedded in a large and heavy positively charged pudding. At about the same time, in Japan, Hantaro Nagaoka devised an atomic model in which electrons circle a large spherical mass forming a system much like the planet Saturn and its rings. Neither this "Saturnine" model nor Thomson's "pudding" turned out to be anything like the real atom. Nature, once again, proved herself to be more

ISOTOPES

Thomson is a truly great scientist because he discovered something else, in addition to the electron. A person who discovers only one thing (like Thomas Crapper, of flush toilet fame) may simply have had a bit of beginner's luck. Two such discoveries are a prerequisite for greatness. Frederick Soddy had observed that radioactive elements sometimes appear in two forms, one more radioactive than the other but each with identical chemical properties. He called them *isotopes*. Thomson went on to show that perfectly stable elements have isotopes as well, which are distinguished from one another by their mass. In particular, Thomson showed that the inert gas neon consists of a mixture of two isotopes. One neon isotope is twenty times heavier than the hydrogen atom; the other twenty-two. There are many more varieties of atomic nuclei than of chemical elements.

Each type of nucleus is characterized by two whole numbers—its electrical charge, symbolized by the letter Z, and its mass (approximately) compared to that of the proton, symbolized by A. Here are the values for Z and A for some nuclei:

Nucleus	Symbol	Nuclear Charge Z	Atomic Weight A
Proton	^{1}H	1	1
Alpha particle	^{4}He	2	4
Nitrogen-14	^{14}N	7	14
Neon-20	^{20}Ne	10	20
Neon-22	^{22}Ne	10	22
Uranium-235	^{235}U	92	235
Uranium-238	^{238}U	92	238

Thousands of different isotopes have been identified. Hydrogen, for example, has three isotopes: ordinary hydrogen has atomic weight 1; heavy hydrogen, called deuterium, has atomic weight 2; tritium, a radioactive isotope of hydrogen, has atomic weight 3.

Today we know that atomic nuclei are made of neutrons and protons, each of which weighs about the same. The nuclei of two isotopes contain the same number of protons but different numbers of neutrons.

imaginative than mere scientists. The true structure of the atom was revealed serendipitously by the careful investigations of Ernest Rutherford, the greatest scientist ever from New Zealand.

In 1911, Rutherford and his colleagues directed a beam of particles produced by a radioactive source against a thin piece of gold foil. In essence, they were shooting bullets at the atoms of gold, bullets that were considerably smaller than the atoms themselves. Occasionally one of these subatomic "bullets" would be deflected by a very large angle, as if it had hit something small, hard and heavy within the atom.

Rutherford was later to write:

> It was quite the most incredible event that has ever happened in my whole life. It was almost as incredible as if you fired a 15-inch shell at a piece of tissue paper and it came back and hit you. . . . (W)hen I made calculations I saw that it was impossible . . . unless . . . the greater part of the mass of an atom was concentrated in a minute nucleus.

Thus was born the modern conception of the nuclear atom: a very small, heavy, positively charged nucleus surrounded by a cloud of orbiting, negatively charged electrons. Most of the atom's mass, about 99.98 percent, is concentrated in the tiny nucleus, which makes up only a millionth of a billionth part of the volume of the atom. The astonishing thing is that the atom is mostly just empty space. So are all the things that are made of atoms. So are you!

Electrons, as we have said, seem to be truly elementary particles. Those found in different types of atoms are interchangeable and identical. However, there are many different kinds of nuclei, from hydrogen to uranium and beyond. The nuclei of different chemical elements have different electrical charges. Indeed, it is precisely the electrical charge of its nucleus that determines the chemical properties of any given atom.

The nucleus of the hydrogen atom carries a single unit of positive electrical charge that is neutralized by a single electron orbiting around it. The biggest naturally occurring nucleus, that of uranium, weighs 238 times as much as hydrogen and has an electrical charge of 92 units; ninety-two electrons orbit the uranium nucleus.

Since the hydrogen nucleus is the smallest and simplest of all, Rutherford gave it a name of its own: the *proton*. It is our second candidate for an elementary particle.

Between 1911 and 1932 a great schism developed in physics. With the development of quantum mechanics, the structure of the atom was understood and the mysteries of chemistry were explained. Max Born, one of the founders of atomic theory, wrote: "Physics as we know it will be over in six months." Things were very different, however, for practitioners of the neonatal discipline of nuclear physics. The atomic nucleus, unlike the atom, didn't seem to make any sense at all.

Only two plausibly elementary particles were known: electrons, which are very light and negatively charged; and protons, which are heavy and positively charged. How could the various atomic nuclei be built out of these particles?

Consider, as a typical example, the nucleus of nitrogen. Since it weighs fourteen times as much as a hydrogen nucleus, the nitrogen nucleus must contain precisely fourteen protons. But that would give it an electrical charge of 14 as well. Since the measured positive charge of the nucleus is only 7, perhaps there are seven electrons within it, canceling the positive charges of seven of the nuclear protons. That is the only possible way that the atomic nuclei could be built up out of electrons and protons.

Model of nitrogen nucleus with seven electrons and fourteen protons.

This approach required the electrons to play two disparate roles in atomic structure. There were "insie" electrons inside the nucleus to make the charge come out right, and there were "outsie" electrons that stayed outside the nucleus in nice, well-understood quantum-mechanical orbits. Uranium-238, to take another example, would be built of 238 protons, 92 "outsie" electrons and 146 "insies." The trouble was that electrons could not be made to behave in this curious fashion. "Outsie" electrons were fine; quantum mechanics described their behavior perfectly. However, "insies" made no sense at all.

One of the most serious difficulties with the concept of "insie" electrons involved magnetism. While electrons and protons both behave like small magnets, the electron is the more powerful by a factor of about 1000. Yet the hypothetical "insie" electrons, inside the nucleus, seemed to lose their magnetism completely. How could this be? How could an electron lose its magnetism merely because it had gone from an "outsie" location to an "insie"?

Another well-known difficulty concerned good old nitrogen, the major constituent of the air we breathe. In the proton-electron model, the nucleus of commonplace nitrogen-14 consisted of fourteen protons and seven "insie" electrons, for a total of twenty-one constituent particles. Yet observations of the nitrogen-14 spectrum, interpreted by the newly discovered laws of quantum physics, showed decisively that this nitrogen nucleus had an *even* number of components. The "insie" electron was in deep trouble.

Rutherford and others began to realize that what was needed was a third elementary particle, one that weighed about as much as a proton but had no electrical charge at all. The *neutron*, as the new particle was named, was first detected in the laboratory by James Chadwick in 1932. Its discovery marked the beginning of rational nuclear physics. It made a lot more sense to build the nitrogen-14 nucleus out of seven protons and seven neutrons than out of fourteen protons and seven "insie" electrons.

Protons and neutrons, the true constituents of nuclei, are known collectively as *nucleons*.

Also in 1932, the American chemist Harold C. Urey discovered heavy hydrogen, an isotope of hydrogen that is twice as heavy as ordinary hydrogen. The nucleus of heavy hydrogen is called the *deuteron*. The only nucleus that contains just two nucleons, it is made up of one proton and one neutron that are stuck together.

This brings us to the general question of how and why nucleons

Nitrogen nucleus as it really is, containing seven protons and seven neutrons.

stick together to form the nucleus. In the 1920s only two fundamental forces were known. *Gravity* controls the motions of the heavenly bodies and holds our atmosphere, our oceans and ourselves firmly to the surface of planet Earth. *Electromagnetism* is the force that holds the atom's electrons in orbit around its nucleus. In fact, electromagnetic force is responsible for all we see, feel, hear, taste or smell. But neither of these forces could possibly hold the atom's nucleus together: gravity is far too weak a force for the job, and electromagnetism would push all the protons in the nucleus away from each other, since all protons are positively charged and similar charges repel!

In the 1930s physicists had to admit that there were two additional forces in nature, forces that they had not considered before. They hated to do this because it made the universe seem more complicated, when they were trying to find its underlying simplicities. Doubling the number of fundamental forces from two to four appeared to be doubling their difficulties. Yet the forces undoubtedly existed. The *strong nuclear force* was necessary to explain how the nucleus held itself together. The *weak nuclear force* was needed to explain the mysteries of radioactive decay.

Yet a third remarkable discovery took place in 1932. The *positron* was discovered by Carl David Anderson. The positron is the antiparticle of the electron. Its existence had been predicted by Paul Adrien Maurice Dirac, an English theoretical physicist, a year earlier. Dirac had proven that antimatter is an inevitable offspring of the marriage between relativity and quantum mechanics. We shall return to the positron presently.

Clearly 1932 was another great year for physics.

It was the year I was born.

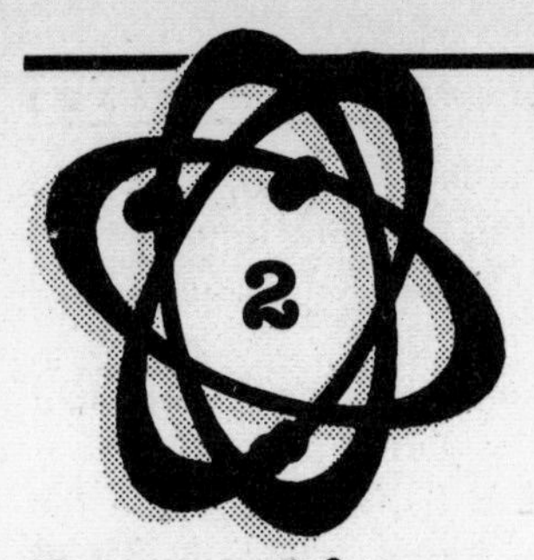

Paradise on Payson Avenue

I WAS BORN AND BRED IN THE Inwood section of upper Manhattan, at the northernmost tip of the island. My parents, my two older brothers and I lived in a two-story red brick house at 65 Payson Avenue, facing the wild and beautiful Inwood Hill Park. The neighborhood, bounded by rivers and parks, was the northern terminus of the A train.

Most of my closest friends were Jewish, but there was a large Irish community in Inwood as well. Although there were occasional fights among the children, the neighborhood was remarkably safe and secure.

I am told that I was an exceptionally curious but ill-mannered little boy. My parents told me later that my first words were, "What's that?" and that my baby-sitters would never return because of my habit of biting them when enraged. The barber came to our home, since it was all but impossible to get me to sit still in the barbershop.

After a year of "Mrs. Stern's Kindergarten," I entered first grade at P.S. 52 in 1938. As I was discovering how to read, Carl David Anderson, the American physicist who first observed the positron, went on to make his second great discovery: a particle we now call the muon. Neither of these remarkable particles was found just lying around a physics lab. Muons are unstable particles that live for only millionths of a second. Positrons are perfectly stable, but when they meet their ubiquitous nemeses, electrons, both the positrons and electrons are annihilated in an instantaneous flash, completely converted into pure energy.

Anderson was not actually searching for new particles. He was studying cosmic rays, which are not "rays" at all but very energetic protons and heavier nuclei that swarm through interstellar space. The earth is constantly being bombarded by cosmic rays, which received their misleading name before scientists realized their true nature. Although they are now known to be subatomic particles, bits and pieces of atoms flitting through the galaxy at incredible energies, the origin of cosmic rays remains a mystery to this day. One thing is sure, though: nature has a *very* powerful particle accelerator at work somewhere in the heavens.

Anderson found positrons, and later muons, among the transient debris produced when a cosmic ray collides with a molecule of our atmosphere. As I have said, the positron is the antiparticle of the electron: it is exactly like the electron in every way, except that its electrical charge is positive and the electron's charge is negative. When particle and antiparticle meet: *poof!* Pure energy.

The muon has many of the properties of the electron, except that it is about two hundred times heavier and it lives only for about two microseconds (two millionths of a second!) before it comes apart. It is an obese electron. Muons play no *known* essential role in the workings of the cosmos. They were the first of the seemingly irrelevant elementary particles to be discovered. These particles certainly do exist (or can be produced in accelerators), but their existence remains an undeciphered clue to the one and true theory of nature.

Soon after the discovery of the muon was made known, I. I. Rabi of Columbia University, who, like most urban American physicists, was fond of Chinese food, exclaimed, "Who ordered *that*?" Rabi was, in a sense, my professional grandfather. He directed the doctoral research of Julian Schwinger, who, in turn, directed mine. All three

A POPULATION EXPLOSION OF PARTICLES

Discovered in 1938, the *muon* was the first of a host of short-lived, apparently elementary particles to be found by physicists over the next several decades. In the 1940s particles called *pions* and a whole set of *strange* particles were discovered among the products of cosmic-ray collisions. These particles are hundreds to thousands of times shorter-lived than muons. With the construction of large particle accelerators ("atom smashers," to the media) that could simulate cosmic-ray conditions in the controlled environment of a laboratory, more and more particles were discovered. By the 1960s the population explosion of "elementary" particles took off. Today well over a hundred of them have been identified. Fortunately, physicists have discovered that most of them are not elementary at all, but are made up of simpler things called quarks. However, the muon that began the population explosion remains as much a mystery as ever. It is not made up of quarks, nor is it itself a constituent of matter as we know it. It seems to be both truly elementary and totally superfluous.

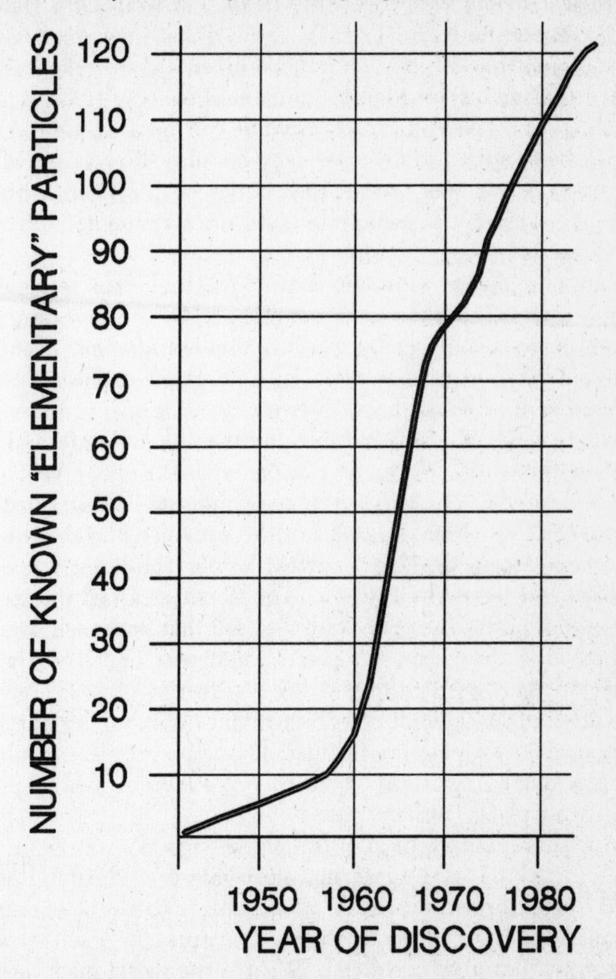

The number of known elementary particles in the past half century has grown. Closely related particles (such as the neutron, the proton and their antiparticles) are treated as a single entry. In 1945 there were four known particles: photons, electrons, nucleons and muons. Today the number is over a hundred and increasing. Spurts in discovery correspond to the development of powerful new techniques: bubble chambers in the early 1960s and electron-positron colliders in the mid-1970s.

of us have Nobel prizes, yet none of us knows the answer to Rabi's question of 1938.

Nuclear fission was discovered in 1938 as well. Otto Hahn and Fritz Strassman, in Nazi Germany, showed that slow neutrons could break uranium nuclei into two roughly equal pieces, releasing a few free neutrons and a considerable amount of energy. Hahn smuggled his findings to his colleague and erstwhile collaborator, Lise Meitner, who had been forced to flee Germany because she was Jewish. The news about fission was carried to America by the Danish physicist Niels Bohr at about the same time that Enrico Fermi fled to America from fascist Italy.

Fermi, together with Leo Szilard (a Hungarian refugee from Nazism), realized that the neutrons released by the fissioning of one uranium nucleus could go on to cause others to fission. A chain reaction could take place that would cause a titanic explosion.

Because fission had been discovered in Germany, there was a real fear that the Nazis would develop the atomic bomb and use it to enslave the world. In October 1939, Szilard, Eugene Wigner and Edward Teller (all Hungarian refugees) induced Albert Einstein to write to President Franklin Roosevelt to convince him that the U.S. government should develop a nuclear bomb. The letter was drafted by Szilard and signed by Einstein. The scientists asked the pacifistic Einstein to send the letter because they felt that Roosevelt would not recognize any other scientist's name. They were undoubtedly right.

Muons and fission, the two big discoveries of 1938, have had vastly different consequences for humankind. Without nuclear fission, neither nuclear weapons nor nuclear electric-power plants could have been developed. The world would be very different today—perhaps a much safer place, perhaps not.

Muons are more typical of the physics I deal with. None of the curious particles I have spent my life trying to understand have as yet led to any direct "practical" application. Particle physicists seek to unravel the most intimate secrets of nature. In a sense, we are overgrown children who still ask, What is the world made of? How does it work and where did it all come from? In our quest we have been led deeper and deeper into the microworld, from molecules to atoms to atomic nuclei to the nucleons themselves and to their constituent quarks. Our great particle accelerators are powerful microscopes with which we observe the very smallest bits of matter. Our search for the most fundamental has led us far away from the workaday world. The logical structure underlying biology, chemistry and engineering technology has been established for decades, and the rules are exceedingly unlikely to change. Not so with particle physics.

Of course, the questions we deal with are the tough ones. While the primary motivation of the particle physicist is to bring us closer to an ultimate understanding of nature, a very practical by-product is the ability to solve what seem to be unsolvable problems. The techniques and technologies we develop are often of great practical importance.

It was the Second World War that got me interested in physics. My brother, Sam, a glider-trooper with the 82nd Airborne Division, patiently explained to me that a low-flying plane must take evasive action after releasing its bombs because if it continued on a straight course, the bombs would explode directly beneath the plane. This puzzled me at first. I had not realized that the bombs would continue to move *forward* as they fell, at the same speed and in the same direction as the plane itself. My brother the dentist-cum-paratrooper got me hooked on classical mechanics—the kinds of problems that Galileo had tackled three centuries earlier.

During the war my literary tastes turned from comic books to science fiction. (But why, oh why, did my parents burn my abandoned comic-book collection? It would have been worth a fortune today!) Higher forms of literature were to wait for my college years.

When the first atomic bomb was dropped on Hiroshima, I was not very surprised and certainly not mystified. Science fiction writers had come eerily close to science fact in the stories I had read. In fact, a story published in the March 1944 issue of *Astounding Science Fiction* magazine, "Deadline," by Cleve Cartmill, described the workings of an atomic bomb so closely that the FBI investigated the author and the magazine's editor, convinced that there had been a security leak in the Manhattan Project. Hiroshima was still more than a year and a half away.

During my years at junior high school I began to realize that I wanted to become a scientist. My father, the plumber, wanted his children to have professions. He could understand that I might prefer medicine to law, but what kind of profession was science? Surely I should become a doctor and do science as my hobby. Nevertheless, he bought me a big Gilbert chemistry set for Christmas. (We were the sort of Jews who attended services once a year and celebrated Christmas, but stopped short of putting up a Christmas tree. We still

GALILEO AND THE BOMBARDIER

Galileo showed that any two bodies, when dropped at the same time from the same height, reach the ground at the same time. Any horizonal velocity of a body is unchanged by gravity and does not affect the rate of fall. This poses a threat to the bombardier, for the bomb he drops will hit the ground and explode when it is directly underneath his airplane.

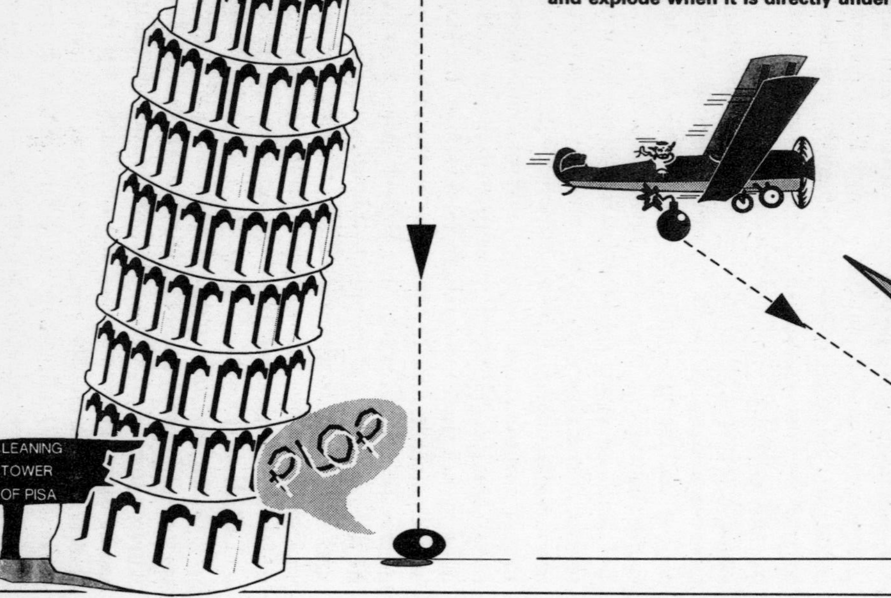

are. One of my Christian friends told me recently that the Christmas tree is of pagan origin, anyway.)

Ah, that chemistry set. In those days, chemistry sets included dozens of wooden bottles and glass tubes filled with exotic chemicals. Most of them were poisonous, and many combinations could be coaxed to explode. Chemistry sets were lots of fun. Nowadays, Big Brother has put an end to all this. Anything you find in today's chemistry sets is safe to put in an omelet. Government regulation has certainly saved a number of fingers and eyes. It has also dramatically reduced the number of child scientists-to-be.

It is one thing to read in a book that many kinds of invisible gases can burn, or extinguish a flame, or smell of apples, horseradish or rotten eggs. It is quite another experience to manufacture these gases all by yourself and to discover that chemistry books are not exercises in abstraction but deal with verifiable properties of the real world. I discovered that science works! What's more, I could do it!

I remember taking a general science course in the seventh grade. We learned that the earth *revolves* around the sun in a year but *rotates* about its axis in a day. A student who claimed the earth rotates about the sun would be marked wrong. We also learned that the moon revolves around the earth in about four weeks, but it always presents the same face to us earthlings—the man in the moon. I realized that this means that the moon must rotate about its own axis in *exactly* the same time as it revolves around the earth. This seemed to me to be a most extraordinary coincidence. I raised my hand and asked the teacher why this was. She immediately recognized that my question was an interesting one and confessed that she didn't know the answer. She was a truly great teacher who could preserve the vulnerable flame of a child's flicker of interest. My question *was* a good one, and it took me years to learn the answer—which Isaac Newton had figured out nearly three hundred years earlier.

The best students in the New York public schools were placed in a special rapid track. Almost all of my chums and acquaintances were part of this somewhat elite group. They were also Jewish, most of them the children of lower-middle-class immigrants. Their parents had placed them firmly upon the intellectual ladder that they felt would (and indeed did!) lead to upward mobility, both socially and economically. Similar enclaves of upward-bound students may be found today in other ethnic communities.

Intellectual curiosity is one of the most contagious of diseases

THE MAN IN THE MOON

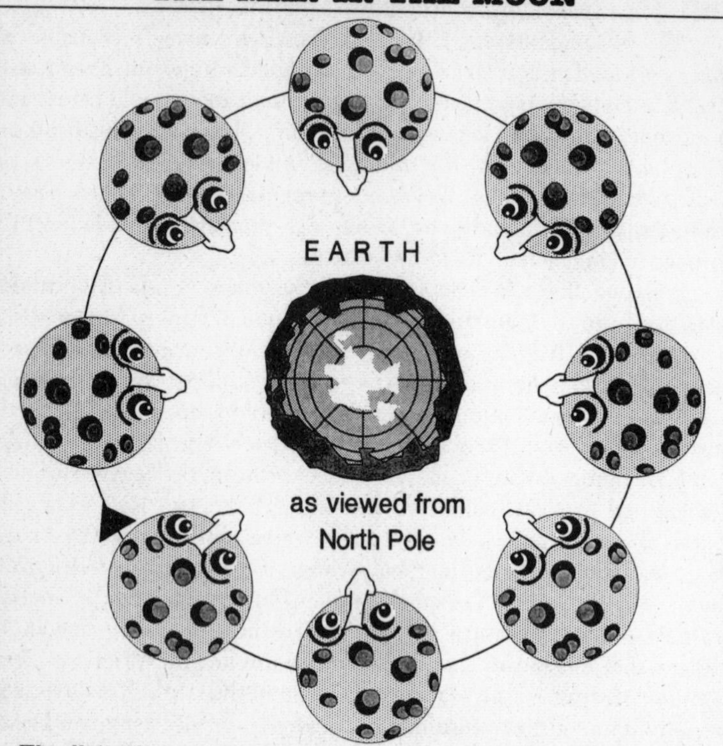

We all familiar with the great tidal force that the moon exerts upon earth. It has even been put forward as a possible solution to the energy crisis. Tidal friction is inexorably slowing down the earth's rotation, so that in some distant future the moon will appear stationary in the sky—our day will equal the lunar month and both will be about fifty-five hours long.

As Newton argued, precisely that has already happened to the moon. It rotates exactly once in each lunar month, so that it always presents the same "face" to the earth.

Recently, similar phenomena have been discovered elsewhere in the solar system. Tidal forces have led to a locked-in relationship between Mercury's periods of rotation about its axis and its revolution around the sun. There are precisely three Mercurian days in each two Mercurian years. This "coincidence" was not known to Newton. It is a recent discovery resulting from radar studies of our solar system.

and it spread throughout my peer group. All through my life, I have felt that I've learned far more from my cohorts than from my teachers.

Once a year, perhaps more often, we were compelled to take some kind of national or statewide scholastic examination. Generally, most of us ended up in the 98th or 99th percentile. To do otherwise was considered a social disgrace. No wonder we viewed New York City as an island of civilization amidst an ocean of boobery.

One spring day, while strolling on the sunny side of Dyckman Street, I (age thirteen) told my friend Henry (age eleven), "I have reached the age of reason. You have not."

Henry responded, "When will I? How will I be able to tell?"

"*I* will know" was my answer.

He claims I said it somewhat airily, and that I never did let him know. Both Henry and I went on to attend the Bronx High School of Science. His autograph in my school album reads, "Best of luck in Science to the only person in 9BR who knows over 80 elements. (I know 79.)—Hammerin' Hank." Evidently science was not his forte: Henry Jay Stern became parks commissioner of New York City.

While still in junior high school I entered an occasional science fair. I won a ribbon once for constructing a simple stroboscope, built with an electric motor I had prized out of a 1920s-vintage Lionel model locomotive that had been my brother Jules's.

We had a somewhat sadistic teacher in the junior high shop course. The first day of class he had all the pupils form a circle and hold hands. So far, so good. Then he sent a mild electrical current through us and gradually turned up the voltage to see who would be the first to break the circuit. My project in shop was to build an electric soldering iron from scratch. It worked quite well, except that it always seemed to give me an electric shock.

Most of the better students graduating from J.H.S. 52 attended one or another of New York City's elite public high schools. I chose the Bronx High School of Science and arranged to take the entrance exam. But an attack of appendicitis hit me. My brother Jules, the doctor, nursed me back to health and let me keep my appendix, which has never bothered me since and to which I am still appended. I took the makeup exam, which I expected to be easy, and it was. Neither my parents nor I ever doubted that I would be admitted to Bronx Science—my gift and my call were apparent. Upon graduation from junior high I was presented with the Ralph E. Horton Memorial Award for "having achieved the first rank in a science sequence."

Summer in New York City was a time of terror for mothers in those days because of the plague of poliomyelitis. My classmate Fred Rein was generally considered the most promising scientist in my junior high graduating class. I was number two. On my first day at Bronx Science I learned from my buddies at the bus stop that Fred had died of polio. Less than a decade later modern science put an end to this horrible disease—too late to save Fred and thousands of others each year.

The first chemistry set my father had given me had helped to make me deeply interested in chemistry, and I convinced my father to convert a large closet in our basement into a relatively sophisticated chemistry laboratory. This was the sort of thing my father could do with ease. When my mother wanted the small porch of our summer home extended around the house so as to produce an arched and tiled portico beneath it, it was done. When my mother preferred the kitchen to be on the first floor, not the second, it was done. When my brother opened his first medical office at our home on Payson Avenue, my father himself built it.

To convert a basement closet to a modern chemistry laboratory was hardly a challenge at all for Pop. It had a workbench, cabinets, a Bunsen burner, sink and a primitive ventilation system powered by a fan driven by the electric motor of my Erector set. My professional brothers ordered the test tubes, beakers, flasks, evaporating dishes, separatory funnels and—most especially—the *chemicals*. As medical professionals, they had access to chemicals that I could not have gotten any other way.

I had all the important concentrated acids (hydrochloric, sulfuric, nitric), powerful oxidizing agents such as potassium chromate, potassium chlorate and even ammonium nitrate (which, at just about that time, was responsible for a disastrous Texas City explosion), and anything else from the Merck catalogue that seemed interesting and not too expensive. I even owned a kilo (2.2 pounds) of bromine. Boy, was I ever going to have fun! It's a miracle that the house survived.

One of my favorite haunts was the New York Public Library on Forty-second Street. Remember, this was in the mid-1940s, long before that part of Manhattan began to fall apart. I copied out procedures for exotic chemical syntheses from ancient textbooks on inorganic chemistry and attempted to carry them out in my basement lab—sometimes even successfully. I get the same joy today on my rare attempts to prepare an exotic dinner for my family.

Making fireworks was particular fun: not rockets or explosions, but wildly colored flames. For some reason that I can no longer remember, I focused eventually on the chemistry of selenium, one of the few chemical elements named after a heavenly body. Only eight of the 109 known chemical elements are so named; helium (the sun), selenium (Greek moon goddess), cerium (after the first known asteroid, Ceres), tellurium (the earth, from the Latin *tellus*), mercury, uranium, neptunium and plutonium, all named after planets of our solar system. Venus, Jupiter and Saturn have somehow been left out.

I began with powdered selenium metal, an exceedingly toxic and noxious substance that smells exactly like horseradish, and synthesized many of its compounds with chlorine and bromine. My most ambitious experiment involved an intricate synthesis of the unstable ring compound selenium nitride. Allowed to dry, this amusing chemical would explode, producing a flash of light, a bang, a puff of smoke and the lingering odor of horseradish. It made a wonderful party favor.

Many of the compounds I produced were placed in sealed glass tubes. I became a collector of interestingly colored, smelly poisons.

My work in my basement laboratory represented an advanced level of curiosity, but it was not yet science. The point was that I simply could not accept chemistry as an abstract rhetorical discipline. Formulas for chemical reactions were of no interest to me unless I could see, hear and smell the chemicals react. For a while it was good fun, but eventually it lost its fascination to me. Chemicals behaved just as they were said to do in ancient mildewed tomes. There seemed to be no profound mysteries lurking unsolved. I was simply following recipes for markedly unpalatable concoctions. I soon forsook the cookbook science of chemistry for biology.

My father had given me an inexpensive microscope with which I retraced the seventeenth-century discoveries of van Leeuwenoek, the Dutch inventor of the microscope. I studied the single-celled creatures that lurk in apparently dead and dry debris from Inwood Hill Park, and identified them as paramecia or blepharisma. Once again, I found that they looked and behaved exactly as my textbooks said they should. The waters of the Hudson River revealed an incredible variety of minute multicellular organisms, mostly microscopic worms. These beasties were not in any book I could lay my hands on. They were undoubtedly human intestinal parasites.

Students at the Bronx High School of Science are encouraged to enter the Westinghouse Science Talent Search, and several of the forty annual finalists are usually Bronx "Sciencites." This is true at other New York high schools as well, since at least a quarter of the finalists have been New Yorkers. Brookline High School in Massachusetts, a highly regarded public school that two of my children have attended, has never produced a Westinghouse finalist. Nor, to the best of my knowledge, has any student from this school ever been encouraged to compete.

Aside from application forms, references and an examination, an original science project has to be submitted to the Science Talent Search. For this enterprise I put together my knowledge of selenium chemistry and my short-lived interest in the life sciences.

Selenium is among the chemical elements that are essential to human life. You will find selenium supplements at your local health-food store. My neighbor the chemist believes that selenium helps to prevent cancer, although why it should do so, neither he nor anyone else can say. The human body needs only minute quantities of selenium, however; too much of it is certainly toxic. Some plants in South Dakota and perhaps elsewhere concentrate selenium from the soil to such an extent that they become hazardous. Cattle that graze on them develop a condition known as alkali disease.

For my Westinghouse science project, I conjectured that the plants were concentrating selenium because the soil in which they grew was poor in sulfur. Selenium belongs to the same family of chemical elements as sulfur. Perhaps the plants were trying to use the available selenium in place of the chemically similar, but unavailable sulfur. I tried to grow tomato seedlings in a hydroponic solution in our small but sunny backyard. Some were fed sulfur, some selenium, and some both. I wanted to learn whether the plants would take up selenium if they could not get sulfur. My results were inconclusive—a way of saying that all the plants died during a weekend I spent at the beach. But the project helped to get me selected as one of the forty Westinghouse Science Talent Search finalists in 1950. We had a wonderful weekend in Washington, during which we shook hands with Harry S Truman and met a lot of big-shot scientists. I was not selected for one of the big money prizes and came away with a mere $100 scholarship. Nevertheless, becoming an "STS'ster" was a tremendous ego boost.

The elements oxygen, sulfur, selenium and tellurium all have similar chemical properties. The same is true for the elements fluorine, chlorine, bromine and iodine. The existence of families of similar elements was a clue to Mendeleev's discovery of the Periodic Table of the Elements.

By 1869 somewhat more than sixty elements had been found. In that year the Russian chemist Dmitri Ivanovich Mendeleev hit on the idea of arranging all the known elements according to their relative atomic weights and placing them sequentially in a tabular array, starting with hydrogen, the lightest element. Mendeleev quickly saw three "holes" in his table—places where an element of a certain weight and with certain chemical properties should exist, but no such elements had been discovered. The table "predicted" that these three elements should be found, eventually, and even foretold their detailed chemical and physical properties. Few chemists took this mad Russian seriously, but within a few years, one after the other, all three elements were in fact discovered: gallium in France, scandium in Sweden and germanium (guess where). Clearly Mendeleev was a genius, after all.

THE NUMBER OF CHEMICAL ELEMENTS

In the eighteenth century the number of known chemical elements began a precipitous increase, continuing to this day. The ancients were familiar with a scant dozen:

lead	tin	carbon
mercury	silver	sulfur
gold	zinc	arsenic
iron	copper	antimony

The thirteenth element, phosphorus, is the first whose discoverer is known. The alchemist, Hennig Brandt of Hamburg isolated white phosphorus in 1669 by distilling human urine. Ironically, his city would be destroyed nearly three centuries later by the very element he discovered: phosphorus-based incendiary bombs dropped on the city by the RAF leveled much of Hamburg in 1943.

The subsequent growth in the number of known elements is shown graphically below. Today we have identified 109 of them.

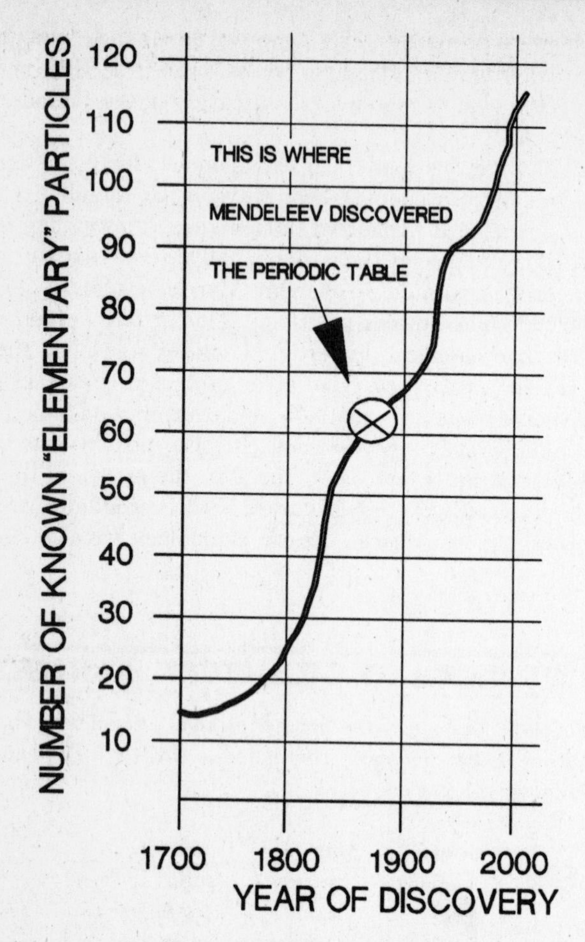

The curve increases inexorably, showing occasional spurts of discovery reflecting technological advances. The invention of the electric battery led to the new science of electrochemistry in the early nineteenth century and permitted many elements to be isolated. The discovery of radioactivity produced a second pulse of discovery in the early twentieth century. The last twenty-one elements, beginning with technetium in 1937, are man-made. They do not occur naturally on earth, but must be synthesized by nuclear reactions. The periodic table was discovered at the point indicated, about halfway up the curve.

MENDELEEV'S PERIODIC TABLE

Dmitri Mendeleev devised his periodic table in 1869. To make it work he had to leave spaces for elements that had not yet been observed to exist. Below, we see a fragment of his table showing several elements (with their discovery dates) properly in place. Mendeleev predicted the precise chemical and physical properties of the three missing elements.

Most chemists ignored his work. Yet, in succeeding years, the elements scandium, gallium and germanium were duly discovered and were found to have just the properties Mendeleev had foretold. The periodic table was then accepted as a major and fundamental contribution to science. It brought order out of the chaos of chemistry and suggested the possibility of atomic structure.

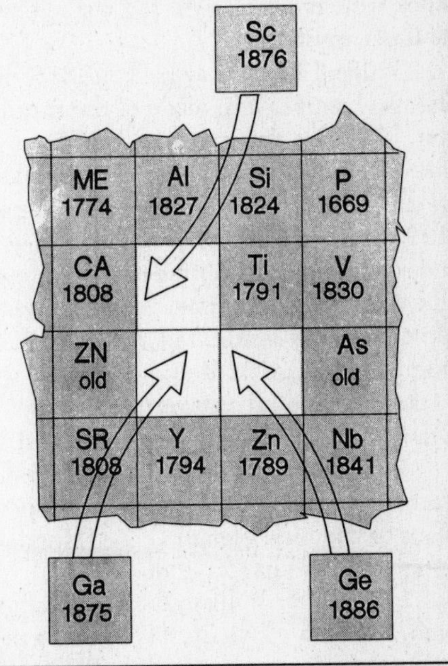

Mendeleev's Predictions

The Periodic Table brought order and sanity to the bewildering welter of chemical elements and helped to convince scientists that each chemical element was composed of its own unique kind of atom. For chemists, the atom was the ultimate chunk of matter. Their quest was (and still is) to understand how atoms form molecules that can interact with one another in chemical reactions. Physicists, however, were not convinced that the atom was nature's ultimate building block. To them, the success of the periodic table in reducing the *chaos* of chemistry to systematic *order* suggested a further level of *structure*. Tantalizing hints that the atom was a composite system appeared in the last years of the nineteenth century. Atoms could vibrate, producing characteristic colors of light. Some newly discovered chemical elements were named for the colors they would impart to a flame: cesium and rubidium, for example, come from the Latin words for blue and red. But what could it mean to say that the atom vibrates? Surely then, the atom must have parts inside that can wiggle and jiggle with respect to one another. It certainly seemed as if the atom was built up out of more primitive things, but what were they? Since we could not even see a single atom, how could we dream of its inner mysteries?

William Prout, way back in 1815, proposed that all of the various chemical atoms were made of the simplest and lightest atom, hydrogen. This idea was suggested to him by the fact that the weights of the different atoms seemed to be whole number multiples of the weight of the hydrogen atom. Could carbon, with an atomic weight of 12, simply be made of twelve hydrogen atoms, nitrogen of fourteen and so on? It was an outrageous yet prescient hypothesis, put forward a century before its time. And in it was a germ of truth. Prout's law ultimately was vindicated in the 1930s, when it was realized that atomic nuclei are built up out of hydrogen nuclei (protons) and their electrically neutral cousins (neutrons). But first, there was a long road to follow, filled with all sorts of almost unbelievable surprises.

By listing the elements in a tabular array, according to their atomic weights, Mendeleev brought seemly order out of the apparent chaos of chemical elements. Yet Mendeleev did not get quite everything right. No one ever does.

In 1894, Sir William Ramsay, a Scottish chemist, and Lord Rayleigh, a British physicist, discovered a new element, argon, a gas that constitutes about one percent of the earth's atmosphere. Argon had no place in the periodic table. Moreover, it had no discernible chemical properties. It was a nonreactive element, an inert gas. The word "argon" comes from the Latin for "lazy one." The discovery of argon was followed by the discovery of neon (the new one), krypton (the secret one) and xenon (the strange one). Here was a whole family of elements that poor Dmitri Ivanovich could not account for. Both Ramsay and Rayleigh won the Nobel Prize for their remarkable discovery: the former in chemistry, the latter in physics.

The columns of the periodic table describe elements of the same valence, a number that characterizes chemical activity. Mendeleev's original table included valence numbers from 1 through 7. The new elements, so chemically aloof that they came to be known as the "noble" gases, exhibited *no* chemical activity—zero valence. A new column had to be added to the periodic table.

At the turn of the century, the success of the periodic table (as amended by the inert gases) was both a triumph and a mystery. The existence of systematic, periodic and predictive *order* among the chemical elements was a welcomed delight. The chemical elements were not a chaotic jumble; they fit into a pattern. But what did the pattern mean? In retrospect we can see quite clearly that the order established by the periodic table was an early indication of *structure*. Each chemical element is composed of its own kind of atom, and different kinds of atoms are built differently, each with a different weight, size, and valence. This was a powerful indication that the atoms are not elementary particles, but are made of simpler things.

While I puttered about with chemistry and biology, my interests in high school tended more and more toward physics. Chemistry, it seemed to me, was child's play. The real question was not how atoms behaved toward one another, but what their innermost structure was. What were the most fundamental rules of nature?

THE PERIODIC TABLE, AMENDED

An unpleasant surprise lay in wait for Mendeleev. In 1894, William Ramsay and Lord Rayleigh discovered a new chemical element making up one percent of our atmosphere: argon. Chemists were appalled, for they could not get argon to form chemical compounds, unlike all the other known elements. It is a chemical element without characteristic chemical affinity. Its discovery was followed by that of neon, krypton and xenon. Helium had been known to exist as a mystery element in the sun since 1868. Ramsay found helium on earth in 1895 and showed that it is the lightest member of the family of inert gases. Radon, the heaviest of the inert gases, is a radioactive element produced in the decay scheme of thorium and uranium. It was discovered in 1900, and found to be so unstable that it lives for only about four days. The figure below shows the first half of the periodic table as it is today.

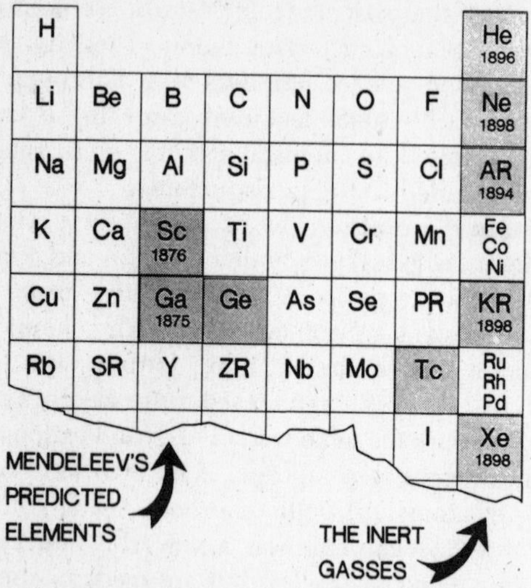

H							He 1896
Li	Be	B	C	N	O	F	Ne 1898
Na	Mg	Al	Si	P	S	Cl	AR 1894
K	Ca	Sc 1876	Ti	V	Cr	Mn	Fe Co Ni
Cu	Zn	Ga 1875	Ge	As	Se	PR	KR 1898
Rb	SR	Y	ZR	Nb	Mo	Tc	Ru Rh Pd
						I	Xe 1898

MENDELEEV'S PREDICTED ELEMENTS

THE INERT GASSES

The so-called inert gases complete a new and entirely unanticipated column of the original periodic table. Only with the amended table in hand were Niels Bohr and his early twentieth-century colleagues able to guess at the mysterious new system of quantum laws governing atomic structure.

Notice, incidentally, element number 43 (Tc) toward the lower right. Mendeleev didn't know about technetium, either. The first of the man-made chemical elements, technetium does not occur naturally on earth.

Our high school physics course was virtually useless. It used an ancient textbook written by Dull, and dull it certainly was. The Bronx High School of Science was not going to teach us what we wanted to know. Gerald Feinberg, Steven Weinberg and I became close friends. Feinberg was to be a Science Talent Search finalist with me. Steve got a mere honorable mention. Each of us was "turned on" to fundamental physics as a career. We hardly thought about anything else. We haunted the library, we learned from books, we learned from each other. We argued physics questions in the New York subways. We had interminable telephone conversations. We learned.

Feinberg and I were among the founding members of the science fiction club at Bronx Science. We published the first high-school science-fiction fan magazine, or fanzine. It sold for a nickel. Its title was explained in the first issue: "In case you are wondering where on Terra we got the name 'ETAOIN SHRDLU,' it isn't Mandark, it is the letter frequency of the English alphabet. . . . [It] is also used by linotype operators who run it off on their machines to see if they are in working order."

A twenty-six-page mimeographed pamphlet printed on pastel paper, ETAOIN SHRDLU contained original works of fiction, articles on science such as Feinberg's "Stellar Yardstick," which dealt with how we know the distances to the stars, and interviews with science fiction authors such as Alfred Bester.

My first published work appeared on page 23 of the first issue of ETAOIN SHRDLU, dated December 1949. It is a factual science article presented with sophomoric whimsy, the sort of thing that might be expected of a student who would, years later, introduce the word *charm* into the adult world of physics, but not the sort of thing to expect in a science fiction fanzine.

My love of science fiction was in part a desperate attempt to make contact with the real world of contemporary physics. My high school physics course and its teacher were blissfully unaware of the "modern" developments of the 1930s. We were taught the five basic tools of mechanics (pulleys, levers, wedges, inclined planes and wheels) rather than the four basic forces of nature. What I knew of atomic bombs and nuclear power I had learned from the factual articles in *Astounding Science Fiction*. In my article "On the Nature of Nothing," I am rather obliquely presenting the original arguments of Paul Dirac that predicted the existence of antimatter in 1931.

Since leaving Bronx Science, Feinberg has become a distinguished author and professor of physics at Columbia University. Weinberg and I have followed one another around the physics world. He and I shared the 1979 Nobel Prize in physics with Abdus Salam, a remarkable Pakistani physicist whom we shall meet again.

As our high school years approached their end and the time came to apply to college, we became concerned about our lack of organized sports activities. Guidance counselors gravely warned that we must

ON THE NATURE OF NOTHING

As a famous philosopher once said (I forgot his name), "Nothing is something of which we know nothing." Truer words were never spoken. It was once thought that nothing, or in more technical terms, the absence of something, is what we call air. However, the Greeks, by discovering the vacuum, disproved this hypothesis. It was postulated by an eminent Greek savant that the vacuum, being the absence of air, was the ultimate in nothingness. This theory lived a mere thirty centuries until it was finally disputed. The glorious postulate of the luminiferous ether was then made. This, proving unsuitable for Einstein, was soon abandoned. A new theory was needed. Practically any schoolchild can recite the verse describing the discovery of nothing:

He sailed the seven seas,
In search of perpetual motion,
Instead he found non-entity,
And sank in Dirac's ocean.

Yes, this marked the beginning of the Nihilist Theory. Dirac, realizing the rarity of nothing (ergo, the abundance of everything), proposed that all space is not, as generally believed, composed of nothing, but consists of a homogeneous field of electrons possessing negative kinetic energy. (This is, of course, obvious.) The absence of one of these electrons, a hole in Dirac's ocean, is, in reality, nothing. But these "holes" act as if they are particles, possessing a positive charge. Dirac's theory is that the positron is the ultimate non-entity.

have an extensive record of extracurricular activities if we hoped to be acceptable to Ivy League schools. I have always regarded the existence of guidance counselors, as a specialized category of humans, as an absolute waste of the taxpayers' money, a mere catch-basin for teachers unable to teach. The role of guidance counselor can be played better by any competent member of the teaching staff.

On the other hand, the only sense in which I was "well-rounded" was in terms of my waistline. Corrective action was needed! I joined the "protozoology squad" and Steve joined the "mold squad." These were merely service groups consisting of a handful of student volunteers who cleaned up the biology labs, but they must have looked good on our records. These "squads" were an ingenious and successful attempt on the part of the principal of Bronx Science, Morris Meister (may his name be blessed!), to contain the school's maintenance expenses. It was a small step toward the system adopted eventually by Taiwan and other nations wherein *all* the cleaning chores are carried out by the pupils, who somehow have time left over to perfect a second language and learn calculus as well.

In the spring of 1950, Steve and I were admitted to Princeton, Cornell and MIT. We were also admitted to CCNY, then referred to as the Circumcised Citizens of New York. Guidance counselors insisted that *all* college-bound seniors apply to CCNY as a safety play. My only refusal came from Harvard University, and it may have resulted from my frank admission to the interviewer that Harvard was only my second choice.

Steve's father took us on a jaunt to all three campuses to help us make up our minds. Mr. Weinberg was a very good driver. Years later, in the mid-sixties, my widowed mother lived in Riverdale and

STEVE WEINBERG AND ME

Not only did Steve and I attend the same high school (Bronx Science) and the same college (Cornell) at the same time, we also spent significant portions of our careers at the University of California at Berkeley, where we simultaneously professed physics for four years during the riotous sixties.

Moreover, we were both members of the Harvard physics department from 1967 to 1981. Each of us spent a year (different ones) at the Niels Bohr Institute in Copenhagen, and both of us were at one time or another affiliated with MIT. On the other hand, Steve has spent time at Princeton and Columbia, and I have taught at Stanford and CalTech. In 1981, Steve forsook fair Harvard to become the Josey Professor of Science at the University of Texas at Austin. I remain primarily a professor of physics at Harvard, but it is curious to note that I too have developed a Texas connection. I spent a sabbatical leave at U.T.'s arch-rival Texas A&M and enjoy a continuing part-time affiliation with the University of Houston. Our trajectories through the world of physics are hopelessly intertwined.

often traveled to midtown Manhattan for her shopping excursions (shirts for me, shoes for herself, etc.). She and her cohorts used a convenient limousine service for that purpose (seventy-five cents plus a quarter tip), gossiping as mothers do all the way there and back about "my son the lawyer," "my daughter the neurosurgeon," and (I love you, Mom) "my son the physicist." One time the driver chimed in that his son was a physicist, too. It was, of course, Mr. Weinberg.

As a result of our junket, Steve and I came to identical conclusions. (No wonder that we would share the Nobel Prize.) MIT was too urban and specialized. Princeton, still obliging its students to wear filthy black academic gowns at dinner, was not only elite and effete but did not admit girls. Cornell won hands down. It was not only the beauty of its campus, "far above Cayuga's waters," but the warmth of our reception and the obvious enthusiasm about physics among the graduate students we met. Moreover, our New York State scholarships were supplemented by generous grants from Cornell University. Our choice was clear.

Far above Cayuga's Waters

IT WAS NOT EASY FOR AN ENTERING Cornell student to get from New York City to Ithaca. Robinson Airlines was expensive, unsafe and irregular. The company went out of business when its plane crashed. The Lehigh Valley Railroad (known to generations of Cornellians as the Leaky Valley, it is now mercifully extinct) could in principle get me there, but I was strongly advised not to discover the charms of its practice. Moreover, there was apparently no direct bus service. Since mere freshmen could not bring cars to the Cornell campus, and since I didn't have a car anyway, I had a problem. The dilemma was solved by my parents, who drove me to Cornell and left me on my own for the very first time. The out-of-town college experience is truly the great American divide between infancy and adulthood. Once I had gotten settled in my new quarters I began to worry about what was to become of me.

If in high school I had to contend with such formidable intellects as Feinberg and Weinberg, what would things be like at a major university such as Cornell? Could I survive intellectually to become the scientist I had dreamed of becoming for so long? Or would I end up as a dropout, or worse, a pre-med?

To my surprise, my four years of college were more like rest and recreation than continued intellectual combat. It was a time to learn about bridge, billiards and girls. The average student at Cornell was inferior to the Bronx High School of Science standard and spent more time drinking and partying than studying. Courses were generally no more challenging than those of high school, and the expectations of our professors were modest. Again, it was necessary for us to take things into our own hands and learn from one another. My choice of pronouns indicates that I again had the good fortune to be among a remarkable group of achievers.

The dorm in which Weinberg and I began our college careers

was a wartime hut that had somehow escaped demolition. The walls were paper thin and broke easily. They were, in fact, frequently broken by collisions with tipsy freshmen. Steve's roommate was Daniel J Kleitman, a gifted young physics major from Morristown, New Jersey. Danny is now a well-known MIT mathematician and my brother-in-law, to boot. Our wives, Joan and Sharon, are sisters. A third sister also lives in Boston. She was once married to Carl Sagan, who convinced her to study biology rather than English: Lynn is possibly Carl's greatest contribution to science. She is a famous biologist known for her controversial "symbiotic theory of evolution" in which sex first appears on earth as a result of viral infection, not natural selection. She is Boston University's sole surviving member of the National Academy of Sciences.

My roommate turned out to be someone I already knew: David Lubell, a precocious mathematician who had been a cofinalist in the Science Talent Search. We had met during the STS weekend in Washington. David came from Midwood High in Brooklyn. Today he lives in Levittown, Long Island, and teaches mathematics at Adelphi University.

The noise, the odors, the lack of privacy—in short, the non-quality of life—at the "temp dorms" soon drove David and me to seek accommodations elsewhere. We moved to a rented room at 522 Stewart Avenue in "college town," as the section of Ithaca abutting the Cornell campus was known. I learned a lot of mathematics from David: simple things that are easy to teach but so rarely taught.

Life off-campus was not quite perfect. We had to cope with a nosy landlord and with the brutish floriculture students across the hall who insisted that we fight them for a fin. It was impossible to postpone combat indefinitely. Simply offering the five dollars only enraged them more. While they were out puttering with their orchids or whatever it is that floriculturists do, we slipped into their room, slit open their pillows, and inserted a raw meatball in each. After carefully stitching up the pillows, all we had to do was wait. About two weeks later our flowery friends moved out, complaining of an unbearable stench. We never heard from them again.

Besides our academic courses, gym and ROTC were mandatory. As an Air Force cadet, I was issued a uniform and expected to march about from time to time and to sit through silly courses in map reading (anyone who can find Cornell can read a map), survival (to a streetwise New Yorker?) and aeronautics. The lieutenant *cum* professor would fly his favorite students to Minnesota in an Air Force plane to pick up a refrigerator for his wife, but he didn't know beans about how a plane could fly. I ad-libbed a song and dance about an airplane being like a kite (it isn't), which he and the whole class bought. Of course, an Air Force officer doesn't really need to know physics any more than a medical doctor needs to know biology. They just need to know which buttons (or pills) to push, and when.

Once a year the cadets were required to publicly demonstrate their precision marching skills, an experience that I found noisome. Ripping my fly open in the men's room, I reported to the drill officer that my zipper had somehow come irrevocably undone and was regretfully excused from the ceremony. The real challenge of ROTC was to get as low a grade in "ROTcore" as possible without failing. In my case the challenge was well met.

In striking contrast to all my other freshman courses, my introductory physics course was an absolute delight. The textbook was intended for a two-year physics sequence at Yale. Our professor, Kenneth Greisen, a famed cosmic-ray researcher, zipped us through the book in less than a year. Physics was being thrown at us with tremendous speed.

Calculus, while not a formal prerequisite for the first-year physics course, had to be picked up along the way. In those days, this most basic mathematical tool was simply not taught in high school, not even at Bronx Science. Most of the students took a parallel and not very well taught booster course in calculus from the math department. Dan Greenberger, a high school classmate who now teaches physics at City University of New York, had taught me the rudiments of calculus in the Bronx Science lunchroom. I had also perused a college calculus cram book that I had bought on one of those frequent clothing-shopping jaunts to Macy's with my mother. It was in one of those lunchroom sessions with Dan that I suddenly realized you *could* turn an inner tube inside out through a hole in its side. Things like that were so much more interesting to me than American history or English literature.

Greisen's course was so high-powered that most of the freshman physics majors had to work their tails off or give up in dismay. For those of us with some mathematical sophistication, like Kleitman, Weinberg, and myself, the course was terrifically stimulating. It convinced us that physics was fun, just as we always suspected.

The math department at Cornell was superb, but so also was the extracurricular supplement in group theory provided by my friend Harold V. Macintosh. He was a grad student in physics from Colorado A&M who now teaches in Vera Cruz, and he had a thing about group theory. Since this discipline is especially relevant to modern physics, I was indeed fortunate to have Macintosh at Cornell to get it across to me clearly and beautifully. The incident illustrates again my belief that one learns as much or more from one's peers as from one's officially designated teachers.

GROUP THEORY

Group theory is an elegant and powerful branch of mathematics that is totally unfamiliar to the average well-educated American. It could and should be taught in grade school to show how beautiful, how useful, and how much fun mathematics can be. It emerges from the study of symmetry, a concept central to both physics and mathematics.

A group is a bunch of operations or transformations with the following essential properties:

1. The operations can be performed sequentially. The resulting compound operation must itself be a member of the group.

2. Every operation of the group has an inverse operation that undoes it.

One of the world's simplest groups consists of the numbers ± 1, where the operation is simple multiplication. Multiplication by one (call it I) is called the identity element of the group. The operation I is a trivial operation that does not change what it multiplies. Multiplication by -1 (call it J) is less trivial: -17 is quite different from $+17$, for example.

The group is defined by its multiplication table telling what is the resultant of any two operations. For this simple example, the table is:

	I	J
I	I	J
J	J	I

the inverse operation to J is J itself.

Let's examine a somewhat more complex group. Consider an equilateral triangle with its vertices labeled as a, b and c. There are precisely six rigid transformations that leave the figure of the triangle unchanged:

I. The identity operation. (Do nothing at all.)
R. Rotate the triangle by 120° clockwise.
S. Rotate the triangle by 120° counterclockwise.
A. Flip the triangle over, leaving vertex a in place.
B. Flip the triangle over, leaving vertex b in place.
C. Flip the triangle over, leaving vertex c in place.

These six operations may be performed sequentially. For example, R × A means: Flip the triangle over, leaving vertex a fixed, then rotate it clockwise by 120°. This compound operation also leaves the triangle intact. However, it is not distinct from the listed operations. Try it, and you will discover that R × A = C.

In a similar way, the "product" of any two of the group elements is itself a group element. Moreover, each of the six operations has an inverse operation. Each of the operations I, A, B and C is its own inverse. That is:

$$\mathbf{A \times A = B \times B = C \times C = I \times I = I}$$

The inverse of R is S, and conversely:

$$\mathbf{R \times S = S \times R = I}$$

Thus, our set of six operations form a group. Its multiplication table is:

	I	R	S	A	B	C
I	I	R	S	A	B	C
R	R	S	I	C	A	B
S	S	I	R	B	C	A
A	A	C	B	I	R	S
B	B	A	C	S	I	R
B	C	B	A	R	S	I

So far, so good. It all looks a little bit like addition or multiplication, but it's not quite. Notice that A × B = S while, on the other hand, B × A = R. In group multiplication, unlike ordinary arithmetic, the order of steps is crucial. Group multiplication is not always commutative. Groups, like this one, are examples of mathematical systems that can be far richer than the elementary operations of ordinary arithmetic and algebra. They can also be immensely useful.

The ozone molecule consists of three oxygen atoms in the shape of an equilateral triangle. The equations governing the motions of such a molecule are complicated, but the symmetry of the system, as described by the group we have discussed, can simplify them. Group theory is a powerful tool with which to study the dynamics of complex but symmetrical systems.

The world itself is such a system. The laws of physics do not depend on the place the experiment is done, nor the time, nor the spatial orientation of the observer. Moreover, they do not depend upon the speed of the observer. Experimental results obtained by observers in different states of motion are related to one another by group operations. The big question is, "What is the group that relates these observations to one another?"

In 1905 Einstein put forward his special theory of relativity, which revolutionized our understanding of space and time. From a purely mathematical point of view, he simply replaced one group by another. His group describes nature correctly, its predecessor did not. From the point of view of mathematics it was only a small step. However, the physical consequences of Einstein's theory are enormous.

The essential property of quantum theory is that experiments affect one another. The sequence in which things are measured matters very much. Quantum mechanics is inherently noncommutative. Its natural language is not ordinary mathematics, but group theory. This mathematical discipline, with which very few American high school teachers are familiar, is at the heart of all major development of twentieth-century physics. For me to elucidate modern physics to an audience lacking this tool is rather like eating a plate of spaghetti with a spoon but no fork. I shall do my best.

Throughout the war my parents had maintained a 1942 Oldsmobile, one of the first American models with automatic transmission. It was the car in which I learned to drive on the then-quiet and deserted streets of Manhattan east of Nagle Avenue. In the summer of 1951 my brothers gave it a spanking new gray paint job and presented it to me. With a physics major friend and an enormous load of luggage, I set off to Ithaca for my sophomore year with my very own car. We never made it. I totaled the car near Binghampton, finding out the hard way what the signs saying "Soft Shoulders" were trying to tell me.

Carless, I finally got to Cornell for my second year. I knew enough to avoid the dormitories, and I wanted nothing to do with the drunken fraternities. That left College Town. I shared a spacious apartment with Danny Kleitman, Kent Gordis, Danny Filson and "Frost Heaves," our bluepoint Siamese kitten, named after another common roadside sign. Kent and Danny Filson (a.k.a. Mouth and Bush) had been classmates at Bronx Science whom I had known only as back-of-the-room bigmouths who would complicate our English and social studies classes with comments about dialectic materialism and other mysteries. They lived in Greenwich Village, products of "the little red schoolhouse." They, too, were physics majors and we soon became fast friends. Much later, Kent made and lost his fortune as Bernie Cornfield's computer hand with Investor's Overseas Services, notorious for having its assets purloined by Robert Vesco. Today he owns and operates a software company in Geneva, Switzerland. His son, Kento, recently graduated from Yale. Danny Filson died in an automobile accident while he was a promising graduate student of physics at UCLA.

Cornell permitted undergraduates who lived off-campus to throw parties, providing two conditions were met. There had to be separate toilet facilities for the two sexes, and at least one more girl present than there were rooms in the apartment. Few parties met these criteria. It was generally believed that violations of the rules were detected and reported by spies from the Women's Christian Temperance Union. Despite these handicaps, one of our lady friends received her expulsion notice at our address.

Others in my crowd included Gerald Sachs, now a professor of mathematical logic jointly at Harvard and MIT; Gerry Neugebauer, now director of the Mount Palomar Observatory; Laurence Mittag, who teaches engineering at Boston University; Ed Wasserman, a big-shot chemist at Allied Chemical; and Tema Ehrenreich, who married her (and my) physics instructor at Cornell, Henry Ehrenreich, now my colleague in the Harvard physics department. I could go on—as I said, I had joined a crowd of achievers.

As an upperclassman I took a number of graduate-level courses, contrary to the advice and wishes of many members of the Cornell physics department. They seemed to feel that undergraduates should take only undergraduate courses. I felt otherwise. Nowadays at Harvard, and at Cornell as well, we strongly encourage our physics concentrators to take advantage of our higher-level courses if they are able. Consequently, a promising senior today knows a lot more physics and math than I did at a corresponding age. Of course, they have to know a lot more than we did, since it appears that we've solved all the easy physics problems.

One of the graduate courses I took dealt with classical electromagnetic theory and presented the great synthesis of James Clerk Maxwell. The ancient Greeks knew of electricity and magnetism. They knew that a piece of amber rubbed with fur would attract small objects. The word *electricity* derives from *electra*, meaning "amber," both in the sense of resin and as a woman's name. (Forever Electra? Mourning Becomes Amber?) Lodestones are natural magnets, and many of them were found in the ancient Greek region of Magnesia, which lies near Ephesus in what is now modern Turkey: hence the word *magnet*. (The same region yielded the names of two chemical elements, magnesium and manganese.)

Maxwell, a shoo-in to anyone's Physicists' Hall of Fame, unified the theories of electricity and magnetism in 1865. That is, he showed that the two phenomena are different manifestations of the very same fundamental force. Most amazingly of all, he showed that light is one variety of electromagnetic energy, and that many other varieties must exist. Thus he blazed the trail for the discovery of radio, television, radar and the entire "electronics revolution" of today.

In homage to this great man, physics majors throughout America wear T-shirts with Maxwell's equations emblazoned in DaGlo colors:

MAXWELL AND HIS EQUATIONS

$$\triangle \cdot E = \rho \qquad (1.)$$

$$\triangle \cdot B = O \qquad (2.)$$

$$\triangle \times B = J + \dot{E} \qquad (3.)$$

$$\triangle \times E = -\dot{B} \qquad (4.)$$

The first equation describes mathematically how electrical charges (denoted by ρ) produce an electric field. In English, it says that the divergence (or, source) of the electric field is the electric charge. This equation codifies the experimental results of the eighteenth-century French scientist Charles Augustin de Coulomb.

The second equation describes the fact that there is no magnetic analog to the electrical charge. No experiment has ever found a fundamental "chunk" of magnetic charge, such as a magnetic equivalent to the electron or proton. This equation is not linked to a particular scientist's name, since it refers to a negative result. (Who was the first person *not* to see a magnetic monopole, as the basic bit of magnetic charge is known?) Some scientists are still hoping that the second equation will ultimately turn out to be wrong, and that a magnetic monopole will turn up in their searches. Every decade or so a diligent physicist will think he has spotted a monopole, only to discover that he has made an error or had a stroke of bad luck. The last reported sighting was on Valentine's Day of 1982, at Stanford University, by Blas Cabrera. A year later, on Valentine's Day, I sent Blas a telegram:

Roses are Red
Violets are Blue
The time is now
For Monopole two.

I got no answer, since the original observation was probably a fluke. It has never been confirmed. I'll bet on the validity of Maxwell's second equation.

The third equation (except for its last term) describes how an electric current, J, produces a magnetic field, $\dot{B}$. The surprising discovery that moving electrical charges—a current—can generate magnetism was the first experimental link between electricity and magnetism. It was discovered by the Danish physicist Hans Christian Oersted in 1819.

The last equation incorporates the observations of a brilliant Scotsman, Michael Faraday. In 1831 he discovered that a *changing* magnetic field can produce an electric field and generate an electrical current. The term $\dot{B}$ in this equation signifies the rate of change of the magnetic field. This discovery made possible the development of electric motors, which convert electricity into motion, and electric power generators (Faraday called them "dynamos"), which convert the energy of motion into electricity. It was Faraday, rather than Edison, who electrified the modern world.

When Faraday was asked by a middle-aged matron of what use was his laboratory discovery, he replied, "Madam, of what use is a newborn baby?"

In a related vein, Robert R. Wilson, designer of America's largest atom smasher, was once asked by Senator John Pastore, "Of what use is your proposed facility for the national defense?"

"Of no use at all," said Wilson. Then he added, ". . . but only to make the nation worth defending."

Maxwell's equations were mostly descriptions of physical processes that were well known from earlier work. But he saw that these descriptions were incomplete and that, as they stood, they were mutually incompatible. His crucially important contribution was the introduction into the third equation of a key extra term $\dot{E}$, which he called "the displacement current." It is for this seemingly innocuous modification that the the equations today bear Maxwell's name.

The most unexpected bonus to come from Maxwell's work was his realization that light is a form of electromagnetic wave.

Maxwell regarded electricity and magnetism as stresses or deformations in a hypothetical elastic medium he called the ether. Imagine, for a moment, a somewhat more tangible elastic medium: a metal bar. It can be bent, stretched or twisted. These deformations are analogous to steady electric or magnetic fields in Maxwell's ether. The bar can also be struck so that it rings like a bell. Such oscillating deformations, or *waves*, may be initiated and transmitted through any elastic medium. Familiar examples include sound waves in air, ocean waves, and seismic waves in the solid earth. Maxwell saw that his equations could describe the existence of electromagnetic waves in his imagined all-pervasive ether. He was able to calculate the velocity at which these waves would travel from the results of simple electrical and magnetic observations. To his astonishment, he found that their velocity was precisely equal to the speed of light, which had been measured to a good accuracy only recently. He concluded that light is simply an oscillatory manifestation of electricity and magnetism. His conclusion was absolutely correct, even though he had no evidence to back it except his own mathematics.

For Maxwell, the ether was a real physical entity with measurable electromechanical properties. In this respect, he was wrong. Careful observations in the late nineteenth century showed that the concept of a material ether simply made no sense. What we call light consists of electromagnetic oscillations, periodically changing electric and magnetic fields, but they do not need ether or any other medium to propagate themselves. They are perfectly capable of propagating through empty space.

Maxwell's extra term, in physical terms, implied that a changing electric field produces a magnetic field. Faraday's work showed that a changing magnetic field produces an electric field. Even in a perfect vacuum with no charges, or currents, or particles about, an "electromagnetic disturbance" could be maintained by a sort of bootstrap effect, where the electric field is produced by the magnetic field that is produced by the electric field, etc. Light, which travels freely through the void of space, is precisely such a self-sustaining electromagnetic wave.

The equations of electromagnetism have the logical consistency and the uniqueness that appear to the physicist as the elegance and beauty of revealed truth. Electricity and magnetism are shown to be so closely intertwined that they may be regarded as different avatars of the same fundamental physical phenomenon.

What could seem more different than rubbing a balloon on one's hair so that it sticks to the wall (electricity) and finding one's way out of a forest with a compass (magnetism)? Maxwell's fundamental equations revealed the unity underlying these disparate phenomena. His was a *unified theory* of electricity and magnetism, the paradigm for today's search for a theory that will unify *all* physical phenomena: the holy grail of contemporary elementary-particle physics.

Maxwell's theory led to something even more exciting, to a theory that incorporated and explained the most commonplace act in nature's repertoire: light. In showing that light is just another electromagnetic phenomenon, Maxwell became the only mortal who could claim for himself the phrase "Let there be light!"

When white light (such as sunlight) goes through a prism, its waves get sorted out by their wavelengths, which the eye sees as colors. Isaac Newton showed this to be so. Red light is the longest of all visible light waves—its wavelength from crest to crest is about 30 millionths of an inch. Violet, lying at the other end of the rainbow, has the shortest wavelength, about 18 millionths of an inch. Ultra-

THE ELECTROMAGNETIC SPECTRUM

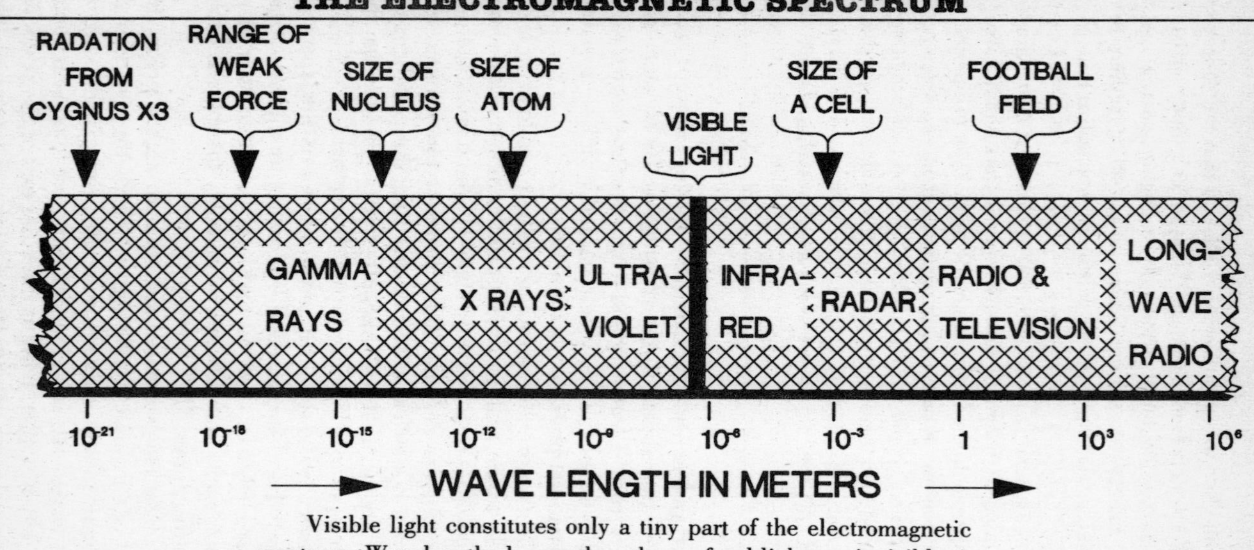

Visible light constitutes only a tiny part of the electromagnetic spectrum. Wavelengths longer than those of red light are invisible to the eye. The longest useful electromagnetic waves are many miles long. They provide a worldwide standard for measurement of time. Wavelengths shorter than those of violet light are also invisible. The shortest electromagnetic waves ever detected come from a mysterious and distant stellar system known as Cygnus X-3.

violet rays, shorter in wavelength than violet, are the invisible component of sunlight to get tan by. Infrared rays, longer than red, are the invisible heat-conveying part of sunlight to bask in. All are forms of electromagnetic radiation, obeying Maxwell's equations. And there are more!

Shorter even than ultraviolet rays are X-rays, discovered toward the end of the nineteenth century and immediately put to use as a valuable tool for medical diagnostics. *Longer* than infrared is the domain of radio waves. Two decades after Maxwell published his theory, the young German physicist Heinrich Hertz produced and detected this new form of radiation whose existence was demanded by the mathematical equations. He sent a radio signal across the width of a room.

The Italian Guglielmo Marconi immediately saw the technological and financial possibilities implicit in this new bit of "pure science." By 1918 he had succeeded in transmitting a radio signal halfway across the world, from Wales to Australia. Only two years later commercial radio broadcasting began, ushering in the age of electronic mass-media communication. Yet the world was not finished with the impact of Maxwell's equations: television, radar, lasers, microwave ovens and much more was to come—and still is.

While I was at Cornell studying electromagnetism, I eventually acquired a 1947 Plymouth with which to escape the tedium of Ithaca (or the tedia of Ithacans). We off-campus physics proto-beatnik types established a "city-a-week plan" and began to discover the joys of upstate New York: the gastronomy of Skeneateles (a restaurant serving unlimited goodies at a fixed price), Cortland State Teachers College (pubescent teachers-to-be), Syracuse (where Danny F. found his wife-to-be), Summerhill (birthplace of the thirteenth president of the nation: let's hear it for Millard Fillmore), and the twin cities of Athens and Sayre at the Pennsylvania border (why not?). One time we got as far as Boston and crashed a Harvard party. It was my first time in Beantown. I liked it a lot and I wanted to return. Harvard had the style that my dominantly agricultural university did not, although it could not boast of the female bullwhip champion of the country, as Cornell could.

In addition to graduate-level courses, we were forced to take physics courses aimed at the average mediocre undergraduate. In an example of unparalleled absurdity, I was obliged to take in my senior year an elementary course in electrodynamics even though I had already received an A in the *advanced* electrodynamics course! Fortunately for me, the instructor was Professor Giuseppe Cocconi, who, with Philip Morrison, was the first to point out the feasibility and importance of a search for extraterrestrial intelligence. Proving that some intelligence exists even on earth, Cocconi wasted no time on the rules and regulations of Cornell—he told me that there was clearly no good reason for me to attend his lectures, do the homework, or take the final examination. He promised to give me a good grade, nonetheless, for I clearly knew the material. Not surprisingly, Cocconi left Cornell soon afterward to be among the first scientists to join the European Center for Nuclear Research (CERN), where he does excellent research and teaches not at all. In the time slot opened by Cocconi's reasonability, I attended the eclectic lectures of Phil Morrison, who taught his students how to think like physicists.

I was, in addition, forced to take an additional course in solid-state physics, or what is now known as condensed matter physics. This was a perfectly sensible requirement. Most of physics, and most physicists by far, do not deal with isolated atoms or individual particles as I do. They deal with material substances. Why is a diamond so hard? they ask. Why does ice expand when it freezes? Why is copper red and a good conductor of electricity? Why is iron, but not zinc, subject to being magnetized? Why does table salt form cubical crystals? This is the physics of innumerable atoms or particles interacting with one another. It is the physics that has led to the transistor and the silicon chip, and to the promise of revolutionary new technologies for a brighter tomorrow.

Sadly, the course I was forced to take was taught by an absolute nudnick in the engineering school, and I was forever alienated from an exciting and important domain of physical science. The problem sets he gave us were exceedingly tedious, the sort that take a lot of time to do but teach you nothing. I neglected the course and ultimately I failed it. I never did learn about the physics of solids and liquids. Not only is there a blank spot in my understanding of physics, there was also a more immediate and direct cost to me. What graduate school would accept a physics student who had *failed* a physics course? It also meant that I did not receive a National Science Foundation fellowship to support my first year of graduate study. Thank God my father's plumbing business was doing well!

My professors and I were agreed on one subject that I and every other physics student had to master: *quantum mechanics*. This is the system of laws describing the motions of material bodies that supplanted the classical mechanics based on Newton's laws. For objects of "reasonable" size—larger than the smallest microorganism—quantum mechanics and Newtonian mechanics yield the same answers in almost every instance. But there have been surprises even there. The phenomenon of *superconductivity*, in which a substance loses all resistance to the passage of an electrical current, is a rare example of a macroscopic system whose behavior can be explained only by quantum mechanics, and not at all by Newtonian physics.

Until very recently, superconductivity was an exotic phenomenon that could be observed only at exceedingly low temperatures. In early 1987 the situation changed dramatically. Researchers at Houston and Alabama produced new materials that became superconducting at the relatively tropical temperature of liquid nitrogen (−196°C, −320.8°F). One day soon room-temperature superconductors may be discovered. Quantum mechanics will have invaded the human environment!

The laws of quantum mechanics first became apparent in the study of the microworld inside the atom. Since its discovery the electron has been known to play a pivotal role in the structure of the atom. In the 1920s, as physicists sought to understand the electron's role, physics underwent a revolution, the revolution of quantum dynamics. Newtonian physics, which regarded the universe as a kind of gigantic mechanical clock, was overthrown and physics took on a completely new look. To many people who are not physicists, modern physics seems to have left the solid world of understandable here-and-now to enter a weird realm of uncertainties and strange, ephemeral particles that have whimsical names and dubious existence. What has actually happened is that physics has gone far beyond the point where ordinary, everyday experiences can provide a kind of human analogy to the things that the physicists are studying. It is a problem of language. The vocabulary and syntax of human language evolved to describe the workings of the everyday world, the world we can see, hear, touch, taste and smell. Words were simply not intended to describe things unimaginably great or incredibly small, far beyond the range of our unaided senses, where the rules of the game are changed. The true language of physics is mathematics.

Even though we know that protons and electrons are so tiny that no microscope can actually see them, we tend to picture them as little round balls. The redoubtable Lord Rutherford, who did as much as any man to uncover the structure of the atom, said, "I was brought up to look at the atom as a nice hard fellow, red or grey in color, according to taste." In Newton's clockwork universe a physicist could—in principle—learn the exact position of every particle in existence and, from that, predict the exact future of all creation. In principle.

The quantum mechanics revolution of the 1920s changed all that, and most people have never forgiven physicists for abandoning the comfortable, predictable world of Newton. Yet the change had to come. Physicists knew that atoms had structure, and soon they discovered that even the particles composing the atoms had structure. Nothing seemed to be *fundamental*: the particles within the atom began to take on the look of those oriental puzzle boxes that contain a smaller box inside each box you open up.

Worse still, uncertainty became the order of the day. Werner Heisenberg's *uncertainty principle* alleged that you could not precisely measure the position of a particle and its velocity at the same time. The more precisely you measure its velocity, the less precisely you know its location, and vice versa. Any dynamical system is *intrinsically* unpredictable. For example, given a small sample of radioactive atoms, there is no way to predict which nucleus will be the first to break down. The universe is not a clockwork orange.

RADIOACTIVE DECAY

All chemical elements past bismuth, the 83rd entry in the periodic table are radioactive. The nuclei of these heavy elements are unstable. They spontaneously emit particles and energy, transforming themselves into lighter elements as they do so. Uranium, for example, undergoes a sequence of fourteen radioactive decays before it becomes a stable isotope of lead.

Natural radioactivity was discovered serendipitously by Henri Becquerel in 1896. Scientists soon found that there are three kinds of radioactivity: alpha, beta and gamma.

In the *alpha* decay process, a heavy nucleus, in its search for a stable configuration, spits out a helium nucleus that consists of two protons and two neutrons. The expelled helium nucleus is called an alpha particle. In this process the weight of the parent nucleus is decreased by four units and its charge increased by two.

In *beta* decay, one of the neutrons in the nucleus becomes a proton. At the same moment an electron and a neutrino are created from the energy made available and are expelled from the nucleus. Electrons of nuclear origin are called beta particles. In this process the weight of the nucleus is decreased by one unit while its charge is increased by one unit.

Often an alpha or beta decay leaves the decaying nucleus in an "excited state" in which the constituent nucleons are wiggling and jiggling about. It falls to its normal quiescent state by emitting a very energetic photon. These photons, which carry far more energy than X-rays, are known as gamma rays. This kind of nuclear de-excitation, in which neither the nuclear weight nor charge is changed, is called *gamma* decay.

Each of the three forms of radioactivity produces radiation that can be hazardous, and each has played a pivotal role in the elucidation of nature's laws.

The occurrence of alpha decay presented physicists with a paradox reminiscent of the famous Zen puzzle: How do you extricate a live duck from a narrow-necked bottle without damaging either the duck or the bottle? Quantum mechanics offers a possible solution. A "quantum fluctuation" could free the duck and leave the bottle intact. Such a happening is a genuine quantum possibility, but its likelihood is infinitesimally remote. Alpha decay is a faithful analog to this question, with the alpha particles playing the role of the duck and the rest of the nucleus the bottle.

Large atomic nuclei contain many neutrons and protons that sometimes gather together to form alpha particles inside the nucleus. According to classical reasoning, these alpha particles simply cannot escape from the nucleus. The electrical forces between the alpha particle and the rest of the nucleus present an absolutely impenetrable barrier. Yet, sometimes the alpha particle *does* escape. It is by just such a mechanism that an ordinary uranium atom decays. The process of alpha decay is a quantum fluctuation. In the case of uranium, it is an exceedingly unlikely event. The average uranium nucleus waits an average of five billion years for things to fluctuate just so. The poor duck would have to wait much, much longer.

This bothers many people. Even Einstein felt repelled by the idea that God plays dice with the universe. Remember, though, that the uncertainty principle comes into obvious effect only when we are looking at things the size of atoms or smaller. We can still predict the behavior of atoms *in the aggregate* well enough to forecast ocean tides, eclipses of the sun and some other macroscopic events with astronomical precision, though not yet, it seems, tomorrow's weather.

Quantum theory emerged when scientists attempted to learn the rules of atomic structure. Rutherford's experiments had revealed the atom to be a small heavy nucleus surrounded by orbiting electrons. The negatively charged electrons were held in orbit by the electrical attraction of the positively charged nucleus. The atom appeared, at first, to be a solar system in miniature, with electromagnetism binding the system together instead of gravity.

The electrons, though, could change their orbits, moving closer or farther away from the nucleus. However, it soon became evident that the electrons could not move indiscriminately from one orbit to another; only certain discrete orbits are permitted, and an electron moves from one orbit to another when it absorbs or emits a *photon*. The photon is the basic unit of electromagnetic energy, including light.

It follows that each kind of atom can absorb only certain wavelengths of light, which are as particular to that atom as a fingerprint is to an individual human being. Sunlight, for example, when it is dissected into its component colors or wavelengths by a prism, displays thousands of dark lines. Each line corresponds to the absorption of light by a specific kind of atom in the cooler gases of the solar atmosphere that lie above the glowing sphere of the sun.

Every one of the lines in the solar spectrum has been associated with a known chemical element. In fact, the lines corresponding to one element were observed in sunlight decades before the element was discovered on earth. That is how helium got its name.

Atomic electrons can be coaxed to hop to bigger and more energetic orbits by the effects of heat, light or electricity. When they fall back down they *emit* light of the same characteristic wavelengths that they normally absorb. Thus, a hot or excited material will give off light of certain discrete wavelengths that depend on its chemical composition. That is why the stove's flame turns yellow when the soup boils over. The table salt in the soup contains sodium, whose dominant spectral lines are bright yellow.

In the laboratory, the spectrum of an unknown material can be studied carefully and its constituent elements thereby identified. *Spectroscopy*, as the procedure is known, is a powerful tool for chemical analysis. Several of the chemical elements were discovered by first seeing their characteristic bright lines. Rubidium is so called because its characteristic photons lie in the red, and cesium comes from the Latin word for "sky blue."

None of this behavior could be understood in terms of the classical physics of Newton and Maxwell. The atom seemed to follow a different set of laws than any other physical system yet encountered. Niels Bohr, the founder of quantum theory, proposed an ad hoc set of rules and regulations that electrons should follow in order for their properties to be as they were observed. This first primitive quantum theory was almost mystical in nature. The rules were simple, but seemingly quite arbitrary. They did not seem to follow from any coherent and consistent theory, but they worked.

What Bohr had done was to write a sort of guidebook for the behavior of the atom. If *this* happens, then expect *that* result. It was all based on observation, with no real understanding of *why* the atom behaved as it did. While Bohr's "guidebook" helped physicists to deal with the atom's behavior, what was really needed was a logical system that would allow physicists to truly understand why the atom behaved as it did, so that they could successfully predict not only the atom's observed behavior but new phenomena that had not yet been observed.

Physicists call such a logical system a *theory*. This word is much misunderstood among nonscientists. When a physicist talks about a theory, he does not mean a hunch, guess, or unproven hypothesis. He means a logical system of ideas that ties together a large number of observations of the real world into a coherent and understandable pattern.

A theory of quantum mechanics was finally devised in 1926 by Heisenberg, Erwin Schroedinger and others. It was a mathematical system from which Bohr's quantum rules could be logically deduced and in which the uncertainty principle was implicit.

One quantum notion that mystifies the novice is "wave-particle duality." Does light consist of a beam of particles or is it a wave phenomenon? The question is hundreds of years old. Newton thought light was probably a stream of particles. Maxwell seemed to answer the problem decisively by showing light to be an electromagnetic wave. Yet Einstein, in 1905, demonstrated that under some circumstances light behaves as if it were a beam of discrete particles, which are now called photons.

Einstein examined the photoelectric effect, which is now so well understood that it is used to open the doors of supermarkets and elevators when you step through a beam of light. In 1905 it was still a mystery.

When a beam of light strikes a metal surface, it causes the ejection of electrons from the atoms of the metal. Einstein showed that the process can be correctly described in terms of *particles* of light (photons) hitting electrons and driving them out of the metal, somewhat like a billiard shot. The effect simply cannot be understood in terms of light behaving like an electromagnetic wave. This did not mean that Maxwell's theory of light was all wet. Many processes, like refraction and diffraction, can be understood *only* in terms of light behaving like a wave.

This schizoid behavior of light was symptomatic of the failure of classical theory. Einstein was awarded the 1921 Nobel Prize for "his discovery of the law of the photoelectric effect."

The wave-particle duality of light was a key step in the development of quantum mechanics, which was, ironically, a theory Einstein could never accept. To the intrinsic randomness of quantum phenomena, Einstein responded, "God does not play dice." For once, he was wrong.

Light consists neither of particles nor of waves. Rather, it obeys the laws of quantum mechanics and is sometimes wavelike and sometimes particulate. Mere words cannot suffice to describe a light beam correctly. Neither category is quite right.

The electron, and any other "particle" as well, suffers the same fate. Sometimes the electron behaves like a wave. The electron microscope, our most powerful tool for viewing the submicroscopic world directly, depends upon the wavelike aspects of the electron. The wave nature of the electron was first conjectured by the French physicist Prince Louis de Broglie in 1923, and it was verified in the laboratory soon afterward.

THE WAVELIKE NATURE OF THE ELECTRON

Imagine the following idealized experiment, illustrated below.

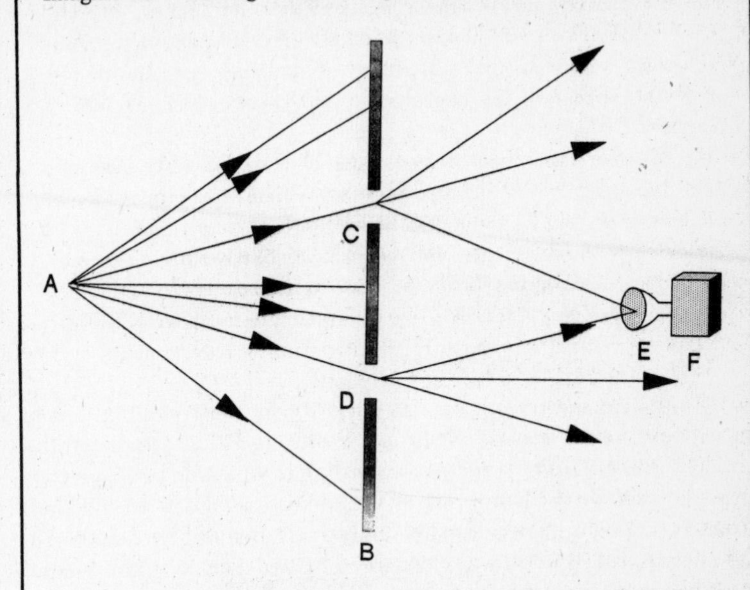

A is a source of electrons, perhaps a hot filament. B is a thin sheet of material that is opaque to electrons. In it are two small holes, C and D. The electrons will pass through the holes, being deflected in many directions. (This itself is an essentially wavelike phenomenon, but it is not what we are focusing upon.)

Downstream, an electron detector, E, is placed. Each time an electron hits the detector, an electronic device, F, produces an audible click. Suppose that the device clicks at an average rate of once per second.

Now, place your finger over hole C, thus making it impossible for any electrons at all to pass through that hole. If electrons were particles, the click rate of the detector would necessarily go down, since only those electrons passing through D could arrive at E.

However, the actual result depends upon the precise position of the detector. At some locations the click rate will decrease. At other locations the click rate will *increase* when one of the holes is obstructed! This is impossible, from a particle point of view.

The mystery is explained in terms of the wavelike nature of the electron. When both C and D are unobstructed, the detector sees a sum of two waves, one coming from C and the other from D. At some sites the two waves will add up, at others they will partially or completely cancel one another out. At any of the latter sites, closing one of the holes undoes the cancellation and causes the click rate to increase.

This one experiment demonstrates both the wavelike and particle-like nature of the electron at the same time. The propagation of the electrons from A, through C and D, to E is clearly wavelike. In the detector, however, the electron behaves like a particle: to each particulate electron the device produces a discrete click.

Paradox? Not at all. We live in a quantum-mechanical world.

Another quantum-mechanical curiosity we must mention has to do with systems of several electrons. Wolfgang Pauli introduced the notion of the *exclusion principle*, according to which no two electrons may play exactly the same role at the same time. Each orbital shell around any atom's nucleus can hold only a certain number of electrons: the closest orbit is filled when occupied by two electrons, the second orbit has room for eight electrons, and so on. The exclusion principle explains why atoms do not collapse, with the electrons spiraling down into the smallest possible orbit. No more than two electrons can be in that orbit at any given time.

THE ATOMIC SHELL GAME

The atomic number, Z, denotes the place to which a chemical element is assigned in the periodic table. It also signifies the magnitude of the electric charge of the atomic nucleus. Thus, an electrically neutral atom with atomic number Z consists of a nucleus surrounded by Z orbiting electrons.

In the figure below, we indicate the structure of a few of the chemical elements, with the heavy nucleus in the center. The several electronic shells are shown as concentric circles upon which the requisite number of electrons are drawn.

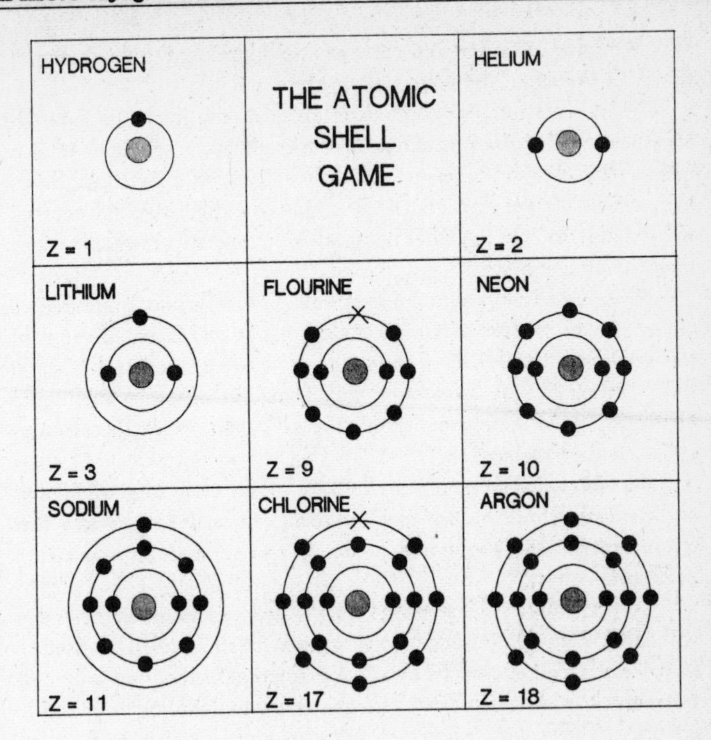

One can begin to see the periodic table emerge. Neon and argon have completely filled shells. They wish neither to gain nor lose electrons and are consequently chemically inert. The elements fluorine and chlorine each lack a single electron in their outer shells. They would do almost anything to steal an electron so that they could complete their electronic shells. Consequently, they are exceedingly reactive elements. On the other hand, lithium and sodium have each just one electron in their outer shells, and would like to find other atoms that could borrow an extra electron. They, too, are very reactive elements.

One atom of sodium could (and does) combine with one atom of chlorine to form a stable and neutral compound: sodium chloride, common table salt.

Thus, the rules of chemistry are seen to emerge from our underlying theory of atomic structure. By the mid-1920s, the triumphs of the periodic table had been explained—by physicists.

We have so far ignored one essential ingredient to the quantum theory of the atom: the notion of electron spin.

Much as a pitched baseball is given a spin about its axis, so also does the electron spin about. However, the microscopic quantum-mechanical electron behaves very differently from a baseball. Baseballs can be old or new, clean or dirty, and can differ from one another in myriad ways. On the other hand, all electrons are absolutely identical to one another.

There is a subtler difference as well. Baseballs may spin rapidly, slowly, or in the case of a knuckleball, not at all. Every electron in the universe (about 10^{80} of them!) is spinning at exactly the same rate. The magnitude of electron spin is an intrinsic and immutable characteristic of the electron. Only the axis about which the electron spins can be changed.

Its spin axis can point up, down, or in any of an infinite number of directions. However, any electron at all can be described as a *sum* of an electron whose spin axis points up and an electron whose spin axis points down.

At this point, the reader who is not a licensed quantum mechanic will balk. What is being added together to make this sum? The electron is described mathematically by a "wave function" rather than a precise trajectory in space. The wave function is a solution to Schroedinger's equation. The wave function for a sideways-spinning electron is gotten by adding together the wave functions corresponding to up-spinning and down-spinning electrons.

While my words fail to describe the elegance of the underlying theory, the essential point is this: There are only two *independent* spin orientations of the electron. That is why just two electrons can snuggle into the innermost shell of an atom, one spinning up and the other down.

In 1926 many unconventional and seemingly conflicting ideas converged into a coherent theory: Bohr's quantum rules, de Broglie's wave-particle duality, Heisenberg's uncertainty principle, the notion of electron spin, Pauli's exclusion principle, the Schroedinger equation, and more. Quantum mechanics emerged as a magnificent synthesis that replaced classical mechanics as a theory of the microworld.

More precisely, quantum mechanics is a *generalization* of the classical theory that works so well for macroscopic physics and astronomy. Any real advance in fundamental theory simply must encompass what is already known, as well as plunging onward to explain something more.

The mysterious success of Mendeleev's periodic table was completely understood and deduced. Thanks to quantum mechanics, scientists could see *why* the different kinds of atoms that composed the chemical elements behaved the way they did. Physics had explained, in principle, the workings of chemistry. A science that had been considered "fundamental" was seen not to be so fundamental at all, but rather a branch of applied physics.

Just as chemistry is ultimately reducible to physics, so is biology ultimately reducible to chemistry. This is the result of Watson and Crick's discovery of the double helix, the structure of DNA, which is the molecule of heredity. Nonetheless, chemistry and biology are still flourishing sciences. Their practitioners will produce for us better and better materials for skis and tennis rackets, and perhaps cures for cancer and AIDS as well. The basic underlying rules of chemistry and biology are well understood. Chemical and biological forces at their most microscopic level result from the electromagnetic forces between electrons and their nuclei, behaving in accordance with precise quantum-mechanical laws. We are essentially electromagnetic creatures inhabiting an electromagnetic planet.

With the atom well understood, physicists turned their attention to more exotic phenomena. For example, they studied our strange visitors from outer space known as cosmic rays. As we have seen earlier, these are not "rays" at all, but very high-energy particles, mostly protons, that flit about the stars at velocities very close to the speed of light. When such a particle hits an atom in the earth's atmosphere it produces a shower of particles, the study of which has provided food for thought for generations of physicists. Physicists also studied carefully the process of radioactivity, in which nuclei change their identity at the same time that new particles are created out of the energy being liberated.

Modern elementary-particle physics is founded upon the two pillars of quantum mechanics and relativity. I have made little mention of relativity so far because, while the atom is very much a quantum system, it is not very relativistic at all. Relativity becomes important only when velocities become comparable to the speed of light. Electrons in atoms move about rather slowly, at a mere one percent of light speed. (The earth in its orbit moves even more slowly: a mere hundredth of a percent of the speed of light.) Thus it is that a satisfactory description of the atom can be obtained without Einstein's revolutionary theory. On the other hand, the study of cosmic rays and of radioactivity involves rapidly moving "relativistic" particles. For their understanding it was necessary to produce a theory that incorporated both relativity and quantum mechanics.

To begin with, a mathematical framework had to be developed that could deal with electrons of any energy, even those moving at close to light speeds. This challenge was met by Paul Dirac in 1928. His relativistic equation for the electron is known as the Dirac equation. It was immediately clear to Dirac that his equation had two kinds of solutions. One set of solutions offered a correct and consistent description of how an electron should behave in an electromagnetic field, including the newly discovered phenomenon of electron spin. The other solutions were puzzling—they called for particles carrying negative energy, a concept that made no sense at all.

Yet no one knew how powerful a tool mathematics can be better than Dirac. If his equation predicted the existence of something that had never been observed in nature, he was confident that the equation was right and the observations were lacking. Dirac stated the ultimate faith of the theoretician when he wrote, "It is more important to have beauty in one's equations than to have them fit experiment."

We come to the point of my high school essay, "On the Nature of Nothing." Dirac made use of the Pauli principle by assuming that all the negative-energy states in his equation were occupied by unobservable electrons. To Dirac, the vacuum was not empty at all; it was a sea made up of multitudes of invisible negative-energy electrons. If, somehow, an electron in the sea were knocked out (say, by collision with a photon), then a hole would be produced, a hole that would behave like a particle with an electric charge opposite to that of the electron. At first Dirac believed that the hole could be the proton, and that his equation described the two then-known elementary particles: electrons and protons.

Soon, however, Dirac realized that there was a complete symmetry in his equation between positive-energy states and holes in the negative-energy sea. The positively charged particle had to have exactly the same mass as the negatively charged electron. It simply could not be the proton, which is about two thousand times heavier than the electron. By 1931 Dirac was reluctantly forced to predict the existence of an entirely new kind of particle, one just like the electron but oppositely charged. The new particle was discovered in that wonderful year of 1932 in the debris of cosmic-ray collisions, and soon afterward, as a product of some radioactive decays. It was called the *positron*, and it was the first observed form of antimatter. It took Dirac a few years to comprehend the true significance of his equation and realize that the mathematics *demanded* the existence of antimatter. This led him to remark that his equation was more intelligent than its author.

Not only does the electron have an antiparticle, but so does every particle or system of particles. When a particle and its antiparticle meet, they annihilate each other on contact, producing an instantaneous burst of energy. Matter is converted to energy with 100 percent efficiency in particle-antiparticle annihilation. The antiproton (antiparticle of the proton) was not observed in the laboratory until 1955, when a particle accelerator of sufficient energy to create one was built at Berkeley.

Antimatter seemed like the stuff of science fiction. (Indeed, the good ship *Enterprise* of *Star Trek* fame is propelled by an "antimatter drive.") It was a surprise that delighted physicist Harold P. Furth so much that he composed a poem after listening to a lecture by Edward Teller on the subject of antimatter. The poem is titled "The Perils of Modern Living":

Well up above the tropostrata
There is a region stark and stellar
Where, on a streak of anti-matter,
Lived Dr. Edward Anti-Teller.

Remote from Fusion's origin,
He lived unguessed and unawares
With all his antikith and kin,
And kept macassars on his chairs.

One morning, idling by the sea,
He spied a tin of monstrous girth
That bore three letters: A.E.C.
Out stepped a visitor from Earth.

Then, shouting gladly o'er the sands,
Met two who in their alien ways
Were like as lentils. Their right hands
Clasped, and the rest was gamma rays.

Many legends illuminate the lives of great scientists. Dirac was married to the sister of Nobel Laureate Eugene Wigner, who was to play a key role in my decision to attend Harvard Graduate School. Wigner always modestly presented Dirac as "my famous brother-in-law." He is the source of a famous physics riddle, "What is the question whose answer is 9W?" The correct response is, "Professor Wigner, do you spell your name with a V?" Well, it's famous among physicists.

Dirac's greatest work was done during the many years he spent at Cambridge University. His textbook on quantum mechanics, written in the 1930s, is still useful today. He was often asked why he simply read from his book, instead of giving "live" lectures to his classes as most other professors do. His reply: "Because the book is perfect. I could not say anything more or better than I have written."

Incidentally, Dirac's book was rejected for publication by the Cambridge University Press, a permanent embarrassment to that prestigious institution, although not too rare an occurrence in book publishing. But that is probably not the reason that Dirac spent his last years at Florida State University, near the location where Tarzan movies were filmed. My thesis supervisor, Julian Schwinger, and my classmate and colleague, Steven Weinberg, have repaired to southern California and Texas, respectively. It seems that many great physicists, as well as nonphysicists, choose to spend their later years in a sunny clime.

Some physicists are born teachers. Dirac was not. In fact, he showed little interest in his doctoral students. "Have you ever *had* a research student?" he was once asked. According to the legend, his reply was, "I think I had one once . . . but he died." Nor was he much of a social butterfly. Seated next to the irrepressible Richard Feynman at a dinner party one evening, Dirac remained taciturn. Feynman tried to bring him out by asking him how excited he felt upon discovering the famous Dirac equation. "Good," said Dirac. Then, after a long pause, he added, "Are you working on an equation, too, Mr. Feynman?"

In the 1930s, while I was busy growing up in upper Manhattan, physicists realized that Dirac's theory of electromagnetic interactions between electrons and positrons was beset by difficulties. In its first approximation, the theory gave plausible answers that were in good agreement with the experimental data then available. For example, the electron behaves like a small magnet. Dirac's prediction of the strength of this magnet was correct to two-decimal-point precision.

As experimental data became more precise, it became clear that the theory did not provide a perfectly accurate description of nature. In 1947 two quantum states of hydrogen that should have had *exactly* the same energy according to the Dirac theory were shown to have a measurable difference. The gauntlet had been thrown. Could the exact value of this "Lamb shift" be deduced from the theory of relativistic quantum mechanics?

The challenge was linked to a long-standing disorder of the quantum theory of electromagnetism. In an atom the electron's motion is determined by the electromagnetic field that it encounters, which includes not only the field of the nucleus but the electron's very own "self-field." In attempting to take into account the effect of the electron upon itself, physicists found that the theory went completely haywire and produced wild answers instead of plausible small corrections. Something was very wrong.

Soon after the Lamb shift was announced at the Shelter Island conference in 1947, Julian Schwinger and Richard Feynman developed the modern theory of *quantum electrodynamics*. (You see, Feynman *was* working on an equation!) Related work had been done during World War II by Sinitiro Tomonaga in Japan. All three scientists shared the Nobel Prize in 1965 for the development of this theory. The crazy results of the older theory were replaced by perfectly sensible predictions that could be tested by laboratory experiments. This procedure, called renormalization, was never accepted by Dirac as a friendly amendment to his theory. Moreover, in recent years, Schwinger has renounced his own brainchild in favor of his latest creation, source theory, or "sorcery," as he sometimes puts it. Feynman himself is not entirely smug about modern quantum theory. In what surely must be poetry, he writes:

> We have always had a great deal of difficulty
> understanding the world view
> that quantum mechanics represents.
> At least I do,
> because I'm an old enough man
> that I haven't got to the point
> that this stuff is obvious to me.
> Okay, I still get nervous with it . . .
> You know how it always is,
> every new idea,

it takes a generation or two
until it becomes obvious
that there's no real problem.
I cannot define the real problem,
therefore I suspect there's no real problem,
but I'm not sure
there's no real problem.

Despite all this quibbling, quantum electrodynamics, or QED as it is familiarly known, has become one of the most successful theoretical accomplishments in the history of science. Theoretical calculations, using giant computers, have produced more and more precise *computations* of atomic and electronic properties. At the same time, parallel developments in experimental technology have led to more and more accurate *measurements* of these same quantities. The calculations and the measurements are in perfect agreement with one another. In the case of the magnetic strength of the electron, experiment and theory have converged upon a value 1.001596524 times the original estimate of Dirac. This means that the theory is accurate in its predictions to at least ten decimal places.

By the time I had become a senior at Cornell, QED had become the very paradigm of a successful theory. Some theorists dreamed of building an equally successful quantum theory of gravitational force, and so they still dream today, three decades later. Others hoped that QED could serve as a guide to the construction of a correct and consistent theory of all subatomic phenomena. This is exactly what has happened, and it is the gist of the tale I am spinning.

Richard Feynman, who was renowned even before winning the 1965 Nobel Prize as a dazzling lecturer and daring bongo drum player, taught for several years at Cornell before accepting a professorship at CalTech, where he remained until his death in 1988. Although we overlapped for one year, I did not meet him at Cornell. As a freshman, I had never even heard of him. I began to learn of the Feynman legend when I was a senior enrolled in an advanced graduate course in quantum field theory taught by a young instructor, Sam Schweber, who now teaches the history of science at Brandeis University.

In the spring of 1954 the only quantum field theory that worked was QED. It offered a seemingly perfect description of some of the particles and forces of nature. It was a theory that dealt with electrons, positrons, and their electromagnetic interactions.

The Pictorial Language of Quantum Electronics

I. Being

Quantum electrodynamics (QED) deals with three kinds of particles: electrons, positrons and photons. The following diagrams indicate the behavior of such particles when they are isolated and free of interaction with other bodies.

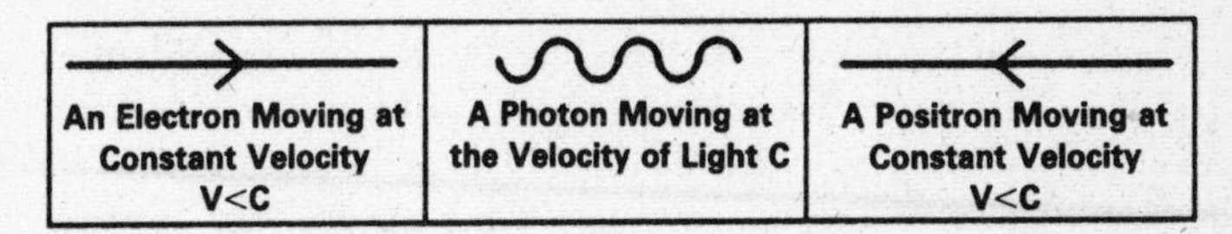

In each case, the particle may be imagined as proceeding from a source to a detector. Its state of motion is completely defined by the magnitude and direction of its *momentum* and by the orientation of its intrinsic *spin*. Nothing happens until or unless several of these particles encounter one another.

II. Becoming

There are six fundamental processes by which electrons, positrons and photons may interact with one another:

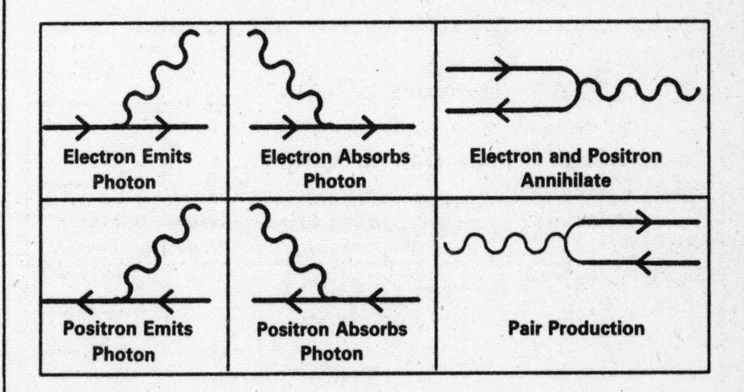

Note that each of these six processes can be obtained from a single diagram by the appropriate bending of legs of the figure:

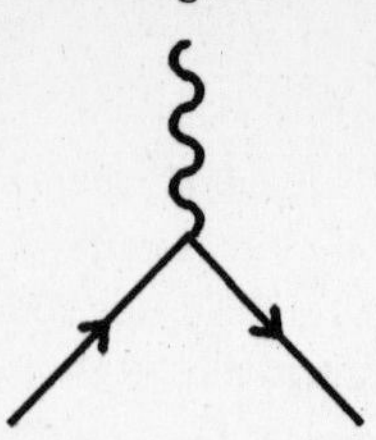

The Single Unified Act of Becoming

Of course, bending a line converts an entering electron into a departing positron. That is why it is sometimes said that a positron is simply an electron moving backwards in time. None of the six avatars of the fundamental act of becoming may take place as such. For example, an isolated electron *cannot* simply emit a photon. Nor can a single photon suddenly convert itself into an electron-positron pair. Such processes would contradict the laws of conservation of momentum and energy.

III. QED Processes

Electrons, positrons and photons do interact nontrivially through *iterations* of the fundamental act. Energy and momentum are conserved overall. Transient violations of conservation laws are permitted by the quantum-mechanical uncertainty principle. Here are some examples of allowed processes:

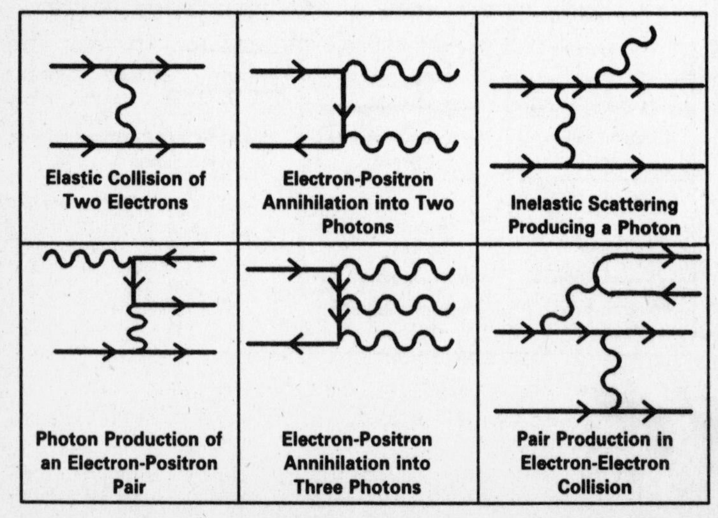

Elastic Collision of Two Electrons | Electron-Positron Annihilation into Two Photons | Inelastic Scattering Producing a Photon

Photon Production of an Electron-Positron Pair | Electron-Positron Annihilation into Three Photons | Pair Production in Electron-Electron Collision

These and more complex electromagnetic phenomena result from multiple iterations of the fundamental act. QED is a paradigm for the construction of more ambitious theories involving other particles and other forces.

There I was, the only undergraduate student in a course whose material was so new that the instructor himself barely could follow it. Our "textbook" was a set of mimeographed notes from a course given in 1951 at Cornell by Freeman J. Dyson, of the Institute for Advanced Study at Princeton. According to its introduction, the text would "discuss how one can make a relativistic quantum theory . . . using the new methods of Feynman and Schwinger."

Finally I had found a course that really was too advanced for me at the time. Schweber could grade the graduate students "pass" or "fail." But I was an undergrad and had to be given a numerical grade, according to the immutable Cornell protocol. "How would 85 be?" Sam asked me at the end of the course. "Not good enough," said I, angling for something closer to an A. Sam then asked me a series of reasonable questions about the material of the course, *none* of which I could answer correctly. "How would an 85 be?" he asked once more.

I settled for an 85.

I applied to several graduate schools, including Princeton and Harvard. Once again, I visited Princeton and was interviewed by two members of the physics department, Arthur Wightman and Eugene Wigner. After the interview, and in my presence, Wigner said to Wightman that I knew the language of physics but not its content. Since that time I have rarely had occasion to visit Princeton University.

My graduate school applications worked out exactly the opposite of my college applications. This time I was rejected by Princeton and accepted by Harvard. Two other physics majors from my Cornell class joined me at Harvard: Danny Kleitman and David Falk (now a professor at Maryland). Steve Weinberg went for a year to the Niels Bohr Institute in Copenhagen, thence to Princeton. In the summer of 1954 my brother Sam bought me my first new car, a blue Ford Crestline sedan, and I was ready to set forth on my professional career.

4 Veritas

Before the Second World War, Harvard's physics department was quaint, quiet and very good. It graduated about five Ph.D. students each year, compared to today's twenty-five. Edwin Kemble was responsible for bringing the new European discipline of quantum mechanics to America—he was our first quantum physicist. He hired two young and prominent American theoretical physicists in the 1930s, the midwesterners Wendell Furry and John Van Vleck, both of whom were to be among my finest teachers at Harvard. Furry delighted in mathematical puzzles, and we would spend many wonderful hours together solving them.

Van Vleck was forever fascinated by railroad trains (all of whose schedules he knew by heart) and by baseball (he once explained to me the feasibility of an unassisted triple play). His work on the magnetic properties of matter earned him the Nobel Prize in 1977—late, but not too late. He died in 1980, an active researcher to the end.

Prewar Harvard was into elementary-particle physics as well. Ken Bainbridge was the world's expert at "mass spectroscopy," a technique of weighing and separating nuclear isotopes. Ken was to

FURRY'S BRAIN TEASERS

Physicists love puzzles, and for many years Wendell Furry was the puzzle master of Harvard. An old favorite involves thirty-one dominoes, each of which is the size of two adjacent squares of a chessboard. Can you cover the entire chessboard except for two opposite corners? Most puzzlers know the answer is no. Each domino covers one white square and one black one. Since opposite corners are of the same color, the deed cannot be done.

Furry invented several wild variations of this puzzle. Suppose you have twenty-one "triominoes," each of which can cover three squares in a row. Can you cover all but one square of the chessboard? And which square can be skipped? The diagram below shows that the answer to the first question is yes.

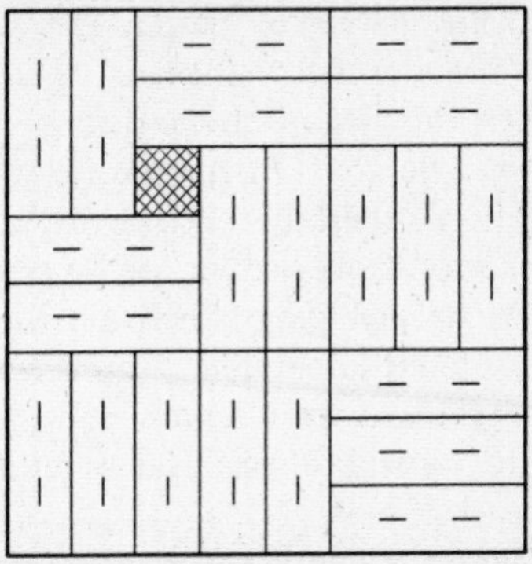

Covering the chessboard with linear triominoes.

It is more difficult to prove that the omitted square must be one of four possible choices.

Consider the same problem with bent triominoes that are L-shaped. The pattern shown below demonstrates that it is possible to cover all but one square of the chessboard with twenty-one of these pieces. This time, however, it is possible to find a solution skipping any square at all.

Covering the chessboard with bent triominoes.

Furry even invented a three-dimensional variation of this problem. Suppose you have 27 bricks, each measuring $2 \times 4 \times 8$ inches. Can you put them together to form a one-foot cube? Good luck!

play a key role in the Manhattan Project at Los Alamos. Jabez Curry Street, a southerner specializing in cosmic rays, was the codiscoverer of the mysterious muon in 1937. By the end of 1939 these guys had built the first Harvard cyclotron for the purpose of carrying out their pure researches into the nature of matter. It was one of only forty such instruments in the world at that time.

The war changed the face of Harvard forever. Many of its scientists joined the war effort at Los Alamos (to which our cyclotron and Ken Bainbridge were secretly sent); at the MIT Radiation Laboratory, where radar was developed for use on ships and planes; and at Harvard-based efforts including the Radio Research Laboratory and the Underwater Sound Laboratory. Those who remained at the Harvard physics department were largely occupied by the school's successful attempt to provide specialized scientific training to our military personnel.

Here is an intriguing example of "reverse spin-off." By working on these very applied military projects, physicists developed skills they didn't know they had.

Julian Schwinger was a precocious, brilliant and chain-smoking nuclear physicist doing his wartime stint at the Radiation Laboratory. At the end of the war he joined the Harvard faculty, where he remained until 1972. At the Rad Lab he was engaged in forefront problems of electrical engineering: how to generate and direct a beam of microwave radiation, and how to detect and interpret the faint reflections from distant aircraft. At first, such problems of radar engineering seemed far removed from his own specialty of nuclear physics. However, he soon began to think of nuclear physics in the language of electrical engineering, an approach that led him to important discoveries, including the so-called effective-range approximation now found in all textbooks on nuclear science.

Research on microwaves led Schwinger to consider the possibility of large electron accelerators in which intense beams of electrons could be accelerated to huge energies. Such a beam, he realized, would itself be an intense source of electromagnetic radiation. In a remarkable tour de force, he computed the exact nature of this "synchrotron radiation." Today, all over the world, electron accelerators are built for the sole purpose of generating this kind of high-frequency light. "Synchrotron light sources" play an essential role both in pure science and its commercial applications.

Schwinger was one of the founders of relativistic quantum theory, or quantum electrodynamics, a self-consistent picture of the interaction of electrons with light that forms the logical basis upon which today's understanding of elementary-particle physics depends. He writes that his interest in the problem began with his work on radar, when he realized that the behavior of an electron depends not only upon the electromagnetic fields that are imposed on it, but also upon the fields generated by the electron itself. The understanding of the self-interaction of an electron was the key step in formulating his theory. Were it not for his war-related activities, Schwinger might never have become one of the World's Greatest Physicists.

He was to become my research supervisor at Harvard.

Ed Purcell and Bob Pound, my colleagues at Harvard today, worked together at the Radiation Laboratory during the war. As a direct consequence of their experience with radar, they were to become the inventors of nuclear magnetic resonance techniques, which have become the highest of today's high technologies. NMR offers, for example, noninvasive imaging procedures for medical diagnoses. The NMR scan is somewhat like the CAT scan, but it has a very great advantage over its predecessor. No X-rays are involved and the patient is not exposed to any potentially hazardous radiations. In addition, the pictures obtained with an NMR scanner are more detailed and often more useful than those obtained with a CAT scanner.

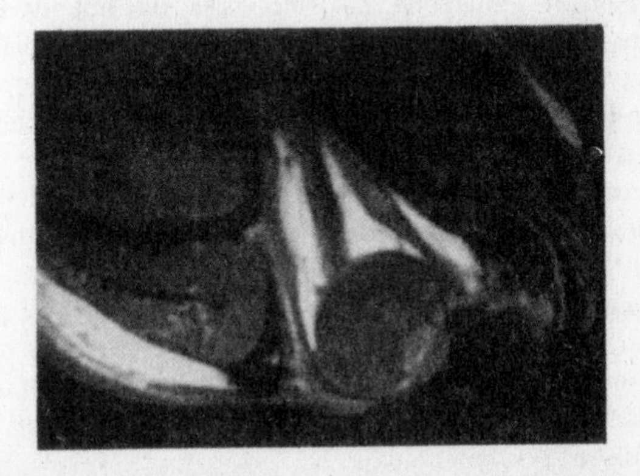

"Image of the eye showing globe, lens, optic nerve, and muscles which turn the eye. The NMR technology has two great advantages over the CAT scan: Soft tissue structures are visualized, and the patient is not exposed to any potentially harmful radiation." William A. Edelstein, et al., Journal of the American Medical Association. Volume 253, page 828, (1985).

Of the influence of his war work, Purcell has said:

> [Our experience at the Radiation Laboratory] provided us with the tools, not merely the hardware, but with a basic understanding . . . of what you have to do to detect a signal in noise. So in all future experiments, whether by radio astronomers or microwave spectroscopists, people knew how to deal with the problem of the ultimate sensitivity of their apparatus. . . .

The war work on radar led directly to the explosive postwar development of radio astronomy. Without it we may never have learned of those fabulous denizens of deep space—pulsars, quasars and black holes.

By the early 1950s, Harvard had become one of the leading centers of physics in the world. It had won its first Nobel Prize in physics in 1946, awarded to Percy Bridgman for his studies of the behavior of matter under extremely great pressures. The second, in 1952, was earned by Ed Purcell for his studies of the magnetism of atomic nuclei.

Professors Bob Pound and Norman Ramsey, together with Ed, formed an incomparable group for the precision study of atomic processes. For the study of sub-atomic physics, Harvard had built its second cyclotron, a machine that still works, but not any longer as a research tool. It has become a unique facility for the medical treatment of certain types of tumors.

To be honest, when I chose Harvard as my school for graduate studies, I was totally unaware of its grand history. I came to Harvard largely because Harvard admitted me—Princeton had been more choosy. What I knew of the place was simply this: (1) It had a snotty reputation. (2) The undergraduates threw great parties. (3) Harvard was near Boston, not quite New York but a lot more of a city than Ithaca, and a place where some of my father's brothers had settled as young immigrants. (4) Harvard was Julian Schwinger's stomping ground, and Schwinger was as godlike to me as Michael Jackson is to my children.

So far, Harvard scientists have won eight Nobel awards in physics, and we are going quite strong. Since joining the university I have come to know more about Harvard, and to love it, warts and all.

I arrived in Cambridge in the autumn of 1954 and settled into a graduate dormitory (the ancient and venerable Perkins Hall) with a roommate who has since become a Silicon Valley scientist. The first thing I discovered was the absence of a central university facility for rest and recreation. There was nothing like the "game room" of Cornell's Willard Strait Hall, no convenient place to shoot pool or play poker. But there were compensating amenities. There were, for example, restaurants in the area serving concoctions more appealing than the Salisbury steaks and Welsh rarebits of Ithacan high cuisine. Wah Yuen's, in Boston's tiny Chinatown, was my first introduction to "real" Chinese food, and the *baba ganooze* of the Red Fez is still second only to my own. Not to speak of the noble Maine lobster.

High Culture accompanies Good Food. I soon inherited from a fellow physics graduate student (Bill Wright, recently the chairman of the physics department at the University of Cincinnati) a pair of front-row center balcony seats to the Celebrity Series at Symphony Hall, whose magnificent acoustics were Harvard's gift to Boston. Wallace Sabine, a member of the physics department in the early twentieth century, founded the science of architectural acoustics. He designed for Boston a concert hall that could rival the best in Europe.

"Mona" and I enjoyed many a fine concert there. There was a real Mona, but here her name signifies a string of girlfriends I knew as a graduate student. New ones I met at Radcliffe's "Jolly-Ups," as dances for newcomers were known. Old ones I imported from Cornell, where I had met them during my undergrad years.

The graduate curriculum was little different then from today. For about two years of full-time study, lecture courses were required. Competence had to be demonstrated in four fields of physics. For the theoretical physicists among us, a laboratory course was required to ensure that we would not become so abstruse and highfalutin as to forget the origins of our science in the behavior of the real world. Moreover, a reading knowledge of two foreign languages was required.

It had not become clear to the Harvard authorities that English and only English is the language of physics. China and Russia, over the past decades, have done more than we to ensure the supremacy of our language. Thank God, since I have never succeeded in learning a "foreign" language, though I have studied from time to time, French, Russian, Danish and Italian. English monoglot I proudly be.

With the coursework done, it was time to begin one's doctoral research. There were no general written exams. A cursory oral examination allowed the professors to determine whether a candidate was adequately prepared. The next step was called "Find an Advisor." The students would scurry about to find a professor who would undertake to become one's research supervisor. It was a little bit like a marriage. Professor and student both willingly entered into a covenant " 'til graduation do you part." Sometimes a student and professor would fall out, but such divorces were shabby and shameful affairs. Finding an advisor was a solemn step.

Then the hard work began. The student, with the advice and assistance of his supervisor, had to initiate and complete a piece of useful and meaningful original research. This could take years. Harvard chose and still chooses its graduate students wisely. Almost all of them succeed in doing substantial research and earn Harvard doctorates. The average period of study is five years, although I know of one student who took twelve and one who did it in only two. I settled for four.

Only a very few of our students fail to complete their theses. They are "sent down" from Harvard with a Master's degree as a consolation prize. Even some of those who didn't get their degrees managed a modicum of success. Edwin H. Land went out and founded Polaroid Corporation, for example.

I had already taken several graduate-level courses at Cornell. Nonetheless, I enrolled in some of these again at Harvard. Different professors have different perspectives, and physics—like good music—improves with repetition.

Because the material was already somewhat familiar to me, I had the time to take advantage of other Harvard offerings. I listened to a fascinating course on the writings of Proust, Joyce and Mann taught by the brilliant Professor Harry Levin, and I attended a marvelous series of lectures and concerts on classic Japanese music. I was catching up on the liberal arts education I had scorned at Cornell. Almost against my will I was joining the ranks of educated men and women (as it is solemnly pronounced at Harvard commencements).

Among my more professional obligations were courses corresponding to each of the four fundamental forces of nature: electromagnetism, gravity, and the strong and weak nuclear forces. Each of my professors specialized in the one or the other of these forces, at least for a time. At Harvard, I enjoyed the opportunity to study with a master of each.

A course in electromagnetism was inevitable, as it is a most important force. Its primary effect is to hold the atom together. The electrons and the nucleus have opposite electrical charges, and are attracted to one another. It is this attraction that keeps the electrons in orbit around the nucleus, much as gravity keeps a satellite in orbit around the earth. Electromagnetic attraction also acts between atoms and gives rise to chemical interactions. Even the forces between molecules that give matter its form and variety are electromagnetic. Everything that we sense comes ultimately from the electromagnetic force.

I have often said from the podium that although it is gravity that holds my feet to the ground, it is the electromagnetic force that stops me from falling through the ground. Electromagnetism binds the atoms together and puts a solid floor beneath my feet.

This brings up the heavenly force of gravity, my second course. It is the force of gravity that makes the oceans and the atmosphere cling lovingly to our planet. Earth's revolutionary flight about the sun and the motion of our solar system through the galaxy are caused by the force of gravity. All of the motions of the heavens are under its sway.

The course in general relativity—which is to say, Einstein's theory of gravitation—was taught by Professor Paul C. Martin, a Stuyvesant High School graduate who had been an undergraduate at Harvard and had continued as a graduate student of Schwinger's. It was the first course that Paul had ever taught. Since that time, Paul has become a very distinguished physicist at Harvard. He is today the dean of the Division of Applied Sciences, Harvard's equivalent of a school of engineering.

Now, *general* relativity is a very different kettle of fish from *special* relativity. Both are essential tools of the modern physicist. Special relativity is rather like a screwdriver—we use it all the time in many disparate realms of physics. General relativity is more like a highly specialized tool used only under very unusual circumstances. It was of paramount importance during the brief course of the big bang that was the birth of our universe. It governs the behavior of black holes—those ultimately collapsed stars that may exist in nature but have never been seen directly. As for the phenomena studied in laboratories on earth, the effects of general relativity are usually far too small to be detectable. This is because general relativity is concerned exclusively with the weakest of the four forces, gravity. Of course, gravity does exist, and a complete theory of physics simply must deal with it as well as with the remaining three forces. However, the incorporation of general relativity into a framework including quantum mechanics remains one of the great unsolved problems of theoretical physics.

Einstein's 1905 *special* theory of relativity has to do with the relationships among measurements made by uniformly moving observers: that is, measurements made by observers who are moving at a steady speed and direction, as if they were in automobiles driving along an absolutely straight and level freeway on cruise control.

Since Galileo's time, physicists had tacitly assumed that the laws of physics deduced by experiments must be the same whether the experiments were done in a laboratory building or a speeding train, or on a spaceship, or on the planet of a distant star. Indeed, when Newton arrived at his theory of gravitation, he called it *universal* gravitation because he was certain that the force of gravity behaved the same way anywhere and everywhere in the universe.

This principle was intrinsic in the classical or Galilean world view; known as the principle of relativity, it is a cornerstone of all physics. However, toward the end of the nineteenth century the principle did not appear to be compatible with Maxwell's very successful theory of electromagnetism.

In retrospect, the contradiction is easy to spot. Maxwell believed that light consists of electromagnetic vibrations of a hypothetical but real "ether" that permeated all of space. After all, he reasoned, since light is a wave phenomenon, it needs a medium in which the waves can propagate. Water waves cannot exist without water; Maxwell concluded that light waves in outer space could not exist without an "ether."

Since the earth must be moving through this ether, it seemed reasonable to ask what the velocity of our planet is, relative to the ether. Yet Maxwell's equations did not make any direct reference to this velocity. Moreover, careful experiments could detect no sign of earth's motion through the ether. Light seemed to travel at the same speed regardless of the state of motion of the observer. If Gertrude

Stein had been a physicist she might have said, "The velocity of light is the velocity of light is the velocity of light."

To resolve this seeming paradox, Einstein had to change the very meaning of our notions of space and time. The consequences were so amazing that it took a decade or more for physicists to fully accept them.

Einstein's special relativity shows that moving objects appear to shrink (the Lorentz-Fitzgerald contraction), that moving clocks slow down (time dilation), that distant events which appear simultaneous to one observer will not appear so to others, and that energy and mass are interconvertible—the most famous equation in all of science, $E = mc^2$. All this is the stuff of Einstein's 1905 theory, *special* relativity. It does not pertain at all to gravity.

By 1916 Einstein was thinking about how the world would look to observers who are not constrained to move in uniform motion. He saw no reason for the laws of physics to depend in any fashion on the state of motion of the observer. However, motion does produce force. Consider an observer who is in an elevator that is constantly accelerating upward—not merely rising, but accelerating, moving faster with each passing second.

You have felt such a force yourself whenever an elevator starts up. You feel heavier for a moment. Then, as the elevator reaches a steady speed, the feeling of extra weight fades away. But if the elevator accelerated all the time, you would feel that force constantly. Conversely, when an elevator starts going down you feel momentarily lighter, as if gravity had been weakened.

If the elevator were somehow placed in deep space, away from any massive body with a discernible gravitational field, the sensation of its acceleration upward would be indistinguishable from gravitational force. If the elevator were big enough to contain a Galileo and a Leaning Tower of Pisa, the sage would discover that objects dropped from the top of the tower fall the same way they do on earth. The force of gravity and the inertial response to acceleration are, in Einstein's view, equivalent phenomena. There is no way to distinguish one from the other simply because they are consequences of the geometrical structure of space and time.

According to the *general* theory of relativity, a massive body causes the distortion of space in a region that extends well beyond the material limits of the body itself. A gravitating body is a little like a heavy weight sitting upon a flat rubber sheet. In the vicinity of the mass, the sheet is stretched and its geometry is changed. A rolling ball bearing, coming within the range of the depression caused by the mass, will be deflected from its path even though it may not collide with the mass. The analogy is imperfect, but we can visualize the distortion of the two-dimensional rubber sheet because it is embedded within our three-dimensional space. The force of gravity may be regarded as a consequence of the curvature of three-dimensional space.

CURVED SPACE

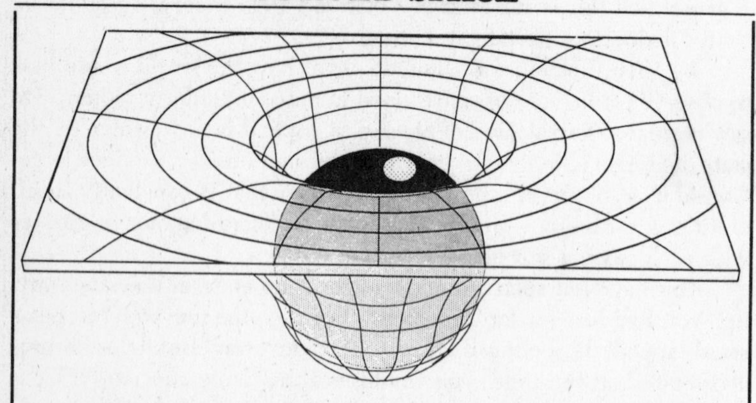

According to Einstein, a massive body causes the space about it to become curved. The figure shows a two-dimensional analogy. A bowling ball is placed upon a flexible rubber sheet, thereby distorting the sheet in its vicinity. A marble rolling along the sheet would be deflected from its path even if it did not collide with the bowling bowl. This is the way in which the force of gravity acts. A massive body affects the geometry of space, which, in turn, affects the motions of other bodies.

For many years, mathematicians knew that there were many possible geometries, of which Euclid's was only one example. According to Euclid, for instance, only one line can be drawn through a given point and parallel to another line. In other geometries there can be many such lines, or even none at all. While Euclidean geometry is certainly a good approximate description of the space we inhabit, Einstein proved that it isn't exactly so. Euclidean geometry is only the limiting case that is valid in the absence of matter. *Matter affects space*, Einstein showed, and creates its own geometric laws.

Newton's theory of gravity had reigned supreme and unchallenged for more than two centuries. Einstein's explained everything that Newton's did, and more. It was subject to three famous tests, which it passed with flying colors.

TESTS OF EINSTEIN'S THEORY OF GRAVITY

No matter how compelling or elegant it is, a theory of physics must be subject to experimental verification or it differs little from medieval theology. At the time of its invention, Einstein himself explicated three ways in which the consequences of his theory differed from Newton's. These have become known as the three classic tests of general relativity.

1. The precession of the orbit of Mercury

The planets in our solar system do not move in perfect ellipses because of small gravitational influences of one planet upon another. The orbit of Mercury does not close upon itself, but moves as shown below in a greatly exaggerated fashion.

The long axis of Mercury's orbit is said to precess about the Sun, completing each turn in about 20,000 years. Almost all of this effect can be explained in terms of Newton's theory of gravitation. In 1859, however, the French astronomer Urbain LeVerrier pointed out that a small inexplicable residue persists, ". . . some as yet unknown action on which no light has been thrown—a grave difficulty worthy of attention by astronomers." The anomaly is a mere 43 seconds of arc per century, corresponding by itself to a turn about the Sun in three million years. In 1915 Einstein saw that this curious motion of Mercury was hardly anomalous, but was *demanded* by his new theory of gravity, which, he wrote, "explains . . . quantitatively . . . the secular rotation of the orbit of Mercury discovered by LeVerrier . . . without the need of any special hypothesis. . . . For a few days, I was beside myself with joyous excitement." Here was the first hint that Einstein's theory was a genuine threat to Newton's.

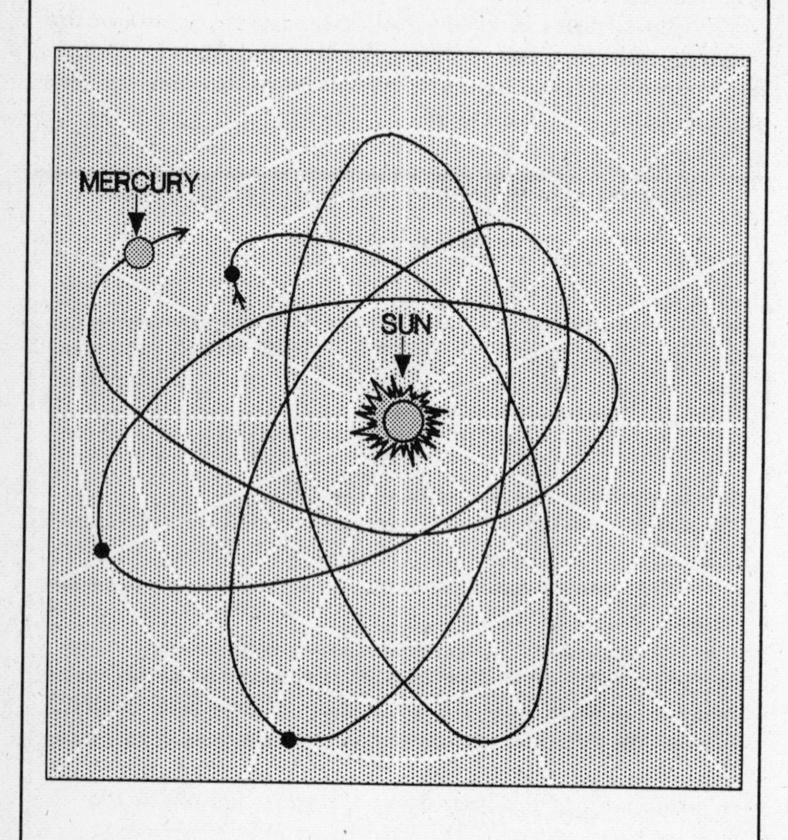

Figure 1. The orbit of Mercury is almost an ellipse. However, the axis of the ellipse rotates about the Sun once every 20,000 years. This precession is greatly exaggerated in the figure.

2. The bending of light by the Sun

The first test of general relativity was "retrodictive" rather than predictive. The theory solved a known observational difficulty about the orbit of Mercury. More impressive is Einstein's prediction of an absolutely new and unexpected effect. He predicted that light waves grazing the Sun are deflected by its gravitational field. There are two contributions to this effect. Half of the bending could have been predicted by Newton, had he realized the equivalence of mass and energy. The other half reflects the curvature of space that is integral to Einstein's theory. The total effect is a mere 1.74 seconds of arc, an angle as small as an incline that rises one inch in two miles.

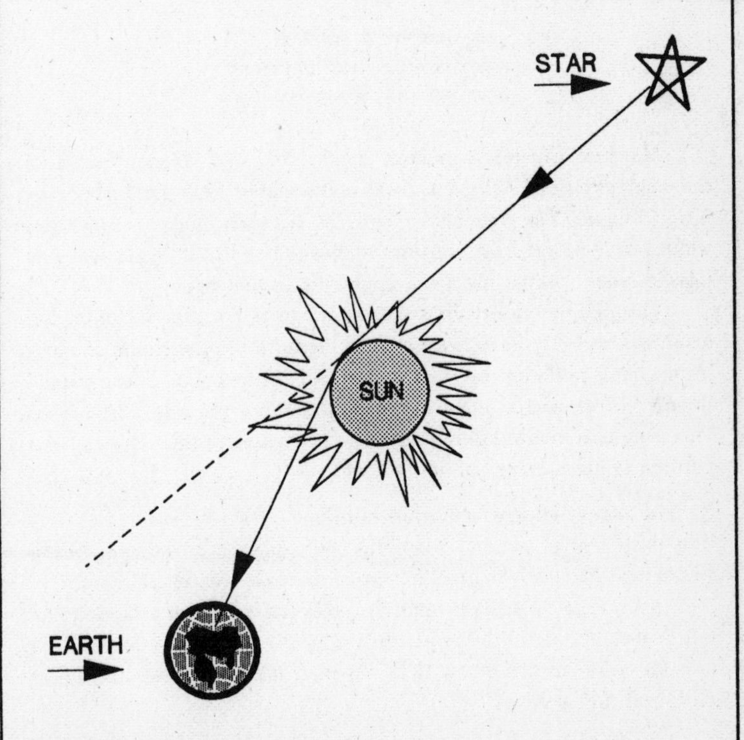

Figure 2. Light from a distant star is deflected very slightly by the gravitational field of the Sun. Because the Sun is so bright, the effect can be observed only during a total eclipse of the Sun.

Since it is difficult to see stars near the Sun, the experimental verification of Einstein's prediction depended on difficult observations done during—and at the site of—a total solar eclipse. At some points in his writings Einstein seems to be well above mere experimental questions: "I do not doubt any more the correctness of the whole system whether the observations of the solar eclipse succeed or not." On the other hand, frustrated at the difficulty in carrying out scientific experimentation while a world war was raging, he wrote, "Only the intrigues of miserable people prevent the execution of the last, new, important test of the theory." On November 6, 1919, the results of a successful British eclipse expedition to South America were revealed to the world and the headlines of the *London Times* next day read:

REVOLUTION IN SCIENCE
NEW THEORY OF THE UNIVERSE
NEWTONIAN IDEAS OVERTHROWN

Many of Einstein's greatest discoveries were accomplished in the magical year of 1905. Yet it was in November 1919 that he became a world figure. *The New York Times*, for example, makes no mention whatever of Albert Einstein prior to November 1919. Not a year has passed since without his name appearing in that paper.

Through the observation of radio waves rather than visible light, scientists recently have been able to measure the deflection of electromagnetic radiation by the Sun to a great precision. These experiments utilize several giant radio antennas in a procedure known as very long baseline interferometry, and Einstein's prediction has been verified to an accuracy of one percent.

3. The effect of gravity upon clocks

The third test of general relativity could not be performed during Einstein's lifetime—the technology of the time was insufficient.

According to general relativity, the rate at which a clock ticks is affected by gravity. Two identical clocks, one placed in the attic and the other in the cellar, will not read the same time: the lower clock will fall behind.

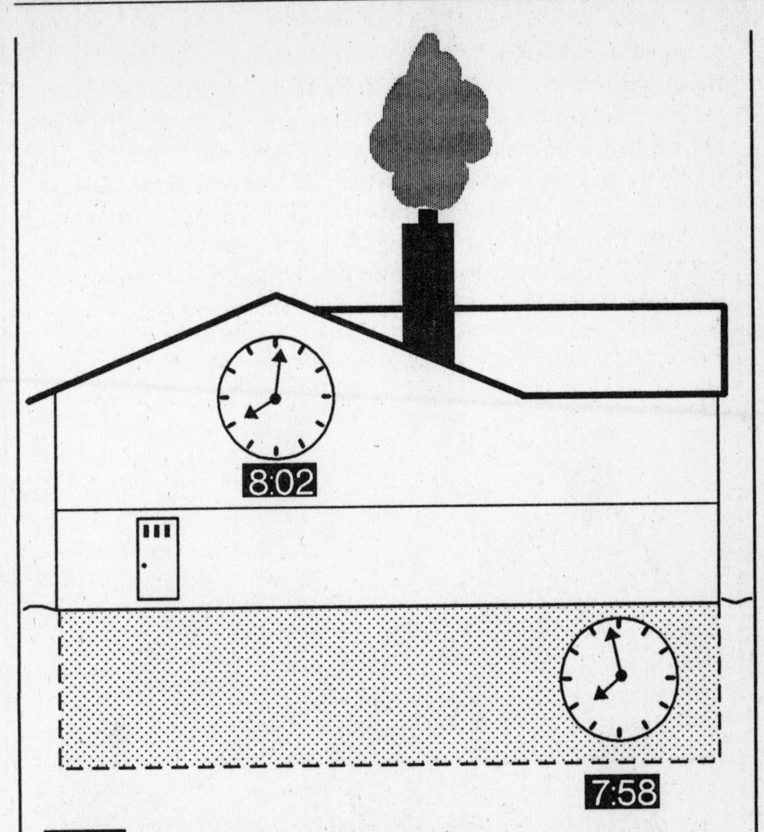

Figure 3. Two identical clocks are synchronized. One is placed in the attic, the other in the basement. The attic clock will tick faster and the basement clock will fall behind. In billions of years the effect will be as large as this figure illustrates.

The effect is very small. If the distance from attic to cellar is seventy-five feet, it would take ten million years before the accumulated error amounted to one second. Yet, just this experiment was performed successfully in the Jefferson Laboratory at Harvard by Professor Robert Pound and his co-workers in 1959. Eventually, they were to measure Einstein's effect to a precision of one percent. Today, far more sensitive observations have been accomplished with spacecraft that compare clock rates at Earth's surface with those at an altitude of thousands of miles. So far, Einstein's theory of general relativity has held up perfectly.

4. Signal retardation

Among the pleasures of student life at Harvard were the parties thrown by Irwin Shapiro, then a graduate student a bit more advanced than I. Irwin is now director of the Harvard Smithsonian Center for Astrophysics and has less time to entertain. In between, he devised and carried out a *fourth* test of general relativity unforeseen by Einstein.

The transit time of a light signal between two "fixed points" is changed if a massive body is placed nearby. In the experiment, the fixed points were Mars and Earth, and the interposed body was the Sun.

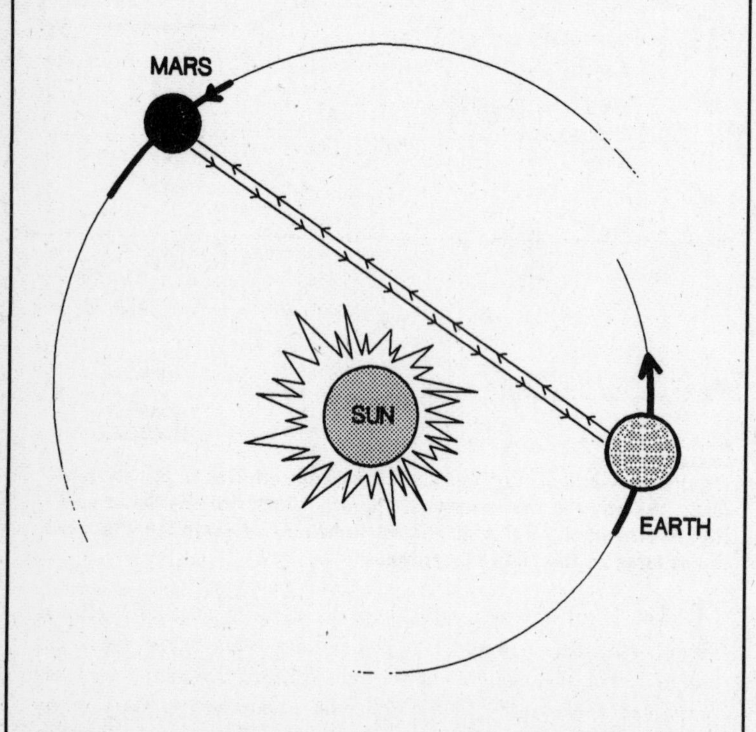

Figure 4. A radar signal from Earth is sent to Mars. A reflected signal is received on Earth, and the transit time is measured. When the Sun lies near the signal path, the transit time is lengthened measurably.

Since Mars and Earth are not really fixed, precise data about planetary orbits was an absolute prerequisite to interpret this experiment. The most precise version of this experiment made use of the two Viking spacecraft that landed on Mars in 1976. Transit times of radio signals between Earth and the Viking landers were measured with the greatest of care with the Sun both near the line of transmission and distant from it. The effects of general relativity were clearly identified. Indeed, the most accurate verification yet obtained of Einstein's theory has resulted from the Viking voyages to Mars. To me, this discovery is at least as important as finding no clear indication of Martian life.

5. Gravitational waves

Electromagnetism makes waves: X-rays, light and radio are examples. Should not gravity make waves as well? The general theory of relativity does indeed demand the existence of gravitational waves and their production in astrophysical events. Unfortunately, the intensity of the gravitational radiation in which we are bathed is minute, and it remains well below the threshold for experimental detection. Nonetheless, far more sensitive devices designed to detect faint gravitational signals are being designed and deployed. Perhaps, in a decade, the frontiers of astronomy will be extended from electromagnetic to gravitational radiation.

Indirect evidence of the existence of gravitational radiation has come from the discovery of a "tight binary pulsar." The pulsar, which sends us radio signals at the rate of seventeen pulses per second, circles a nearby star every eight hours. According to Einstein's theory, this system ought to send out so much energy in the form of gravitational waves that its orbit should be changing rapidly enough to be seen. And indeed it is! Experiments carried out over the past decade on this curious object (called PSR1913+16 by the astronomers) appear to have proven that gravitational waves behave according to the theory of general relativity, and offer yet a fifth quantitative test of the theory.

Conclusion

Just because the theory has passed all its examinations thus far does not mean that it must be accepted as true, any more than a child who does well in first grade can be expected to graduate from Harvard. Far more precise tests can, should, and will be done. Although a very fine theory, general relativity is neither a quantum theory nor one that is even compatible with the rules of quantum mechanics. At some level, general relativity is most certainly *wrong*, and the sooner we find out how it is wrong, the sooner we will come upon its successor.

According to my old buddy Irwin Shapiro, "For the long term we can confidently expect the appearance of experimentally verified deviations from the predictions of the general theory of relativity. The ultimate synthesis of gravitational and quantum phenomena will surely point to experimental consequences not envisioned or predicted by Einstein's theory. Einstein himself would surely have welcomed such progress."

Since the advent of space science and radio astronomy, far greater sensitivity and variety have been achieved in the tests of Einstein's theory. Recently, C. M. Hill wrote, "So far [general relativity] has withstood every confrontation, but new confrontations in new areas are on the horizon. Whether general relativity survives is a matter of speculation for some, pious hope for some, and supreme confidence for others."

It is clear that Einstein was of the latter category in the final three decades of his life. He wrote in 1930, "I do not consider the main significance of the general theory of relativity to be the prediction of some tiny observable numbers, but rather the simplicity of its foundations and its consistency."

Einstein believed that gravity and electromagnetism were "all ye know on earth, and all ye need to know." He considered them to be the fundamental forces of nature and thought that a theory which unified them could explain all the phenomena of nature. He even wrote a paper arguing that the force of gravity might be responsible for holding the atomic nucleus together. Einstein spent the last three decades of his life fruitlessly seeking his "unified field theory."

Today we know he was doomed to fail for two reasons. He never really accepted quantum mechanics, even though his early work lies at its very foundation. In particular, Einstein did not accept the statistical nature of quantum theory. The second reason he could not succeed is that Einstein simply did not know enough physics! He never cared to learn about the two other forces that exist in the universe, beyond gravity and electromagnetism. In the 1930s, as experimenters probed the atomic nucleus, it became apparent that two new forces were at work inside the nuclei of atoms. The new nuclear forces could not be ignored, and the more I learned about these forces the more I became fascinated by them.

Paul Martin's course convinced me that I would not specialize to become a "general relativist." I was much more concerned with the mysteries of the nuclear forces, where the confrontation between experiment and theory was immediate and developing rapidly.

I took a course in nuclear physics in 1955 which led to my first professional publication, an article written jointly with my professor, Walter Selove. It was a theoretical study of a certain type of nuclear reaction in which a proton strikes a nucleus, picks up a neutron, and emerges as a deuteron.

PICKUP AND STRIPPING

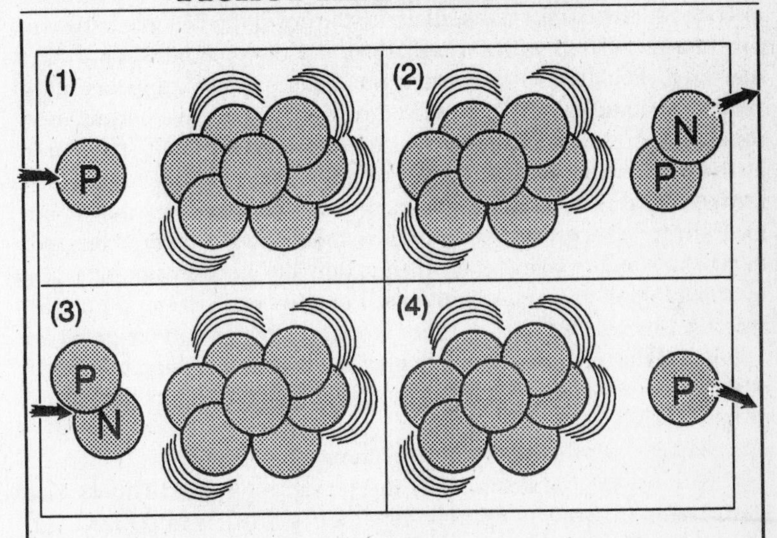

In the pickup reaction, a proton collides with a nucleus (1) and picks up one of its neutrons, to emerge as a deuteron (2).

In the stripping reaction, a deuteron collides with a nucleus (3) and loses its neutron to the nucleus, to emerge as a proton (4).

In the jargon of the trade this is known as a pickup reaction. The converse process, in which a deuteron loses a neutron to become a proton, is called a stripping reaction. I prefer not to comment on the etymologies of these expressions.

What forces govern processes such as these? The force of gravity is far too weak to be even remotely relevant. Earth's gravity seems to be strong because our planet contains so very many neutrons and protons that act on us in concert. But you can pick up a pin with a ten-cent magnet, despite the whole earth's gravitational pull. Gravity is by *far* the weakest of the four forces.

Electromagnetism cannot govern our proton nucleus collision. It works the wrong way. Nuclei contain lots of protons, all positively charged. They repel each other, and the nucleus would fly apart if another force was not holding it together.

This "other" force is called the *strong* nuclear force. It acts only over the extremely short range of the atomic nucleus. In this respect, the nuclear force differs greatly from the other forces. Gravity and electromagnetism are said to be forces of infinite range. The electric force between two charged particles or the gravitational force between two masses falls off with the square of the distance between them; the force dwindles with distance but never quite disappears altogether. The nuclear force, on the other hand, is effective only at very short ranges, comparable to the size of the nucleus itself. Beyond a distance of a trillionth of an inch or so, the nuclear force is completely ineffective. But what it lacks in range, the nuclear force makes up for in brute strength. About a million times more energy is needed to take an atomic nucleus apart than to defy the electromagnetic force by stripping off an atom's electrons. For the same reason, a nuclear reaction liberates millions of times more energy per unit weight than a chemical reaction. That is why a nuclear weapon weighing hundreds of pounds produces the explosive equivalent of hundreds of thousands of tons of TNT.

Where does the strong nuclear force come from?

To solve this riddle, in 1934 the Japanese physicist Hideki Yukawa boldly injected an entirely hypothetical particle into the world of physics. His particle, now called the *pion*, could be emitted or absorbed by nucleons (that is, protons or neutrons) in the same way that photons were known to be emitted and absorbed by electrons.

Once upon a time, physicists wondered whether light was an electromagnetic wave or a beam of particles. Quantum mechanics revealed that this question is simply not meaningful since neither answer is quite correct. Light, and all other forms of electromagnetic radiation, sometimes displays particle-like properties as photons, and sometimes behaves like a wave. The same is true for the electron, which is particulate as it produces a flash of light on the TV screen and wavelike as it passes through an electron microscope. In everyday life, when a pebble is thrown into a pond, the pebble is the particle and the ripple is the wave. In the quantum-mechanical microworld, there is no such clear-cut distinction. The wave-particle duality is a universal attribute of material systems.

Another aspect of quantum theory blurs the distinction between a *particle* and a *force*. Two electrons repel one another. One may say that an electron, being charged, produces a field of force about it that acts upon the other electron. Equivalently, one may say that the electric force is produced by the exchange of virtual photons between the two electrons. Every particle in nature can play a dual role: as a "real particle" that can produce an observable signal in a detector such as an eyeball or a Geiger counter, and as a "virtual particle" that acts to generate a force or interaction among other particles.

In one sense, the photon is the particle of light and other forms of electromagnetic radiation ranging from radio waves to gamma rays. "Real" photons can be absorbed or emitted by electrons or other charged particles. They correspond to observable particles or waves. On the other hand, it is by the exchange of "virtual" photons that charged particles exert forces upon one another. When a proton attracts an electron, they are exchanging "virtual" photons. In this role, the photon is not observed as a "real" particle, for it does not emerge from the region of interaction to impinge on a detector—it is consumed in the act of producing the electromagnetic force.

Yukawa's argument was based upon an analogy with electrodynamics. He saw the need for a new interaction in nature: a strong force that acts selectively on the protons and neutrons within the nucleus, and nowhere else. He recognized the inalienable association between forces and particles: the price of a new force is the existence of a new particle. And although the particle exerts its force in its role as a virtual particle, it must become observable—and be observed —before physicists will accept its reality.

Yukawa invented his pion for the purpose of mediating the nuclear force. He imagined that the nuclear interactions could be understood in terms of the exchange of virtual pions between nucleons in exactly the same way that the electromagnetic force is understood in terms of the exchange of virtual photons. If the properties of these hypothetical particles were stipulated correctly, their exchange between the nucleons within the atomic nucleus could produce the strong force and bind the nucleus together. According to Yukawa, his theory could be formulated in terms of a single "fundamental act of becoming," just like quantum electrodynamics. Instead of the emission or absorption of a photon by an electron, the fundamental nuclear act consisted of the emission or absorption of a pion by a nucleon.

NUCLEAR FORCE ACCORDING TO YUKAWA

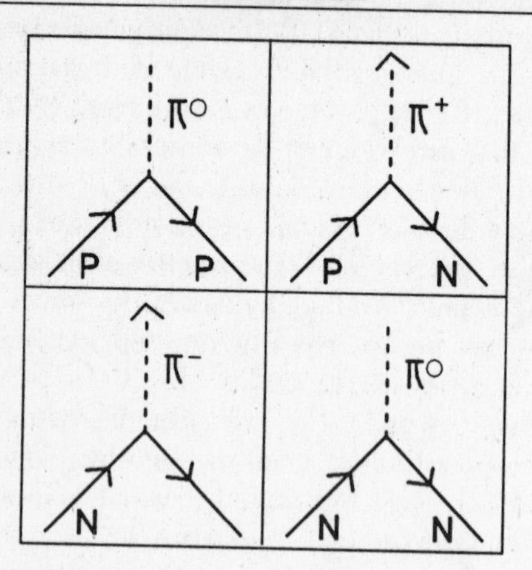

The electromagnetic force arises from the exchange of virtual photons. The fundamental act is the emission or absorption of a photon by a charged particle. Yukawa suggested an analogy between QED and the strong nuclear force. The strong force would arise from the exchange of virtual pions. The fundamental act, shown above, is the emission or absorption of a pion by a nucleon. Unlike the photon, the pion is a massive particle weighing a seventh as much as the nucleon. It comes in three varieties: positively charged, negatively charged, and neutral.

The neutrons and protons that make up the atomic nucleus are so close together that they touch one another. How is it that the nuclear force is exerted only over very short distances, while the electromagnetic force is long-range enough to span the distances among the far-flung electrons and beyond? Yukawa saw that he could obtain this result if his hypothetical particle was not massless like the photon, but was several hundred times heavier than the electron. The exchange of such a heavy particle between nucleons would seem to violate the law of conservation of energy, for where does the energy come from to create a virtual pion?

The curious laws of quantum mechanics permit such departures from the law, but only for very short times and over minute distances, such as are allowed by Heisenberg's principle of uncertainty. The intrinsic uncertainty in time corresponding to the energy of a virtual pion is about equal to the time it takes light to traverse the nucleon, about a trillionth of a trillionth of a second. Thus the virtual pion has an exceedingly fleeting existence.

It is through Heisenberg's principle as well that the mass of the particle determines the range over which the force may act. The massless photon acts over an unbounded range. A charged particle produces electric forces everywhere in space. The pion, however, has a significant mass, about one-seventh that of the nucleon. Consequently, it gives rise to a very short-range force with an effective radius of one fermi, or 10^{-13} centimeter. (A trillion fermis equal the width of lead in a mechanical pencil.)

The range of the nuclear force is about the same as the observed size of the proton. Indeed, the nuclear force produces the effect that we regard as the nuclear size. The heavier the pion, the shorter is the range over which the nuclear force may act.

Three other properties had to be determined to specify the pion theory of nuclear forces. One had to do with the "spin" of the particle, the measure of its internal degree of rotation. Another had to do with its "parity," or its behavior when viewed in a mirror. The final property was the specification of its electrical charge. It turned out that pions exist in three different charge states: positive, negative and neutral.

The role of the pion in mediating nuclear forces is a passive or virtual one. However, according to the rules of the game, the pion must also be capable of being produced as a real particle in a nuclear collision that is sufficiently violent. More precisely, the energy of the collision must be at least as large as the energy equivalent of a pion mass. Only then could a real pion be produced and observed. Not until the late 1940s, more than ten years after Yukawa's brilliant insight, were accelerators developed that could search for the new particle. However, there is a natural source of energetic pions: cosmic rays.

$E = MC^2$

The law of conservation of energy says that energy can neither be created nor destroyed, but only changed from one form to another. The law of conservation of mass says that mass can neither be created nor destroyed, that the mass of the reactants in any natural process precisely equals the mass of the end products. Einstein showed that neither law is correct, because energy and mass are interconvertible: mass can become energy and energy can become mass. There are not two independent conservation laws, but only one. The famous formula expresses the energy equivalent E of a mass m, and c stands for the speed of light.

To illustrate the interconvertibility of energy and mass, consider the energy that could be obtained from one pound of heavy hydrogen. I choose this exotic fuel because it can be used in several ways.

1. Combustion

Burn it! That is, chemically react it with an appropriate amount of oxygen (about four pounds) to form heavy water. This process releases energy that could be used to generate ten kilowatt-hours of electricity, worth about fifty cents. The mass of the heavy water is almost equal to the mass of the reactants. Almost, but not quite! The heavy water is lighter than its constituents by about one part per billion. The mass loss is the exact equivalent of the energy released, but it is far too small to notice.

2. Nuclear Fusion

Two heavy hydrogen nuclei can be coerced to unite to form one helium nucleus, with a consequent release of energy. This process is much more efficient than burning. One pound of fuel can generate sixty million kilowatt-hours of electricity, worth about $3 million. This is why scientists are trying to develop a practical system for controlled nuclear fusion. The mass of the helium produced by the fusion is measurably less than the mass of the fuel consumed. About one-half percent of the original pound of heavy hydrogen is converted into energy.

3. Annihilation

Suppose that *all* of the fuel can be turned into useful energy. It could be put in contact with an equal mass of antimatter, if we had some.

Or, by some unknown mechanism, we could turn half the fuel into antifuel. In either case, none of the fuel survives as mass. Matter and antimatter totally annihilate one another. One pound of matter (any stuff at all) has locked within it an astonishing 12 billion kilowatt-hours. This is enough energy to supply a fair-sized city for a year, and is worth about $600 million!

One way or another, when the energy supply of a fuel is tapped, some of its mass is converted into energy. The end products weigh less than the original material. A little mass goes a long way because the speed of light "c" in Einstein's formula) is very large.

Conversely, it takes a lot of energy to create even a tiny bit of mass. That is why particle physicists need giant accelerators to produce beams of very energetic particles. In the collisions of these particles, some of their energy is converted into the mass of newly created particles. For example, if a proton is to be energetic enough to produce a pion in a collision, it must have a speed half that of light. If it is to produce an antiproton, it must be so energetic as to move at 96 percent of the speed of light. The more energetic a particle is, the closer it approaches light speed. It can never reach the speed of light, let alone exceed it. That is another of Einstein's discoveries.

Back in the 1930s physicists discovered two new particles that are produced in the collisions of cosmic-ray protons with air molecules. One was the positron, the electron's antiparticle. The other was the muon, a particle that is now known to play no essential role in the atomic nucleus. The muon is simply a particle much like the electron but about two hundred times heavier. It has about the right mass to be Yukawa's predicted particle, and many physicists assumed that it was. It was tentatively christened the Yukon in honor of Yukawa (not the northwestern region of Canada).

The trouble with such an interpretation first showed up in experiments done in Italy shortly after World War II. Three Italian scientists showed that negative muons that are allowed to come to rest in carbon are not absorbed by the carbon nuclei, as they must be if they are the particles mediating the nuclear force. The muon was not the pion after all.

The mystery was solved in 1947. A young Brazilian physicist, Cesare Lattes, working in Bristol with Cecil Powell and Italian physicist Giuseppe Occhialini, brought a stack of photographic emulsions to a meteorological station in Bolivia situated 18,600 feet above sea level. Here, much closer to the region where primary cosmic rays make their first collisions, is where Yukawa's pions might hang out. The behavior of these particles would be recorded on the photographic film for later study and analysis at Bristol. The experiment worked! The tracks of some thirty pions were captured on the film. Yukawa's particle, as it was expected to be, was unstable. It decays, or comes apart, after only a hundred-millionth of a second. The big surprise was that it decays into a muon, the particle that for a decade had been taken to be the pion.

At the end of 1947, Lattes left England for Berkeley, where the world's largest cyclotron had just been put into operation. He was the first to show that pions could be artificially produced by the high-energy particle beam of the cyclotron. And why not? The protons in the machine's beam behave just the way that similarly energetic cosmic-ray protons do. Within a scant few years beams of pions were being routinely produced and studied at atom smashers throughout the nation. Yukawa's brilliant insight had been vindicated and he was awarded the 1949 Nobel Prize in physics "for his prediction of the existence of [pions] on the basis of theoretical work on nuclear forces." Cecil Powell, for his development of the photographic techniques making the discovery of the pion possible, captured the prize in 1950.

With the discovery of the pion the elementary-particle zoo had just six species: neutrons, protons, pions, electrons, muons and photons. Not so simple as earth, air, fire and water, but not much worse. Simplicity was not to endure. The hunting expedition for new particles produced by cosmic rays made its first catch in Manchester in 1947. In succeeding years many species of "strange particles" were discovered: charged and neutral kaons, lambda particles, three subspecies of sigma particles, etc.

They were called "strange" particles because of their seemingly paradoxical properties. Their relatively long lifetimes could not be reconciled with their copious production by cosmic rays. While these particles were subject to the strong nuclear force, they seemed and still seem to play no essential role in the structure of atomic nuclei.

By the time I entered graduate school, the list of known elementary particles had grown considerably, and it was to grow much longer.

Taxonomy of Elementary Particles (Ca. 1954)

	HADRONS This is the name given to particles that are subject to the strong nuclear force. ↓	**NOT HADRONS** No word has been coined to designate particles that are not subject to the strong nuclear force. ↓
FERMIONS → Any particle that satisfies the Pauli exclusion principle is called a fermion, in honor of Enrico Fermi.	**BARYONS** These are particles that are both hadrons and fermions. The nucleons (neutrons and protons) are baryons. In 1954, there were also six known species of strange baryons.	**LEPTONS** This is the name given to fermions that are not subject to the strong nuclear force. In 1954, electrons were the only observed leptons. Neutrinos had been conjected by Pauli, but not yet detected in the laboratory.
BOSONS → Any particle that does not satisfy the Pauli principle is called a boson in honor of the Indian physicist Satyenda Bose.	**MESONS** This is the name given to particles that are both hadrons and bosons. In 1954 this category included the three *pions* and four strange mesons called *kaons.*	**OTHER BOSONS** In 1954 *photons* were the only known bosons not subject to the strong nuclear force. Since 1983 they have been joined by *W-bosons* and *Z-bosons*.

The classification of elementary particles is complicated because there are two essential but distinct divisions among them. Do they, or do they not, satisfy the Pauli exclusion principle (fermions versus bosons)? Are they, or are they not, subject to the strong nuclear force (hadrons versus others)? A similar complexity besets the taxonomy of life-forms if we attempt to classify them as animal versus vegetable, and at the same time as landlubbers versus seafarers.

The time has come to introduce some technical jargon into our discussion of the elementary-particle bestiary. Remember the Pauli exclusion principle? No two electrons can do the same thing at the same time. The same is true for protons and neutrons. However, not all elementary particles satisfy the exclusion principle. Those that do (such as nucleons and leptons) are called *fermions*. Those that do not (such as pions and photons) are called *bosons*. Every elementary particle belongs to one of these two broad categories.

Some elementary particles, like nucleons and pions, are subject to the strong nuclear force. Others, like electrons, muons, and photons are not. My close Russian friend and colleague Lev Okun' coined an important new word when he wrote, in 1962, "In this report, I shall call strongly interacting particles HADRONS." Today, *hadron* (from the Greek word for "thick and bulky") is the accepted term for any of those seemingly elementary particles that interact with one another through the strong force. Pions and nucleons are hadrons. So are all the various strange particles. Well over a hundred hadrons have since been discovered.

Some hadrons satisfy the Pauli principle. Thus they are included in the general category of fermions. There is a special word for hadrons that are fermions: they are called *baryons*, and their antiparticles are called *antibaryons*. Neutrons and protons (generally called *nucleons*) are examples of baryons. There are very many others.

Other hadrons do not satisfy the Pauli principle. They are included in the category of bosons. The special word for a hadron that is a boson is *meson*. Pions are mesons. So also are the strange particles called kaons. Again, there are many other known mesons.

All this Byzantine complexity becomes much simpler in the language of quarks. Unfortunately, I didn't know about quarks during the 1950s. Neither did anyone else. The concept was invented by Murray Gell-Mann and independently by George Zweig in 1963. They were appalled at the uncontrolled expansion of the elementary-particle zoo. Perhaps, they argued, hadrons were not really elementary particles after all but were made of simpler things. They were absolutely right! Quarks are now believed to be the truly fundamental building blocks from which all the hadrons are constructed.

Here are the rules that must be followed to build hadrons from quarks:

1. There are several different kinds (or "flavors," as we say) of quarks. Any combination of flavors may be combined together to

make a hadron. Thus, a wide variety of hadrons can be made up out of a small number of quark flavors.

2. Three quarks stick together to form a baryon, such as a proton or neutron.

3. Three antiquarks stick together to form an antibaryon.

4. One quark can be attached to one antiquark to form a meson. If the flavor of the quark is different from the flavor of the antiquark, the constituents cannot annihilate one another, so the meson will be long-lived. It does, however, eventually decay by means of the *weak* nuclear force. When the quark and antiquark are of the same flavor they rapidly annihilate so that the meson they form (for example, the neutral pion) is exceedingly short-lived.

5. No other combination of quarks sticks together to form a particle.

6. *Strange* particles are hadrons that contain one or more strange quarks or strange antiquarks. Strange quark is one of the original flavors that Gell-Mann introduced. Today five flavors are known and more are almost certainly on the way.

7. The quark is a fermion, as is any system containing an odd number of quarks. Systems containing an even number of fermionic constituents are necessarily bosons, such as the mesons, and in particular the pions.

These rules work very well, and they greatly simplify our understanding of the hadron bestiary, Yet they are no more than a system of arbitrary rules: they hardly constitute a coherent theory. The rules leave unanswered such questions as: What is the origin of the force holding the quarks together? And why do three quarks stick together to form a hadron, but not two? It was not until the mid-1970s that these questions were answered by the theory of quantum chromodynamics.

What about those particles that are not hadrons, are not made of quarks, and do not partake of strong nuclear force? Such weakly interacting particles that are fermions are called *leptons*, from the Greek for "weak." The electron is a lepton. So are the muon and the neutrino. Today there are exactly six known species of lepton. Each of the six seems to be a genuinely elementary particle. As yet, there is no indication that leptons are built up out of tinier or simpler building blocks.

The neutrino had been predicted by Wolfgang Pauli in 1930 on the basis of indirect but compelling evidence, but it was not until 1956, when I was a graduate student at Harvard, that the elusive neutrino was actually observed.

Strange particles were an unanticipated discovery of 1947. One of the great challenges of the fifties was to find some systematic pattern into which all the hadrons, including the strange particles, could be fit.

History of the Pion

1932 The discovery of neutrons makes possible a rational theory of nuclear structure. But what produces the forces that hold neutrons and protons together?

1934 Yukawa predicts the existence of then-unseen particles, about two hundred times heavier than electrons, whose exchange produces strong forces between nucleons. He called his hypothetical particles mesotrons, but they are now called pions.

1937 Jabez Curry Street at Harvard, and Carl Anderson and Seth Neddermeyer elsewhere, demonstrate the existence of a new particle in cosmic radiation. This discovery led to a period of confusion since it (now called the muon) was not Yukawa's predicted particle.

1947 Conversi, Pancini and Piccioni present compelling evidence that the muon is not the pion. The muon penetrates through matter too easily because it does not interact strongly with the nucleus. It is simply a by-product of the radioactive decay of Yukawa's pion.

1947 Cesare Lattes, Giuseppe Occhialini and Cecil Powell discover a second new particle that *is* Yukawa's expected particle. Pions are readily produced in the collisions of primary cosmic-ray protons with the nuclei of air atoms. Charged pions decay in a hundred-millionth of a second to produce muons.

1949 Hideki Yukawa is the first Japanese to win a Nobel Prize.

1950 Powell of Britain wins the Nobel Prize "for his development of the photographic method of studying nuclear processes and his discoveries regarding mesons with this technique."

Thus began a period in which *pions* and *nucleons* were regarded as the ultimate building blocks of matter. The meson theory of nuclear forces was analogous to quantum electrodynamics, with electrons replaced by nucleons and photons by pions. It was a useful but wrong model. Pions and nucleons are two of more than a hundred known "hadrons" with equal claim to elementarity. Today we know that they are all composite systems made up of quarks.

Pions were understood to be the glue that holds nucleons together. Physicists believed that the exchange of pions inside the nucleus determined the properties of the nuclear force, just as the exchange of photons determines the properties of the electromagnetic force. As the photon is the "carrier" of light and all the other manifestations of electromagnetism, the pion is the "carrier" of the strong nuclear force. What was needed was a computational scheme that would allow physicists to unravel the details of how pions are exchanged from one nucleon to another. Until this mathematical treatment could be developed, the pion theory gave satisfactory *qualitative* understanding of the underlying rules of the game known as nuclear physics. But physicists wanted a *quantitative* theory, one that could predict in detail what was going on inside the nucleus.

During my first graduate year I decided that nuclear physics was no more for me than was chemistry. There were, after all, about a hundred chemical elements (see figure on Growth of the Number of Elements), and the chemical workings of the atom had been essentially completely explained by the quantum theory before I was born.

In the same way, who could be fascinated by nuclei when there were so many different species? It seemed to me that nuclei were not much more fundamental than atoms. They were simply balls of nucleons stuck together by forces arising from pion exchange. True enough, many questions remained unanswered, but they did not bear the cachet of truly fundamental questions. Nuclear physics, especially after my one sortie with Selove, was simply not for me.

I was more fascinated by neutrinos than by neutrons. Elusive almost to the point of undetectability, neutrinos were known to be produced in certain radioactive processes called beta decay.

The beta process is essential to the *thermonuclear fusion* mechanism by which the Sun produces energy. In the Sun, and all the other stars as well, hydrogen nuclei are forced together to become nuclei of helium, releasing heat and light in the process. For this to take place, protons must be converted into neutrons by a process

THE NUCLEAR POPULATION EXPLOSION

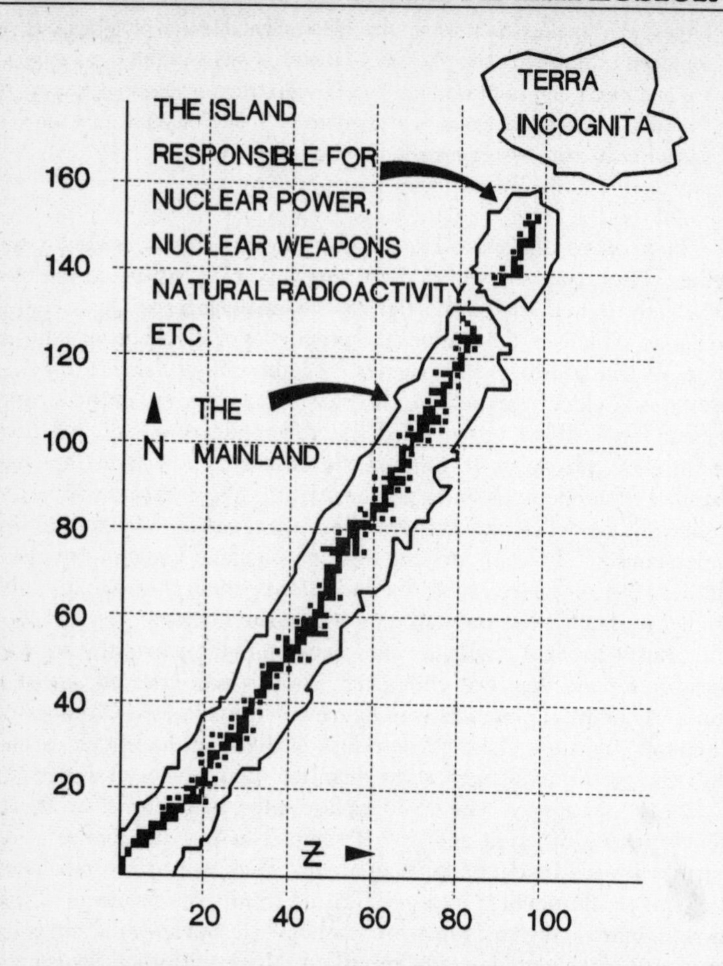

While only 109 chemical elements are known, there are thousands of different isotopes. In this plot we show the 399 nuclei that have a lifetime of at least one year. Each point corresponds to a definite number of neutrons, N, and protons, Z. There are so many different nuclear species that I could never regard nuclear physics, any more than chemistry or atomic physics, as a truly fundamental science. Yet nuclear physics could be fun! Some physicists expect there to be an island of stability, as yet undiscovered, lying in the region marked Terra Incognita. Perhaps this hypothetical island resembles the real island shown in the figure, which consists of the radioactive elements from radium to californium. The voyage of discovery, the search for superheavy and relatively stable new isotopes, is being pursued by many nations. Wherever you look in the physical universe, you can find exciting and challenging problems.

akin to beta decay. It is thermonuclear fusion that produces the sunlight that allows life to exist on Earth. (Thermonuclear fusion is also the process that makes hydrogen bombs work and threatens to end life on Earth.) Without the neutrino and the beta-decay process, Spaceship Earth, if it existed at all, would be a frozen ball of solid hydrogen.

In 1930, Pauli was forced to conjecture the existence of a new and unobserved elementary particle which is produced in beta decay. In this process, a parent nucleus is transmuted into a different daughter nucleus when one of its neutrons spontaneously becomes a proton. In this process an electron is created and expelled from the nucleus. The emitted electron, however, does not always carry away the same amount of energy. Sometimes it carries off most of the available energy, but at other times only a very small part of it.

How can this be? All the parent nuclei are identical to one another, as are the daughter nuclei. Surely, in the process of beta decay, the electron should always carry away exactly the same amount of energy. Madame Curie suggested the heretical notion that the law of conservation of energy failed when it came to beta decay.

Pauli would brook no such nonsense. He assumed that some other unobserved particle was produced in beta decay along with the electron, and that this mysterious particle carried away the missing energy. The particle had to be electrically neutral and very light—perhaps massless, like the photon.

It may seem odd to believe in the existence of a particle that no one had ever seen. Yet this was exactly the case for decades. When faced with the possibility that the conservation of energy may not hold true, physicists almost unanimously preferred to "explain" the missing energy by postulating an undiscovered particle. It was not that they were embarrassed by the missing energy and invented the new particle to save their reputations. It was that they believed so strongly in the conservation of energy (a bedrock of physics) that they fully expected the particle to be discovered. Fermi dubbed it "neutrino," an Italian coinage for "little neutral one."

It did not seem likely that neutrinos would ever be found in the laboratory. Calculations showed that neutrinos from beta decay hardly interact with matter at all: on the average, a neutrino could penetrate a slab of steel many *light-years* thick without being stopped. It seemed that no one was going to prove Pauli wrong—or right—for a long, long time.

Experimenters, however, are a persistent and ingenious breed. Twenty-six years after Pauli's hypothesis, neutrinos were "seen" at an experiment performed at a nuclear reactor by Clyde L. Cowan and Fred Reines. More precisely, antineutrinos from the reactor impinged on a huge instrumented tank of water. About one in a trillion of the antineutrinos passing through the device struck a proton of the water, turning it into a neutron and a positron, both of which were detected with the apparatus. The ghostlike neutrino, or at least its direct effects, had been spotted at last. It's true that neutrinos very rarely interact with matter. The trick was to work with an enormous number of neutrinos.

The process of beta decay is an example of a "weak interaction," which represents the fourth fundamental force of nature. Although the strong force holds the nucleus together, it is the weak force that permits protons to turn into neutrons, and vice versa.

Soon after radioactivity was discovered, three distinct kinds of radioactive emanations were recognized. Physicists called these alpha rays, beta rays and gamma rays. Subsequently, it was discovered that alpha rays are nothing other than helium nuclei (two protons and two neutrons bound together). The process by which such a particle is emitted from the nucleus is called alpha decay. Then it was discovered that beta rays are simply electrons. In the process called beta decay, the decaying nucleus spits out an electron and a neutrino. The gamma ray was eventually identified to be a very high-energy photon, one carrying millions of times more energy than a photon of visible light. Gamma decay is the process by which a radioactive nucleus emits a gamma ray.

These three processes are characteristic of the three forces that govern the microworld: the strong force, the weak force and electromagnetism, respectively. In the early 1950s theoretical physicists had an enormously successful picture of electromagnetism in terms of the modern theory of quantum electrodynamics developed by Schwinger, Feynman and others. In addition, there was a plausible theory of strong interactions which seemed to be at least partially correct, though it did not explain the existence of strange particles. There was also a quantum theory, of sorts, of the weak force.

The fourth force, gravity, was still the holdout. And it still is.

My first year at Harvard gave me a state-of-the-art knowledge of each of the four forces, but this was only a small part of my graduate education. It was also mandatory in my day to demonstrate a command of two foreign languages. Having taken French in high school and in college, I passed the first exam handily. To prepare for the second language exam I enrolled in an intensive summer course in Russian. (Ironically, although both my parents came to the United States from Russia as teenagers, they never spoke Russian at home; Yiddish on occasion, but never Russian.)

Wendell Furry, who was a fluent and self-taught Russian reader, was the only member of the department competent to administer the Russian exam, which consisted of portions of scientific text that he had photocopied (not Xeroxed—there was not yet such a process) out of the very few Russian texts in the departmental library. The night before the exam was given, I examined these books carefully and found that they could be opened readily and repeatedly to certain pages. Clearly, the photocopying procedure had been hard on the books. I studied these pages well and consequently passed the Russian exam.

Today there are no language requirements in the Harvard doctoral program in physics, not even in proper English. Throughout the world, including China, Japan and the Soviet Union, the language of physics is Pidgin English.

METHODS OF PARTICLE DETECTION

An experiment in high-energy physics begins at an accelerator and ends at a detector. The accelerator produces a beam of energetic particles that is made to collide with a target. In some experiments the target is at rest. In others, it consists of a second beam of particles moving in the opposite direction. In the collisions, many new particles are produced and are observed by the detector. The detector is like the journalist who must determine what, where, when, which, and how?

What is the identity of the particle?

Exactly *where* is it when it is observed?

When does the particle get to the detector?

Which way is it going?

How fast is it moving?

Detector technology is a large and rapidly evolving discipline that overlaps significantly more practical science such as medical imaging, xerography, photography and meteorology.

One of the earliest detector techniques was photographic. When an energetic particle traverses a photographic emulsion, it leaves a trail of ions. When the emulsion is developed, the trajectory of the particle can be seen. While this technique is still used occasionally by cosmic-ray physicists, it is far too slow and cumbersome for accelerator physics.

The cloud chamber and its successor, the bubble chamber, are devices that present us with a picture of the collision event. In the cloud chamber, vapor condenses in a trail along the path of a charged particle. In the bubble chamber, the ionization produced by a charged particle generates microscopic bubbles along its path. In either case, the resulting image is photographed for subsequent study.

Other kinds of detectors simply react to the passage of a particle by producing an electronic signal. Wire spark chambers consist of a large number of parallel wires that are maintained at a high voltage. When a particle passes between two wires, a small spark is produced, generating a signal in the affected wire. These devices can determine *when* and *where* with dispatch and efficiency.

The speed of a particle can be determined by measuring how long it takes to get from one detector to another. Alternatively, the Cherenkov detector can estimate speed. This device is named after the Russian physicist Pavel Cherenkov, who won the Nobel Prize in 1958 for his discovery of the effect that bears his name. He observed that a charged particle produces visible light when it moves faster than light. But didn't Einstein show that particles cannot exceed light speed? Not exactly. In a transparent medium, such as air or glass, light moves more slowly than it does in a vacuum. In such a medium, particles can move faster than the speed of light *for that medium.* Thus, only those particles moving very rapidly will produce a signal in a Cherenkov detector.

When an energetic particle passes through matter it disrupts atoms in its path, producing ions, which are atoms that have lost one or more of their orbital electrons and thus acquired a net positive charge. These ions can be collected electrically and counted. The more energetic a particle is, the more ions it produces. Thus the ionization chamber can determine how much energy a particle possesses.

In certain crystals or organic liquids the ionization produced by the passage of a particle creates a flash of light. These materials form the basis of scintillation detectors, which can be used to detect charged particles, X-rays or gamma rays.

We have only scratched the surface of the art of particle detection, as challenging and exciting an endeavor as may be imagined. Detector scientists work at the verge of the impossible. Consider, for example, the detector problems presented by our dream machine, the Superconducting SuperCollider. Each event will produce about a hundred particles of various kinds moving every which way. Events occur at a rate of a billion per second! The detector must select only the most interesting events, for there are far too many for each event to be studied.

This instantaneous electronic wizard is called the trigger. For each selected event, the detector must determine the identity, energy and direction of motion of each of the secondary particles and record this hard-won data on a magnetic tape. Only much later does a physicist study the result and try to figure out just what happened.

Shown below is the result of a proton-antiproton collision at a world-record energy of 1800 GeV. It is one of the very first events produced by the Tevatron Collider and seen at its Collider Detection Facility, or CDF. The computer reconstruction of the event shows some seventy particle tracks. The collision energy at the Superconducting SuperCollider will be twenty-five times greater, and some events will produce hundreds of particles. An effective SSC detector will cost about $100 million and will require a cast of hundreds of Ph.D.s to design, construct and operate.

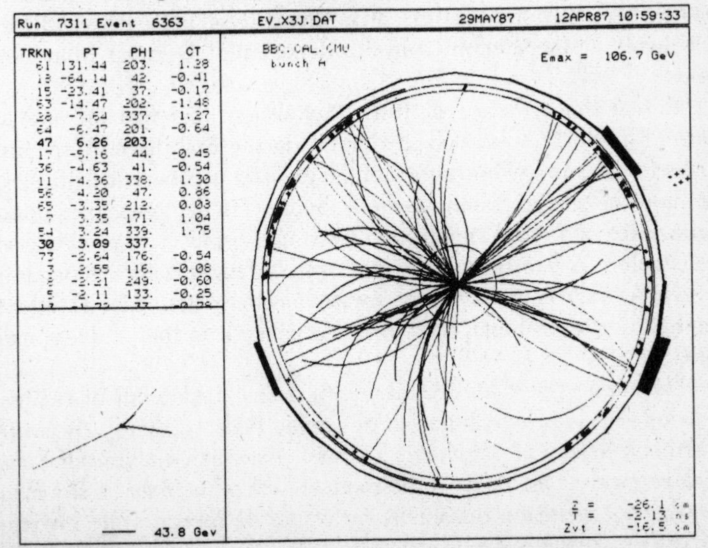

Another requirement for the doctorate was to have a sufficiently broad training in physics. For the aspiring theorist, this ordinarily included some laboratory experiences. Since I was no longer the experimentalist of my youth, I was willing to go to any length to avoid the lab course. Reading the pamphlet on degree requirements, I discovered an option that is rarely exercised: I volunteered to prepare for and submit to an oral examination on the methods used to detect and observe elementary particles. It was a valuable experience. To be an effective theorist, one should understand just what can and cannot be done in the laboratory.

During my second year at Harvard I shared an apartment in the Boston neighborhood of Brighton with Chuck Zemach, who, like Paul Martin, emerged from Stuyvesant High School in New York. The type is easy to spot. Instead of saying "Hi!" or "Hello!" or "Howdy!" they invariably come up with "Greetings!"

Chuck and Paul were both ahead of me, having gotten their doctorates under the guidance of Roy Glauber, who hailed from Bronx Science and from my very own Manhattan neighborhood of Inwood. I remember, or more precisely, I can't remember, a fabulous drunken party that Chuck and I threw for the physics crowd and their groupies. Days later I discovered that somebody at the party had covertly inserted a frankfurter into my flute. Glauber eventually confessed to his sin. I began playing as a graduate student, but I never became a good flutist. I was, nonetheless, an enthusiastic, if not always audible, participant in the Harvard physics department German Marching Band.

It was the presence of Julian Schwinger that had led me to consider graduate education at Harvard in the first place. Mere undergraduates knew of him as one of the greatest in theoretical physics, a founder of modern quantum field theory. His lectures, presented in the voice of a radio announcer, were masterpieces of pedagogical organization. A late riser, Schwinger rarely taught before noon. His lectures would run overtime, so it was almost impossible for a student who had not taken lunch beforehand to make it to the graduate hall afterward.

He often ended his lectures with a chalkboard full of rapidly scrawled formulas (which had to be copied because there was *never* a textbook in any of Schwinger's courses) and poised himself for a quick getaway at the point of the room closest to the door. A shy man who did not welcome questions, he would disappear while his students were still ascribble. It was said that a graduate student once dared to interrupt an otherwise perfect lecture with a question. He drowned soon after at the annual physics department picnic.

The material of Schwinger's course was sheer dynamite. Much of it was his own research, so new that he had not yet had the time to publish it in the journals. Everything was presented in the unique Schwingerian format. It took me years to be able to translate what he had taught into the vernacular of the hoi polloi.

Toward the end of my second graduate year I decided to pop Schwinger the question—that is, I wanted to ask to become one of his graduate students. I screwed up my courage, since Schwinger had the reputation of selecting only the very best students. I was dismayed to discover that ten of us had the same thought at the same time. Schwinger interviewed us together. He gave us each a problem and told us not to return until we had solved it. That way, he thought, he could deal with us one at a time. His mistake was that he gave us all the *same* problem. Of course we collaborated and returned a few days later with the solution in our collective hands. Stuck with his troops, Schwinger assigned each of us a different problem that could serve as a beginning to serious doctoral research. In the end, he accepted us all as research students.

Schwinger's career at Harvard was particularly noteworthy both for the large number and the exceptional quality of his graduate students. One of us, a Spaniard named Luis Garrido, did not succeed in obtaining his doctorate. However, just the fact that he had been a student of Schwinger's was enough for him to secure a position of great power in the Spanish physics establishment.

Among Schwinger's ten new students were Danny Kleitman, my old buddy, and Charles M. Sommerfield. Charley, a New Yorker who had attended Brooklyn Polytechnical Institute, had the most miserable taste in movies. If he liked a flick we avoided it. If he hated it, it was a must! We called him Wrong-Clue Summerfoot, but loved him all the same. Schwinger gave Charley a problem in electrodynamics, Danny a problem in the strong force, and I inherited the weak. Three forces for three friends.

Charley's problem had to do with the magnetic moment of the electron, the measure of just how strong a magnet it is. Dirac provided the first, crude answer to this question with his famous equation of 1930. However, he did not know how to take into account small corrections due to the influence of the electron's very own electric and magnetic fields. One of Schwinger's greatest achievements was to show how this could be done. In the late 1940s, Schwinger had calculated the magnetic moment to a precision of one part in a thousand. More recently, his disciples alleged they had improved the calculation to one part per million precision. However, their result was in disagreement with the latest experimental measurements of the quantity. Charley's task was to redo the calculations using Schwinger's new and improved method.

It took poor Charley three years to complete the calculation and get it right. But his reward was not only his Harvard doctorate but, ultimately, tenure at Yale, where he remains happily ensconced to this day. Occasionally he has worked on even more precise calculations of the magnetic moment of the electron, but today this activity is best performed by a Japanese supercomputer.

Charley found his charming wife, Linda, through a computer dating agency. She studies the physiology of taste and boasts of the largest private pornography collection in all of New Haven. They are indeed obscenely happily married.

Charley and I took two big trips together during our graduate years. During the course of one summer we drove some 12,000 miles through Canada, the States and Mexico. It was the first time we experienced the grandeur of the American West. It was not at all like New York, Boston or even Ithaca! The ostensible purpose of the trip was to attend the first "Latin American Summer School in Physics," where distinguished physicists from all over the Western Hemisphere gathered in Mexico City to lecture us students. This was the first of many such summer schools I was to attend, either as a graduate student or a speaker: a dozen or so in Italy, three in France, two in Scotland, two in Turkey, and one each in several other countries.

Toward the end of our graduate career, Sommerfield and I both contacted the Boeing Company in Seattle about possible employment opportunities. First-class air tickets appeared as if by magic and we were off on our second grand tour. In our brief interview at Boeing we were asked whether we had competence or interest in the subject of missile nose cones or flush-mounted radar antennas. We didn't, and that was it for Boeing. In those days, a transcontinental air ticket permitted an almost unlimited number of free stopovers. We did.

Danny Kleitman was assigned a problem having to do with strong interactions and involving the search for some kind of systematic order among the many hadrons. In the meson theory of nuclear structure, pions and nucleons were regarded as elementary particles. The Pakistani physicist Abdus Salam had shown that this was a "renormalizable" theory, which at that time was a buzzword for a theory's being sensible. Yet all these strange particles were floating around that didn't fit in. Called "strange" because their properties seemed inexplicable, they were produced very copiously by cosmic rays. This meant that they partook of the nuclear force—they were strongly interacting particles, hadrons.

Since they were created by the strong interactions, it was reasonable to expect that they would decay by strong interactions and consequently display exceedingly short lifetimes. But they did not! They had relatively long lifetimes, just as one would expect if their decays proceeded through the weak force.

The concept of "relatively long lifetime" deserves some clarification. Strange particles have lifetimes ranging from a tenth of a nanosecond to ten nanoseconds. (A nanosecond is a billionth—10^{-9}—of a second; there are as many nanoseconds in one second as there are seconds in thirty-two years!) Certainly, strange particles do not have long lifetimes in everyday terms. Yet, it is enough to permit an energetic strange particle to traverse many meters before decaying. To the particle physicist, a nanosecond is an *eternity*. Particles that decay by virtue of strong interactions have lifetimes a trillion times smaller. They barely have time to escape from an atomic nucleus before the time comes for them to decay. You have to spot them quickly! The great mystery was how the strange-particle lifetimes could be a trillion times longer.

The resolution to this seeming paradox emerged in the early 1950s. The Dutch-born physicist Abraham Pais put forward the ad hoc hypothesis that strong interactions simply could not produce or destroy strange particles singly. For unknown reasons, it could produce or destroy only *pairs* of strange particles. The collision of an energetic proton with a nucleus at rest could sometimes produce two strange particles but never just one. Once these two strange particles had separated from one another, the strong force could no longer undo what it had done. It's as simple as 1-2-3. Production of strange particles could be copious, yet the particles could be long-lived because their decays had to invoke the weak force.

Pais's hypothesis was no mere codification of experimental data. He offered the daring prediction that the strange particles, when produced and observed in the laboratory, would never appear singly but only in associated pairs. Pais's conjecture was established as fact by careful cloud-chamber experiments performed in 1954 at the Cosmotron accelerator located at Brookhaven, New York.

Why does the strong force satisfy the curious law of associated production? To answer this question I have to explain what the physicist means by a conserved quantity, or a "quantum number." Quite simply, a conserved quantity is a measurable quantity that does not change over the course of time, unlike my age, my weight, or my net worth. Two good examples are energy and electric charge. In any known process, whether it be a chemical reaction, a physical collision or a radioactive decay, the total electric charge of the initial group of particles is identical to the total charge of the particles at the end of the process. Electric charge is a conserved quantity. Similarly, the total energy (including the energy equivalent of the mass of the reacting particles) is conserved.

Another quantity that appears to be conserved, and thus is said to be a useful quantum number, is called baryon number. This is the number of neutrons, protons and other baryons involved in a process, minus the number of antibaryons involved. If baryon number were not a good quantum number, if baryons were not conserved, then atomic nuclei could decay into electrons and other particles. The whole world would be madly radioactive, if indeed it were still there. Thank God that baryon number is exactly conserved—well, almost exactly. (More on that later.)

Murray Gell-Mann did the obvious thing, just as he has done so many times. He is most famous because he does things first, and things are often obvious only after they have been pointed out. Gell-Mann invented a new quantum number whose conservation law could explain the observed systematics of strange particles. He called his new quantum number "strangeness" simply because it was a property uniquely characteristic of strange particles.

To each hadron he assigned a numerical value called its *strangeness.* To the pions and nucleons, which are not strange at all, he assigned a strangeness value of zero. The kaon, a strange meson, received a strangeness value of +1, while the lambda, a strange baryon, got a strangeness of −1. Pais's hypothesis of associated production was encompassed by the assumption that strong interactions conserved the strangeness quantum number. Strong interactions could produce a pair of strange particles, but not just any pair. They had to produce a pair of particles with opposite and canceling values of strangeness. Strange particles could not decay through the strong force simply because the strong force could not do away with their strangeness quantum number.

With the realizations (1) that there is a strangeness quantum number borne by strange particles, (2) that strong interactions conserve strangeness, and (3) that weak interactions may change strangeness by one unit, much of the strangeness of strange particles was explained. One could still ask, Why is there a strangeness quantum number, and why does it behave as it does? Today we can answer such questions simply. A strange particle is simply a particle containing a strange quark or a strange antiquark. Not even Murray knew this in the 1950s. He had not yet invented quarks.

While Murray's work made strange particles a little less strange to us graduate students, their existence was still entirely baffling. Why do they exist? How do they fit into a theory of the structure of matter? (Today these questions are still with us, but in an altered form: Who needs a strange quark?) Be that as it may, here is my catechism as a graduate student, a table of the strangeness assignments to each of the fifteen hadrons that were known in 1954:

STRANGENESS NUMBERS OF THE HADRONS

Strangeness Numbers of the Hadrons

Symbol	Mass(MeV)	Charge	Strangeness	Baryon Number
Baryons				
p	938	1	0	1
n	939	0	0	1
Λ^0	1115	0	−1	1
Σ^+	1189	1	−1	1
Σ^0	1192	0	−1	1
Σ^-	1197	−1	−1	1
Ξ^0	1315	0	−2	1
Ξ^-	1321	−1	−2	1

Symbol	Mass(MeV)	Charge	Strangeness	Baryon Number
Mesons				
K^+	494	1	1	0
K^0	498	0	1	0
π^+	140	1	0	0
π^0	135	0	0	0
π^-	140	−1	0	0
$\bar{K}^0$	498	0	−1	0
K^-	494	−1	−1	0

Listed in the table are eight baryons, which each have baryon number one, and seven mesons with baryon number zero. In addition to their strangeness assignments, the particle masses (measured in the energy equivalent of millions of electron volts, MeV) and their electric charges are shown. I have left out the eight antibaryons. Each antibaryon has the same mass as the corresponding baryon, but its charge, strangeness and baryon number are opposite in sign.

At the time that Danny had been given the problem of finding some kind of order among the welter of hadrons, most of the information in this table was brand-new, coming from experiments that had just been done.

Here are some examples of the kinds of observed reactions between particles that interact by virtue of the strong force that Danny had to deal with:

$$p + p \longrightarrow p + p + \bar{p} + p$$

A collision takes place between two very energetic protons and yields not only a third proton but an antiproton $\bar{p}$ as well. This reaction can take place only if the collision energy is sufficient to produce the new proton and antiproton. Energy is conserved, and so are charge and baryon number. In the reaction

$$p + p \longrightarrow n + n + \pi^+ + \pi^+$$

two positive pions are produced, while the original protons are converted into neutrons. Again, energy, charge and baryon number are conserved.

Now, in processes involving strange particles:

$$p + p \longrightarrow p + \Lambda^0 + K^+$$

we have an example of associated production yielding two strange particles with opposite strangeness. Energy, charge, baryon number and strangeness are all conserved. Finally, consider a process in which a negative kaon collides with a proton:

$$K^- + p \longrightarrow \Xi^- + K^+$$

In this case, one of the initial colliding particles carries a strangeness of -1. Both of the final particles are strange, one with strangeness -2 and one of $+1$. Once again the strong reactions are seen to obey Gell-Mann's rule that strangeness is conserved. In fact, his rule has been tested under many circumstances and it always works.

Of course, Murray didn't invent the concept of strangeness and its conservation any more than Einstein invented the theory of relativity. He discovered strangeness, he gave it a name, and he guessed (correctly!) that it was conserved by the strong force. As a fundamental law of nature it was always there, waiting to be recognized by mortal man.

Take another look at the table of hadrons and notice something queer. The hadrons often appear in multiplets of two or three with nearly (but not quite) the same masses. The nucleons (neutrons and protons) form a doublet of particles whose masses are the same to about a tenth of a percent. The kaons form another doublet, as do their antiparticles. Both the pions and the sigma particles form triplets. Why is this?

The point is that the strong force is not only strong but blind. It cannot distinguish which is the proton and which is the neutron. For example, you could replace the proton field by an arbitrary mixture of the proton and neutron fields, and the strong interactions would be none the wiser. As far as the strong force is concerned, neither the proton nor the neutron, nor any combination of the two, has any special entitlement. Except for weak and electromagnetic effects, the proton and neutron are entirely equivalent and coincident to one another in all their properties. This symmetry is called isotopic spin conservation, and it was first thought of by Werner Heisenberg in the 1930s.

Isotopic spin symmetry demands that the strongly interacting particles, or hadrons, must come in families with almost identical properties. The proton and neutron form a nuclear family of just two particles. The pions, π^+, π^0 and π^-, form a family of three: the members of a family differ electrically, but not at all in their nuclear properties. *All* observed hadrons are seen to appear in such families. There are singlets (bachelor particles like the lambda), doublets (like nucleons or kaons), triplets (like the pions) and even quartets. Indeed, the first quartet of hadrons, known as the delta, was revealed by experiments in the mid 1950s.

It is a curious fact that no hadron family (more precisely, no isotopic multiplet) has ever been discovered with more than four family members, even though over a hundred hadron multiplets are now known. The theory of isotopic spin admits the possible existence of large families, but the hadrons seem to display remarkable self-control. This was once a fascinating mystery. It, too, was explained in terms of today's quark theory of hadron structure.

What is incredible to me, as I write, is how so many of the great mysteries of matter have been solved since I was a student, and how simply and elegantly.

What is most remarkable about isotopic spin symmetry is that it is only an approximate symmetry of nature, not an exact symmetry. In this respect it is most unlike the laws of conservation of energy or of electric charge. Clearly there is a fundamental difference between the neutron and the proton. The proton is electrically charged, but the neutron is neutral—and a little bit heavier than the proton. The electromagnetic interaction clearly senses a difference between the proton and the neutron to which the strong force is entirely blind. It is as if only the lowly servant could distinguish the sex of the chicks, not the powerful mogul who owns the chicken farm. The strong force respects the isotopic spin symmetry; the electromagnetic interaction does not.

The idea of an approximate symmetry law is not entirely alien to everyday life. To a first approximation, the directions north, south, east and west are entirely equivalent to a person lost in a trackless desert. Fortunately for him or her, the symmetry is not exact. The sun rises in the east and sets in the west. The prevailing winds are in a well-defined direction. The earth's magnetic field acts upon a compass and guides our lost friend to civilization.

However, the notion of approximate symmetry in a fundamental theory of nature is admittedly a bit repulsive. Laws of nature would be more seemly if they were either exactly true or dead wrong. Physics, unlike economics or other sociopathic sciences, is supposed to be an exact science. Well, tough! The laws of physics are what we see them to be in the laboratory. If God, in his wisdom, wanted a symmetry of strong interactions that is not shared by electromagnetism, then that is what he got. Anyway, in today's theory, three decades removed from Danny Kleitman's time of trial, we do understand things quite a bit better.

Once again, the appearance of approximate isotopic spin symmetry is clearly explained by today's theory of quarks and the color force. The idea of a symmetry principle applicable to some forces but not to all was an essential artifice in its time, and it still plays a useful role. Our instincts, however, were correct. Approximate symmetry is not part of the fundamental nature of things, but only a secondary manifestation of an underlying yet remote and inaccessible "reality."

Gell-Mann realized that his strangeness quantum number, like isotopic spin, could be only an approximate symmetry of nature. It is indeed a symmetry of the strong force and electromagnetism, but it could not possibly be a symmetry of the weak force. Eventually, a strange particle does decay. It comes apart into particles that do not carry any strangeness number. For example, the strange kaon decays into two or three pions, thus violating the law of strangeness conservation. It seemed as if there were a hierarchy of symmetry principles. The strong force respects both the isotopic spin and strangeness symmetries. The feebler electromagnetic force conserves strangeness, but not isotopic spin. The weak force has the least symmetry and conserves neither isotopic spin nor strangeness. Could there be a grander level to the symmetry hierarchy, a scheme that included both strangeness and isotopic spin in which the underlying pattern of particles would be revealed?

Perhaps the eight known baryons (which formed two doublets, a triplet and a singlet) could be organized into a single supermultiplet. Perhaps kaons and pions could be related to one another. Perhaps a unifying and predictive pattern of hadrons could be found, an order among the hadrons analogous to Mendeleev's periodic table of the chemical elements. The search for such a "higher symmetry scheme" was a central issue among theoretical physicists of the late 1950s.

Schwinger's entry into the higher symmetry sweepstakes was known as global symmetry, and poor Danny was assigned the problem of exploiting this conjectural scheme. He did a credible job and even deduced a formula that was to relate the various masses of the eight known baryons to one another. However, the formula didn't work too well. In retrospect, this was not surprising. Global symmetry, if the truth be told, was all wet. Schwinger's effort was a noble one, but it was simply wrong. The correct higher symmetry scheme was discovered much later, long after Danny and I had graduated from Harvard.

It was not at all unusual in theoretical physics to spend a lot of time on a speculative notion that turns out to be wrong. I do it all the time. Having a lot of crazy ideas is the secret to my success. Some of them turned out to be right!

Danny's work on global symmetry was a bit weak, in itself, to form a Harvard thesis. His work with Schwinger was combined with some research on nuclear physics done under Roy Glauber's supervision. My own doctoral work involved Schwinger's wild speculation on the true nature of the weak force, and it is the subject of the next chapter.

By the summer of 1958, Charley, Danny and I had completed our theses. Actually, I missed the deadline, which is why I did not officially receive my Ph.D. until 1959. Both Danny and I were awarded National Science Foundation Postdoctoral Fellowships with the astonishing stipend of $283.33 per month. We went off to "finishing school" in Copenhagen together.

Many of the physics-oriented men (and a few women) whom I encountered at Cornell and Harvard were to establish themselves as professional physicists in academia, with industry, or at government laboratories. Eight of the ten of us who began our research with Schwinger followed this route. The other two became mathematicians. However, often a person trained in physics makes his mark in a totally different field of endeavor.

I have known physicists who have become surgeons, lawyers, entrepreneurs, inventors and successful investment brokers. Many switched to biology, where they have made significant contributions. Others work on international relations, or on arms control, for our government. One became a famous French songwriter, another (a student of mine) is an even more famous computer scientist, and one is a staff writer for *The New Yorker* magazine. One who has remained a physicist is founding a public school for gifted students in Los

Angeles. I recently nominated another for the Journalist-in-Space program. Many have written best-selling books.

Do you have a child about to enter college who doesn't know what to major in? Tell him or her to learn physics, whose study clearly builds sound minds in strong bodies.

The Mystery of the Weak Force

TODAY AT THE HARVARD PHYSICS department all of the graduate students are provided with office space. Many of the offices, on the top floor of the hundred-year-old Jefferson Physical Laboratory, enjoy a superb view of the Charles River and the new Boston skyline.

This is not the way things were when I was a graduate student. In my day, only those students who were accepted for research, and were beginning to write their theses in earnest, had a right to a desk. We theorists were relegated to the windowless warren of basement offices known as B-24, which had been, in the distant past, the habitat of Harvard's version of the World's First Computer. There wasn't much of a Boston skyline in those days, anyway.

With no windows, there was no indication of the passage of time. Day and night, winter and summer were all the same. We had at our disposal desks, chairs, antique mechanical calculating machines, and a dusty and neglected collection of useless books and journals. Each of us was expected to produce a magnificent contribution to human knowledge, our thesis. Then and only then would we be released.

It was to B-24 that Sommerfield took his problem on quantum electrodynamics and Kleitman his puzzle on the pattern of the strong force. I repaired to that subterranean world to pursue Schwinger's idea concerning the weak force. The experimental situation was changing very rapidly. It was a time of great discoveries, and a time for me to learn a lot more physics.

I used to say, in those days, that there were two different kinds of theoretical physics: radial and angular. The schism has to do with the solution to a textbook problem in quantum theory, the behavior of the hydrogen atom. As it is usually taught, the analysis neatly separates into two parts. The position of the electron orbiting the proton is described by two angles (the latitude and longitude upon an imaginary sphere centered on the proton) and by a radial separation. The angular part of the problem can be dealt with independently, and can be addressed with elegance and simplicity by the purely algebraic techniques of group theory. The radial problem, on the other hand, involves a messy differential equation needing the methods of advanced calculus for its solution. Much of the rest of physics separates in just this fashion. Some problems are amenable to the trickery of imaginative mathematics, while others need brute force and sheer computational "number-crunching" power.

I have always preferred to work on angular problems, leaving the hard work of radial problems to others. My father the plumber once told me that he had repaired a complex heating system with a single well-placed kick of his foot. Balking at a seemingly exorbitant bill, the owner asked my father why he should pay so much for so little. "The kick was free," said my father. "The charge is for knowing just where to kick." Like father, like son.

Angular physics consists largely of the study and exploitation of the symmetry properties of physical laws. A symmetry is simply an operation under which the system doesn't change. For example, the letter A is symetric under mirror reflection, and the letter Z is symetric under being turned upside down.

A symmetry more relevant to physics is time translation invariance. It says that the laws of physics do not change with time. They were the same yesterday as they will be tomorrow. Dinosaurs, had they been smart enough, could have measured the magnetic moment of the electron two hundred million years ago and gotten the same answer that Charley Sommerfield computed.

It is a curious but important fact that every symmetry principle entails, as a logical consequence, a conservation law for a physically measurable quantity. The time invariance of physical laws is the symmetry principle that implies the conservation of energy, the fact that energy can neither be created nor destroyed. Thus the impossibility of perpetual motion machines follows from the constancy of physical law, from the fact that night will always follow day.

Another fundamental physical principle says that the laws of physics are identical everywhere in the universe. Arthur, the intelligent alien from Arcturus, measures the same value for the electron's magnetic moment as we (or the dinosaurs) do. Physical laws are invariant under space translation. The conservation law for momentum is the logical consequence of this symmetry. From the principles of momentum and energy conservation you can understand the physics underlying the game of billiards. But it may not do much for your game.

Here's another symmetry: Australians observe the same laws of nature as we do despite the fact that they, relative to us, are standing on their heads. The orientation in space of the observer is of no consequence. Physical laws are symmetric or invariant under rotations. This symmetry law is a bit more complicated than space or time translations. Rotations do not "commute" with one another. This means simply that the order in which two successive rotations are done matters very much. An example:

A fellow walked from CalTech to Harvard by going north until he reached Harvard's latitude, then heading dead east. He recorded the distances he walked and wrote of the wonders of his trip to his twin sister back at CalTech. She decided to make the same trip, but since she preferred the warmth of a more southerly route she chose to walk east first and then north. It was a fatal error. Because the earth is round, she ended up at Cornell rather than Harvard. Maybe that's why they put Cornell where it is.

In exactly the same way, the order in which rotations are done affects the final orientation of a body. Again, rotational symmetry implies a conservation law. The quantity that is conserved is angular momentum. When a spinning ice skater brings her arms inward her rate of spin increases. This may be the most familiar demonstration of the law of conservation of angular momentum, the understanding of which is essential for every physicist—and figure skater.

Energy, momentum and angular momentum are conserved quantities associated with symmetries in space and time. There are other conserved quantities, like electric charge, that do not seem to be closely associated to the geometry of space and time. Charge conservation results from mathematically manipulating the fields describing the charged particles without producing any physical consequences. It is a so-called internal symmetry that is not so obviously related to behavior in space and time. We have already encountered other examples of conservation laws that follow from internal symmetries, like baryon number, strangeness and isotopic spin.

The geometrical symmetries of nature have another important consequence in the context of quantum theory. They imply that every elementary particle must possess two inalienable characteristics: mass and spin (intrinsic angular momentum). The mass may be a positive number or, as in the case of the photon, zero. Every particle of a given species must have exactly the same mass. Thus, every single electron weighs just 9.105953×10^{-28} gram, no more and no less. It is as if all the zillions of electrons in the universe were produced in a single factory whose quality control is better even than the Japanese.

Spin, as the word suggests, is a measure of the intrinsic angular momentum that a particle possesses. Its value can be any whole number, or half a whole number, when measured in terms of the fundamental unit of angular momentum. Niels Bohr introduced the idea of a fundamental quantum unit of orbital angular momentum into the world of physics in 1911. He argued that the angular momentum of any system of particles, such as an atom, had to be a whole-number multiple of his fundamental unit—that is to say, angular momentum is quantized.

Bohr did not realize that an elementary particle carried a built-in irreducible angular momentum: its intrinsic spin. The spin angular momentum of a particle can be either a whole number of fundamental units or half a whole number. Particles with integer (whole number) spin are called bosons, and those with half integer spin are called fermions. Fermions satisfy the Pauli exclusion principle; bosons do not. The structural components of matter, nucleons and electrons, are all ½-spin fermions. Force-carrying particles, like the photon, are bosons.

When Yukawa first predicted the pion, in the 1930s, he actually suggested the possibility that both the weak and strong nuclear interactions were generated by the exchange of force-carrying bosons. For a time he believed that the pion itself could mediate both forces. Yukawa was the first to attempt to unify the strong and weak interactions. He soon realized, however, that nature was not as economical as he was ambitious. That idea didn't work at all. Today Yukawa is remembered for his successful prediction of the existence of the intermediate boson for the strong force, the pion.

The credit for proposing an intermediate boson for the weak force is usually given to Italy's successor to Galileo, Enrico Fermi. The many things named after him include Fermi surfaces, Fermi statistics, fermions and the fermi, a unit of a quadrillionth of a meter that is used as a yardstick for measuring nuclear sizes. Fermi was responsible for building the world's first nuclear reactor during World War II, under the University of Chicago squash courts. With pions discovered, Fermi's hypothesis seemed more attractive than ever.

UNSTABLE PARTICLES

Most of the so-called elementary particles are unstable. They live for a short time before decaying into other elementary particles. The decay products, in turn, may also be unstable.

Some particles decay relatively slowly because their decay requires the participation of the weak force. Others decay more rapidly, by means of the electromagnetic force, or even quicker, by means of the strong force.

Decay Scheme	Lifetime	Mechanism
$\rho \rightarrow \pi\pi$	10^{-23} sec	strong force
$\omega \rightarrow \pi\pi\pi$	10^{-22} sec	strong force
$\eta \rightarrow \pi\pi\pi$	10^{-18} sec	electromagnetism
$\pi^0 \rightarrow \gamma\gamma$	10^{-16} sec	electromagnetism
$\pi^+ \rightarrow \mu\nu$	10^{-8} sec	weak force
$K^+ \rightarrow \pi^+\pi^0$	10^{-8} sec	weak force
$n \rightarrow pe\bar{\nu}$	1000 sec	weak force

The table shows the decay schemes and lifetimes of a few hadrons. Notice that strong decays (ρ and ω) take place much more rapidly than electromagnetic decays (η and π^0), which themselves are much more rapid than weak decays. The neutron, because it is not much heavier than the proton, has quite a respectable lifetime in human terms.

All three forces (gravity is always the holdout!) would then be put on a similar footing: photons carry the electromagnetic force; pions, the strong force; and Fermi's hypothetical new boson, the weak force.

Over the course of time, many names were proposed for the conjectural charged particles whose exchange could mediate the weak force. Gell-Mann called it the Uxyl, and T. D. Lee introduced a variant of such a particle that he named the Schizon. Schwinger used the letter designation W, and others used X or Z. Most physicists, if they believed in such a speculation at all, used the phrase *intermediate vector boson*, or its acronym, IVB. The word *vector* was physicists' jargon for the expectation that the weak boson would bear one unit of spin, just like the photon.

UNIVERSALITY OF WEAK FORCES

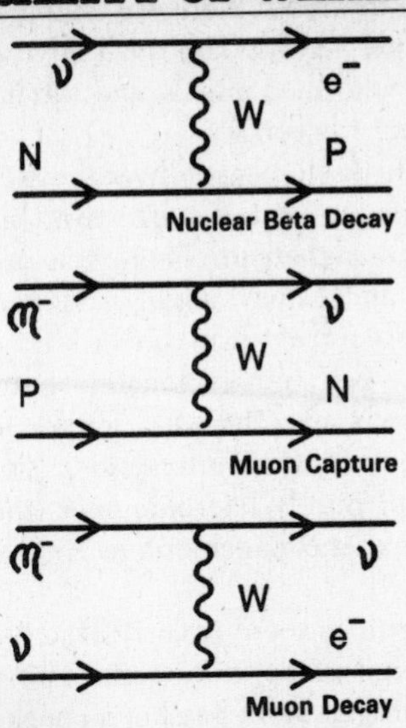

It has been known since the 1950s that the weak force manifested the same strength in three disparate processes: nuclear beta decay, muon decay and the nuclear capture of negative muons. Feynman's diagrams shown here involve three fundamental acts: emission of a W by a nucleon, an electron or a muon. Experimentally, each process is characterized by approximately the same universal strength.

The trouble was that the idea of an IVB did not seem to be very useful. For one thing, no one had ever seen such a particle in the laboratory. It was not to be discovered until 1983! The idea didn't seem to have any decisive experimental tests, either. The concept was elegant but not predictive. Elegance is a nice thing, but by itself it is not, and should not be, convincing evidence of truth. To be valid, a theory must not simply explain the results of experiments that have already been done, but must point the way to new ideas and new observations.

There was another problem. With or without an intermediate vector boson, the theory of the weak force was in deep trouble. It was not "renormalizable," which meant that attempts to calculate a weak process with precision produced results that were infinite and meaningless. Back in the 1950s there was simply no acceptable theory of the weak force, only a crude and qualitative description and codification of the observed phenomena, and the hope for a better theoretical understanding tomorrow.

The intermediate vector boson hypothesis was founded upon a perceived analogy, imperfect but suggestive, between the weak interactions and electromagnetism. There was one big hint from experiment that such an analogy might have something to do with reality, after all. Both forces displayed a kind of universality. The electromagnetic force acts indiscriminately between two electrons, two protons, or one of each. The weak force is just as promiscuous: there are purely leptonic weak interactions (the muon decays into three lighter leptons), purely hadronic ones (the kaon often decays into pions), and those that connect leptons and hadrons together (like nuclear beta decay).

Electrons and protons seem to carry exactly the same amount of electric charge, though of opposite sign. The consequent electric neutrality of the hydrogen atom has been checked in the laboratory to a remarkable twenty-two decimal-place accuracy. As the most precise measurement ever performed, this result qualifies for the *Guinness Book of World Records*. Weak interactions also display a universality of interaction strength. The weak force is observed to be just as strong in nuclear beta decay as in muon decay and in many other disparate weak processes.

Schwinger believed that the weak and electromagnetic forces might actually be two aspects of the same phenomenon. To me he presented the tantalizing possibility of unifying what were then thought to be two basic but unrelated forces of nature. It was a problem that would lead me eventually to the Nobel Prize.

Nonetheless, only a very few grown-up physicists believed in Schwinger's fancy of unifying the weak and electromagnetic interactions. (I, of course, as his novice graduate student, had little choice in the matter.) After all, the weak interactions were so very weak compared to electromagnetism. This could be explained away, but only in a seemingly ugly and artificial fashion. It had to be believed that the W-boson was very, very heavy: about a hundred times heavier than the proton. This meant that it would be awfully hard to produce a W-boson in the laboratory. It's all a question of conservation of energy. If a collision of two elementary particles is to result in the production of a W-boson, then the energy of the collision must be large enough to generate a W-boson's mass.

By the early 1950s the antiproton (with about one percent of the W-boson's mass) had just been successfully produced and seen. There appeared to be little chance that a particle as heavy as the W-boson could ever be studied in the laboratory. The more massive a particle is, the more power an accelerator must have to produce it. No existing particle accelerator was even remotely big enough to do the job. Schwinger's notion appeared to be more like medieval theology than the physics of observable stuff. Who would dream that the European Center for Nuclear Research (CERN), which was just then being established in Geneva, would discover Schwinger's particle two and a half decades later, thus marking the renaissance of particle physics in the Old World?

Most physicists of the 1950s were unwilling to contemplate an elementary particle with such great mass. They argued that a particle so heavy could not possibly be married to the massless photon in a hypothetical electroweak synthesis. They were wrong.

Another apparent obstacle to Schwinger's notion came from experiment. If the W-boson were to be related to the photon, then it had to have the same spin as the photon. Both particles had to be what are called *vector bosons*. The experimental results available in the early 1950s seemed to show that this could not be the case. It looked as if the W-boson, if it existed, had no spin at all. The reason for this was that three of the key experiments had been done wrong. All three were eventually to be retracted. Had Schwinger or I taken these data too seriously I would not have written a thesis about the intermediate vector boson and I would not be writing this book. The idea of unifying the weak and electromagnetic forces was too tantalizing to give up. Sometimes (but not too often) faith in one's own ideas should override mere observation.

As I progressed in my doctoral research, the experimental conflict eventually evaporated. It turned out that the properties of weak interactions were just as one would expect if the boson were, like the photon, a vector boson.

To realize the dream of unifying the weak and electromagnetic interactions, a specific theoretical framework was necessary in which to implement Schwinger's hunch. Just such a mathematical construct was invented in 1954. It has the forbidding name of non-Abelian gauge theory, and it will take a bit of explaining (see box on Gauge Theory).

GAUGE THEORY

Physicists always seek simple and elegant principles from which the minutiae of natural phenomena may be deduced logically. The discipline of geometrical optics, for instance, describes the passage of light rays through complex systems of lenses, prisms and mirrors. This is the science that governs the design of cameras, microscopes, telescopes and even tinted contact lenses.

Way back in the days of ancient Greece, Hero (that was his name, honest!) noted that a light ray proceeding from one point to a mirror and on to a second point invariably follows the shortest possible route. In the seventeenth century the famed mathematician Pierre de Fermat proved a powerful generalization of this result: the path of a light ray is always such as to minimize its time of transit, even when it passes through a complicated set or mirrors or lenses. This curious result can be established from the laws of optical reflection and refraction. Conversely, and more to the point, all the laws of optics can themselves be deduced from this "principle of least time." The science of geometrical optics is embodied in this one simple principle. What an elegant reformulation of optical theory!

In a similar way, all of Newtonian mechanics can be reexpressed in terms of the "principle of least action." Its formulator, the French mathematician Pierre Louis Moreau de Maupertuis, wrote in 1732: "Here then is this principle, so wise, so worthy of the Supreme Being: Whenever any change takes place in Nature, the amount of action expended in this change is always the smallest possible."

Principles of this kind really don't tell us anything new about the world. They simply represent elegant codifications or concise representations of what is already known. Often they may suggest useful calculational procedures. Sometimes, however, they may indicate new and fruitful directions for research. In the case of electromagnetism, it was the generalization of such a principle that led to our present theories of the strong and weak nuclear forces.

Electromagnetic gauge invariance is the simple principle (despite its fearsome-sounding name) that underlies and embodies electromagnetism. In quantum theory, each of the various types of particle in nature is described by a mathematical device called a quantum field. According to the principle of gauge invariance, each of these fields may be redefined in a certain fashion at each point of space and time. Such a redefinition, or gauge transformation, has no effect upon physically observable phenomena.

The contemporary usage of the word *gauge* in physics dates to the work of the German mathematical physicist Herman Weyl in 1918. Weyl called a gauge "any function introduced into the equations of the potentials of a field such that the derived equations of observed physical quantities are unaltered by the introduction." Somewhat more transparently, Arthur Eddington wrote in 1923 that "a change of gauge is a generalization of a change of unit in physical equations, the unit being no longer a constant but an arbitrary function of position."

Imagine that people used as their measure of length the inch in New York and the mile in Los Angeles, with a smooth and continuous gradation in between, the unit getting longer and longer as one went westward. While such a convention—a kind of gauge transformation—would surely lead to chaos, it would not affect the physical distance between New York and Los Angeles. No matter which units were used to express the distance, the East and West coasts would remain the same distance apart.

Maxwell's electromagnetism, unlike the U.S. Bureau of Standards, accepts such peculiar conventions naturally. One may meddle with the mathematical definition of the electromagnetic potential everywhere without affecting the physical consequences anywhere. Gauge invariance is a bizarre, but sometimes computationally convenient, aspect of the theory of electromagnetism.

But gauge invariance is far more than a convenience or a curiosity. Just as the theory of optics follows from the principle of least time, and the theory of mechanics is epitomized by the principle of least action, so does the whole kit and caboodle of electromagnetic theory follow from the principle of gauge invariance. All of electromagnetic theory, including the very existence of the photon, and especially the fact that it is a massless particle with a unit spin, may be logically deduced from the principle of gauge invariance alone. It is strange and surprising that a requirement for symmetry has as a consequence the existence of a force. How wise is this principle, so worthy of a Supreme Being that it sayeth for him, "Let there be light!" And how tempting it is to extend this principle so as to generate all the forces of nature. How wonderful it would be if all of the fundamental properties of matter could be understood as deductive consequences of an overriding symmetry principle.

In 1954 C. N. (Frank) Yang and Robert Mills invented a generalized principle of gauge invariance that led to a different form of gauge theory, richer and more elaborate than electromagnetism. They believed that the strong nuclear force could be coaxed into existence by their generalization of electromagnetic gauge invariance.

Their idea was that the redefinition of fields that constituted a gauge transformation could mix up the identities of neutrons and protons as well. Thus, Heisenberg's notion of isotopic spin symmetry, in which a neutron can turn into a proton, and vice versa, was promoted into a new kind of gauge invariance. As in the case of electromagnetism, this kind of symmetry implies the existence of new particles whose exchange mediates new forces. The new force must be able to change protons and neutrons into one another. To do this, the gauge bosons (the particles carrying the force, analogous to the photon) must be three in number: one positive, one negative and one electrically neutral.

All three of these particles had to be massless, like the photon, according to Yang and Mills's mathematics. Principles of gauge invariance inevitably give rise to massless gauge particles. For precisely this reason, Yang and Mills abandoned their brainchild.

There simply did not exist in nature the massless gauge particles that their theory called for. Or so they thought. (Much more recently it has turned out that a Yang-Mills–like theory is at the heart of the strong nuclear force. The massless gauge particles of the strong force are known as gluons, the glue that holds the nucleus together. They do exist, and serve to hold together the quarks that form hadrons. Patience! That story comes later.)

Thirty years ago, Yang-Mills theories did not seem to have the least relevance to the physics of the real world. Non-Abelian gauge theories were just a sort of theoretical toy, except in the fertile imagination of my very own thesis advisor, Julian Schwinger.

Schwinger sensed that a Yang-Mills theory could form the basis of a unified theory of weak and electromagnetic interactions. He suspected that the neutral member of the triplet of gauge bosons was the photon. Thus, ordinary electromagnetic gauge invariance was inserted as part of the more general scheme that Yang and Mills had created. Schwinger regarded the two oppositely charged gauge bosons as the intermediate vector bosons W^+ and W^-. In this fashion, the electromagnetic and weak interactions were inextricably linked. They were both parts of a unified theory. The symmetry properties of the theory meant that the W-bosons could be interchanged with the photon, or the weak force with the electromagnetic force, and that such an interchange would have no physical effect.

It is as if the three vector bosons pointed in three mutually perpendicular directions in an imaginary three-dimensional space, with the photon pointing (say) downward, and the Ws aligned north and east. If all three directions were exactly equivalent, electromagnetism would be indistinguishable from the weak force.

Clearly, this is not the way things are in the real world. Electrical currents and magnetic fields are very different from the slow radioactive decay of an atomic nucleus. Electromagnetism and the weak force seem to be very different from one another.

The three directions are not exactly equivalent. One of them is different from the other two. One of the imagined electroweak directions is singled out: the photon is massless and the W bosons are not. Something is breaking the underlying Yang-Mills symmetry. What this something is was not known until 1967. At that time, Steven Weinberg and Abdus Salam identified the symmetry-destroying culprit—the vacuum, acting through its agent, the Higgs Boson. But that is still another story, and we will tell it in its time.

Neither Schwinger nor I had the foggiest idea of what produced the W-boson's mass and removed the perfect symmetry between weak and electromagnetic interactions. It was a most nettlesome question, and it probably led Schwinger to abandon his quest for an electroweak synthesis and pursue other fields of research. I didn't have that option. I had a thesis to write.

Someone, someday, would figure out the problem of the boson mass. Little did I dream that it would be my old red-headed high school buddy Steve. I simply put the mass into my calculations "by hand" as a temporary and ugly expedient, and went on to fry my other fish. Could I adjust the details of my electroweak theory to account for the rapidly unfolding experimental developments?

The detailed properties of the weak force were just being revealed by careful laboratory studies of nuclear beta decay. Strange particles were being produced and examined at accelerators, and they showed mysterious properties. For example, the universality of the weak force did not seem to extend to the decays of strange particles. It was not at all clear how to handle strangeness in any potential theory of the weak interactions. Most important of all, was it possible to make the theory "renormalizable"? (*Renormalizable* is a buzzword that means calculations based on the theory will not go haywire and produce meaningless results.)

In the midst of these cogitations a great bombshell hit. I learned in *The New York Times*, of all places, of the experimental discovery of parity violation. It looked as if my thesis was going to be detoured—maybe over a cliff! To understand why, we ourselves must make a detour for a moment.

Electrons and nucleons are particles with spin. They are a little bit like light bulbs or screws that twist around as they move forward. In principle, there are two kinds of light bulbs: right-handed bulbs move forward as they turn clockwise; left-handed bulbs move backward as they turn clockwise. I have never seen a left-handed bulb, but I am told they are used in certain business establishments so that the employees are not tempted to steal them. (How many employees does it take to screw a left-handed bulb into a right-handed socket?)

Handedness plays an important role in ordinary life. The words *dexterity* and *dextral* (meaning, "on the right") have the same origin. The word *sinister* comes from the root meaning "left hand." More people are right-handed than left-handed, and almost everyone's heart is on the sinister side. But physicists once believed that the laws of physics did not distinguish between right and left, in particular, that right- and left-handed electrons would play exactly the same role in natural phenomena. The evident lack of perfect left-right symmetry in biological systems such as people had to do with the accident (or miracle) of the birth of life on earth.

A left hand, when viewed in a mirror, appears to be a right hand, Yet the fundamental laws of physics were expected to be identical in the real world and in a mirror world. This geometrical principle is known as space-reflection symmetry, and its consequence is the law of parity conservation.

Everyone believed in it except two young Chinese-American scientists, T. D. Lee of Columbia University and Frank Yang, then at the Institute for Advanced Study at Princeton. They recognized that the evidence for parity conservation in nature, while good, was not ironclad. They pointed out that there was no good reason to believe that the weak interactions were unchanged by the space-reflection operation. They even went so far as to spell out experiments that could be done to determine whether or not the weak force conserves parity.

Madam Chien-Shiung Wu, an experimental physicist also at Columbia, accepted the challenge and did the relevant experiment. She and her collaborators carefully studied the beta decay of cobalt-60, a nucleus with a spin. They arranged for all the cobalt nuclei to be polarized, which is to say, all spinning in the same direction. They then observed that the decay electrons emerged preferentially in one direction. It was as if you stood for days on end outside the doors of a department store, and everyone you saw coming through the doors turned and walked away from the store in only one direction.

Madam Wu's experiment was clear proof that the weak force violated parity, and that the electrons produced in beta decay were more often left-handed than right-handed. The discovery was soon confirmed by other scientists. Not only did the weak force violate parity, but it did so as much as it possibly could. Parity violation is now seen to be a dominant characteristic and an experimental signature for the action of the weak force. For "their penetrating investigation of the so-called parity laws which has led to important discoveries regarding the elementary particles," Chen Ning Yang and Tsung Dao Lee, then aged thirty-seven and thirty-one, respectively, were awarded the 1957 Nobel Prize in physics.

There I was, writing my thesis on "The Vector Boson in Elementary Particle Decays" while the rules of the game were being changed with abandon. It was like trying to surf in a hurricane. The next development was the construction of what is called the V-A theory of weak interactions. It was the invention of Feynman and Gell-Mann at CalTech and, independently, of Robert Marshak and George Sudarshan at the University of Rochester.

The V-A theory was not so much of a theory as it was a model. It explained some aspects of parity violation by the assertion that the neutrino is intrinsically a left-handed particle. It was exactly the sort of structure that could be generated by a theory involving intermediate vector bosons.

V, you see, stands for vector, and A stands for axial vector. Intermediate vector bosons could be responsible for both of these forms of the weak force. Other forms of weak interaction, called S, P and T, could have been present in principle and could not have been explained in terms of the intermediate vector boson hypothesis. Mercifully, experiment demonstrated that these types of forces were absent. The weak force showed just those properties that an intermediate vector boson could reproduce. It was an encouraging portent.

Physicists immediately strove to incorporate the V-A model into a consistent and comprehensive theory of the weak force. I became convinced that a unified electroweak theory based on the Yang-Mills formalism was an appealing possibility.

My old high school friend Gary Feinberg was at that time a graduate student at Columbia, and we remained in close contact. Feinberg, the first graduate student of T. D. Lee, realized that an intermediate vector meson could lead to a serious experimental difficulty in theories that incorporated just one species of neutrino. I didn't see this as a problem since Schwinger had convinced himself —and his students—that there must be two distinguishable kinds of neutrinos.

Yet Feinberg's calculation was relevant in another way. The one loop diagram he considered was, in general, infinite. This meant that it generally gave meaningless results and was useless for calculating actual phenomena of nature. Yet it turned out finite, usable results under very special circumstances: when the underlying theory put electromagnetism and the weak force together into a system of the Yang-Mills type. Any other theory involving intermediate vector bosons is not only manifestly not renormalizable and hence unacceptable, but it would spoil the renormalizability of good old quantum electrodynamics. It was for reasons such as these that I was led to the central conclusion of my thesis:

> It has been our declared viewpoint, an attractive if not overly plausible one, that the photon and the [charged W-boson] comprise a family. . . .
>
> It is of little value to have a potentially renormalizable theory of beta processes without the possibility of renormalizable electrodynamics. We should care to suggest that a fully acceptable theory of these interactions may only be achieved it they are treated together, in accordance with our identification of the neutral [W-boson] as the photon.

At this point, many of the problems of particle physics were still unsolved. We did not yet understand that pions, nucleons and strange particles were really not elementary systems, and that the pion-nucleon interaction was not nearly as fundamental as we thought. The weak interactions of strange particles were exceedingly perverse—they did not share the universality of other weak processes. George Sudarshan and I worked on this problem briefly, both at Harvard and at Rochester, but made no progress at all. There was also the mystery of how weak interactions could violate parity while electromagnetism did not. All this would have to wait. I had completed an acceptable thesis.

It was not a great thesis, but it would get me out of Harvard. I had not quite risen to the challenge and constructed a fully acceptable unified theory of weak and electromagnetic forces. However, I had come across some hints that Schwinger's idea might be right. Still and all, my thesis was so very speculative that I began it with a quotation from Galileo: *E forze dire che gl'ingeni poetici sieno di due spezie: Alcuni, destri ed atti ad inventar le favole; ed altri, disposti ed accomadati a crederle.* Loosely translated, Galileo finds it necessary to say that there are two kinds of poetic imagination: there are those who invent fables, and those who are disposed to believe them.

While at Harvard I had painfully shed the two (sequential) girlfriends whom I had inherited from Cornell. One ran off and married Ray Sawyer, a Schwinger student with me, who promptly took her to southern California. The other demonstrated her disdain for physics (and me) by marrying a mathematics professor at Cornell who went on to solve one of the most important outstanding problems in group theory. I had tried to convince her to wed me rather than Walter, but she told me it was too late: they had already bought the furniture.

Quite unencumbered, I decided to spend a postdoctoral year as a scientific visitor to the Soviet Union. (Perhaps I wanted to put my hard-won knowledge of the Russian language to some practical use.) It would be professionally useful and lots of fun. Early in 1958 I wrote to Professor Igor Tamm at the famed Lebedev Institute to see if such a plan was workable. Since I had won a National Science Foundation postdoctoral Fellowship, there was no problem about finances. All I needed was the hospitality of the institute and a Soviet visa. A few months later I received a strongly encouraging letter from Professor Tamm.

In the summer of 1958 I followed Schwinger to the University of Wisconsin, at Madison to take my final doctoral examination. Paul Martin was there, too, for he and Schwinger had turned their powerful analytical techniques to the study of condensed matter phenomena—the subject I had failed at Cornell. I had a wonderful week with my friends. Clarisse Schwinger prepared her best beef Stroganoff ever (fitting, if I was to visit Russia) and baked yet another fabulous cake.

The Martins, with whom I was staying, threw such a successful party that the neighborhood children were evidently instructed to avoid our house for the duration of our visit. Late one night, while

Julian and I were discussing physics in his baby blue Cadillac, we were water-bombed by hostile neighbors. One night was spent in a long and fruitless search for a bottle of Indian chutney. The Midwest is still a bit like that.

My examination committee consisted of Schwinger, Martin, Robert Sachs (a onetime collaborator of Schwinger's who was then head of the physics department at Madison) and Frank Yang. Most of the exam consisted of a heated argument between Schwinger and Yang that was inspired by my casual remark that there were *two* varieties of neutrinos in nature—for so I had been taught by Schwinger. Yang did not realize that the two-neutrino hypothesis was testable and meaningful. It did have experimental consequences, as Schwinger pointed out to him.

Three years later, Schwinger's prescience was rewarded as the "Jewish Mafia of High-Energy Physics" (Leon Lederman, Mel Schwartz and Jack Steinberger), working at the Brookhaven National Laboratory, proved the existence of the second neutrino, winning a tight race with European experimentalists at CERN.

I passed the examination. My parents, Lewis and Bella Glashow, neither of them college-educated, finally acquired three sons who were "doctors."

Back in Cambridge, I discovered that nothing new had happened in the Russian visa bit. It was time to activate Plan II. I had written of my circumstances to the famous Institute for Theoretical Physics at Copenhagen, popularly known as the Niels Bohr Institute. Aage Bohr, Niels's son, suggested that I stay in Copenhagen while my Russian plans congealed. At the end of the summer I set out on my first trip to Europe.

I never did get to spend a year in Russia, even though I did get to know the Soviet consul in Copenhagen quite well. In October 1958 I learned with delight that Tamm along with two other Russian physicists had won the Nobel Prize for the discovery and interpretation of Cherenkov radiation. Soon after the ceremonies in Stockholm I received a second letter from Tamm—still quite encouraging. The letter was mailed from the Grand Hotel in Stockholm, a hotel I would learn to love two decades later when I received the prize.

Meanwhile, I was accumulating a thick stack of correspondence from the Soviet Academy of Science, the U.S. State Department, Harvard University and the National Science Foundation. The red tape was flowing furiously. A year later, when I was visiting CERN,

I received my last correspondence in re Sheldon Glashow versus the USSR. It was from the Soviet embassy at Bern: "[W]e must let you know that we have not received authorization to issue you a visa. Should such authorization appear, we shall call you."

In the end, it was Russian intransigence that led me to win the Nobel Prize in physics. The key contribution that I made to the electroweak theory was accomplished during my pleasant visit to Denmark. And, however discouraged I was not to be let *into* the Soviet Union, I thank my stars that I was not in the position of wanting to get *out*.

6 Wonderful, Wonderful Copenhagen

In September 1958 I sailed to Europe aboard the *Flandres*, a French passenger liner bound for England. My ultimate destination was Copenhagen. Thirty-six years earlier the Danish government had created a research institute for the world's best-known Dane, Niels Bohr. The Universitetets Institut for Teorisk Fysik was known throughout the world as the Niels Bohr Institute. Only after Bohr's death in 1962 did that become its official name. During the 1920s and 30s the Bohr Institute was the unquestioned worldwide hub of theoretical physics. In the 1950s it was still a flourishing institution where young and aspiring physicists could spend a postdoctoral year with the great guru of Copenhagen.

I was to spend twenty-seven months, on and off, at the Bohr Institute between 1958 and 1964. Two of my best research papers were written there. James Bjorken and I were to introduce the fourth (charmed) quark to an unbelieving world in 1964. My paper about the electroweak theory, "Partial Symmetries of Weak Interactions," written there in 1960, was to earn my share of the Nobel Prize.

The Danish physicist Hans Christian Oersted was a contemporary of Hans Christian Andersen of fairy tale fame. Andersen always referred to Oersted as Hans Christian the Great. Oersted had shown, in 1820, that an electric current flowing in a wire causes a magnetic compass needle to move. He was the first scientist to establish a connection of any kind between the forces of electricity and magnetism, just as I was among the first to relate electromagnetism to the weak force. Obviously, Copenhagen was exactly the right place for me to be.

En route, I shared a second-class cabin with Jack Schnepps, a young experimental physicist bound for a postdoctoral year in Italy. He is now the chairman of the physics department at Tufts University, just a stone's throw from Harvard. The fare was $180. My parents and brothers treated us to a last-minute champagne party just before departure. We soon discovered, to our delight, that many of our fellow passengers were college girls spending their junior year abroad. Coming from Catholic schools, they were watched carefully by silent nuns.

Playing poker was permissible. We spent every evening drinking Ricard and gambling with the girls till dawn while the sleepy nuns watched silently. Sometimes a world-weary Frenchman would join us. He told us that the only good butter is French and the only real women were Parisian whores. He lost a lot of money, but reneged at the end of the trip. He was very French.

We disembarked at Plymouth. Jack had purchased a Morris Minor to be delivered upon his arrival in England. The car was almost ready, but its radio had yet to be installed. We were to wait briefly in the salesman's private office until the work was done. A grubby note pinned to a grimy wall left unforgettable instructions from management to its help. Short-term automobile insurance policies could not be issued to certain persons: gypsies, Hungarians, jewelers, furriers and Jews. Jack is a Jew and so am I, but it didn't seem to matter. We were Rich Americans. Fully insured, we drove to London to the Cumberland Hotel at Marble Arch.

Late at night we checked into a magnificent room with heated bath towels fit for giants. We were awakened the next morning by a stunningly beautiful young lady bearing menus. We ordered, then hurriedly shaved and showered and returned to our beds to await the damsel's return. There was a knock at the door and in rolled a flower-bedecked table laden with British breakfast delicacies. It was propelled into our room by an ancient and wizened ogre with a knowing smirk.

Jack and I drove to Paris, whose charms had been sung to me since seventh grade. My teachers hadn't exaggerated. This was no New York, Chicago or even London. Paris proves that the city—simply a place where many people live, work and play—can be one of the marvels of human creativity. No wonder that every Frenchman dreams of a life in Paris, that the Germans wanted it so desperately, and that the tourists occupy it every summer—when the Parisians take to the beach. I spent my days wandering about the city, a pleasure that is even more intense today. I met no "real woman," but I did discover that French butter is special.

Too soon, it was time to take leave of the City of Light. My destiny lay in what the Danes optimistically call the Paris of the North, which indeed it is in the sense that Stanford University (as their sweatshirts proudly proclaim) is the Harvard of the West.

I took the train to Denmark with my three overstuffed Lark suitcases. It took twenty-four hours. In Copenhagen I found a bunk in a student hostel filled with boisterous Germans.

One of the highlights of the standard tourist itinerary of Denmark is a guided visit to one of the major breweries, Carlsberg or Tüborg, and a generous tasting of their products. Carlsberg, in those days, supported Danish science and was a generous contributor to the institute, and Tüborg bankrolled the industrial arts. Today, like the forces of physics, Carlsberg and Tüborg have been unified.

I succeeded in doing both breweries in one beery day. Leaving the second brewery in somewhat of a stupor, I met what I believed was a beautiful woman. While her beau was searching for a taxi in the rain, I succeeded in getting her name and phone number.

Months later I found the slip of paper with a name, Birgitta Torstenson, and a Swedish telephone number. I phoned her and arranged to take the boat across the Øresund to her home town, Landskrona, where she would meet me at the pier. Birgitta turned out to be a sixteen-year-old high school student who had just had a spat with her seventeen-year-old boyfriend, Ragnar. I was to take her to the annual prom that evening. Her somewhat bemused parents were impressed with my academic credentials, but they suggested that upon my next visit they would accompany me to Malmö, where I might buy some decent clothing. I was more nerd than fop.

The prom began with an elegant fashion show, but the drunk and disorderly students, forcibly demonstrating their preference for striptease, soon drove the models off in terror. Ragnar appeared with a girl on either arm, but nonetheless seemed deeply upset to find Birgitta dancing with me. Physical combat was narrowly averted, and I managed to escape on the last boat to Copenhagen. I never saw Birgitta again.

I was warmly welcomed to the institute, where I was assigned a desk, given a set of keys and a tour of the facility, and loaned a decrepit bicycle. Life at the institute was not so different from life at Harvard. I was again surrounded by dozens of young scientists like myself who were devoted to physics and wanted to make names for themselves. We had frequent lectures by eminent visitors, a fine library, distinguished resident scientists to consult, and offices (unlike Harvard's) with views.

On the other hand, the city of Copenhagen, like Boston a maritime city, revealed an entirely different conception of Man versus Nature. Copenhagen lovingly makes the sea an integral part of the city, with its promenades, parks, bicycle paths along the seaside and spanking-clean ferries to other parts of Scandinavia near and far. Boston shows only contempt and disregard, perhaps even Puritan fear, of its potentially magnificent but ruined setting. Things have changed for the better in the New Boston, but its filth and disrepair still shock me after a return from its beautiful Danish sister city.

Let me introduce some of the young and multinational postdoctoral students who were to make my life at the Niels Bohr Institute so thrilling. First of all, there was my chum from both Cornell and Harvard, Danny Kleitman. He was still a physicist and my friend, for only many years later would he become a mathematician and my brother-in-law. Danny was not much better at speaking foreign languages than I. One day he tried to buy a used bicycle in a new bicycle store. Having explained his purpose to the owner, so he thought, he was patiently escorted to a bus stop and put on a bus. He had no idea of where to get off. His Danish dictionary had failed him again. He never did find the bicycle store, nor could he figure out just why he was put on the bus.

While he was in Copenhagen, Danny collaborated with a Canadian physicist, Nick Burgoyne, on problems of a very mathematical nature. Nick had a young and exceedingly beautiful German girlfriend and a series of expensive sports cars: a Mercedes Benz 300SL (the one with the butterfly doors) and a Porsche Carrera. The time he won second place in the Danish auto races at Roskilde, he completed the last few laps with a flat tire. The newspaper accounts seemed to upset Niels Bohr, even though he himself had been a motorcycle nut in his youth. Years later, Nick and Christa were married in Berkeley, and I was a guest at their wedding. He, too, has become a mathematician, but now he drives crummy old American cars.

Poland was represented at the institute by Wieslaw Czyz and his lovely wife, Maria. Wieslaw and I wrote several inconsequential papers together, and I spent New Year's Eve of 1960 chez Czyz in Kraków. We visited Auschwitz together and skied in the Tatra Mountains at the Polish-Czech frontier. I gave lectures in Kraków and Warsaw, but it felt a bit strange having my expenses paid by the Polish Atomic Energy Commission.

A Swedish physicist, Torlief Ericson, was also involved in our collaboration. I was invited to attend his Ph.D. examination in Lund, Sweden, where he was interrogated by three professors: one friendly, one hostile and one funny. The four of them were dressed in top hats and tails. It sounds odd, but that's the way it's done in Sweden. Of course, I couldn't understand a word of the proceedings. Torlief seemed to pass, because there was a twenty-one gun salute and then everybody got drunk. Czyz and Ericson have become well-known physicists at Kraków and CERN, respectively.

Ben Siderov was one of my several Russian friends in Copenhagen. The Bohr Institute encouraged visits by scientists from throughout the world. As winter approached, my student hostel went through its annual transformation into a home for the aged and I was evicted. Ben Siderov and I rented rooms in Fru (i.e., Mrs.) Moltke's house on Moltkevej, three or four miles from the institute. Fru Moltke was related to the Prussian count von Moltke, one of the great military strategists. The bed in my room was kind of short, so that I slept with my feet in the linen closet at the end of the bed.

Fru Moltke had very international boarders, so much so that she had to label each bathroom appliance with a picture indicating its proper use. Ben and I were close friends in Copenhagen. It was through him that I met the Soviet consul in Denmark, whom I continually pestered about that visa I was waiting for but never got. Each time I saw him, he would say, "Is coming tomorrow!" and hand me a bottle of Stolichnaya vodka.

I've seen Ben a number of times since those glorious days in Copenhagen, both in America and in Russia. Today he is quite a big cheese in Akademiigorsk, the Science City of Soviet Siberia. Brr!

I had many other colleagues from many other countries, including Finland, France, Norway, Spain, Yugoslavia, Italy, China, Denmark, Holland, Great Britain, and the States. Strangely enough, there were no young German scientists. All of us had been lured to Denmark by the reputation and the person of Niels Bohr.

The man who had invented the very rules of quantum theory was still an active scientist and a visible presence. At seminars, no one dared choose the last seat on the right in the front row: it was exclusively Bohr's. He attended many talks and always asked pertinent questions. I had several conversations with Bohr in the institute lunchroom, but I can recall little of what he said. Sometimes I couldn't tell whether he was speaking Danish or English. Even when I could follow the words, the precise sense of what he was saying often escaped me. "Never express yourself more clearly than you think," was one of his sayings. "He utters his opinions like one perpetually groping and never like one who believes to be in possession of definite truth." So Einstein had written of Bohr, and so I learned for myself.

No matter that his speech was clumsy, nor that his principal interests had turned to grand questions of peace, war, philosophy and politics—to have known Bohr is one of my cherished experiences. Whenever I teach a new class of freshmen about the history of physics and of its Greatest Masters, Galileo, Newton, Maxwell, Einstein and Bohr, I am always astounded to recall that I once knew such a man.

By the time I arrived in Copenhagen, the central concern of most members of the institute was the discipline of nuclear physics, not elementary particles. Great things were being done in that field by Niels's son Aage, together with Ben Mottelson, an American expatriate living and working in Denmark. They were to share the 1975 Nobel Prize for their work in determining the structural shape of the atomic nucleus, a subject in which I had not (nor have now) the least interest. Neutrons and protons are the stuff of nuclear physics: just two of the many seemingly elementary particles. They are special only because we, our artifacts (such as nuclear reactors) and the world we inhabit are mostly made of them. Nuclear physics, in this light, is as anthropomorphic a science as human genetics or terrestrial geology. I was interested in a bigger picture, the how and why of the whole gamut of particles and forces of nature. I could never understand those around me who would seek their fame and fortune in the study of the peculiarities of isotopes of lead. Only a few of my colleagues at Copenhagen shared my views. We were tolerated, but mostly left to ourselves.

Nonetheless, it was the spirit of the institute that was the key to its success and to mine. A visitor to Copenhagen is said to have remarked to Bohr, "It seems that at your institute nobody takes anything seriously." Bohr responded, "That is true—and it even applies to what you just said."

Hunkering down to work, I produced a paper by November 1958 summarizing the important parts of my doctoral thesis. I wrote:

> Recent advances in the study of elementary particle decays have encouraged speculation concerning the nature of the [weak] interactions responsible for these processes. Those to whom renormalizability seems a desirable attribute of a quantum field theory may object to such a

> theory because of the alleged unrenormalizability of most vector-[boson] couplings.

I then dared to predict that a unified electroweak theory based upon a gauge principle, with the W-boson mass put in by hand, would be renormalizable. That is, the results of calculations based on the theory would be useful and sensible rather than infinite and meaningless. I went on to predict the mass that the W-boson had to have.

I got the wrong answer, but this is not surprising since I was dealing with the wrong theory. Schwinger's proposed unification of the photon with the charged W-bosons was not quite the right idea. Yet I knew that "to obtain a realistic theory of the decay processes, we must [produce one] which contains both symmetric parity violation and the known strangeness selection rules. . . . The discussion of these points is reserved for a more speculative [!] sequel."

Right on, young Shelly! The parity problem I would solve in a year, the strangeness problem in a decade.

Danish winter was coming and I decided I needed a car. Bicycling through the slush to and from the institute was too Scandinavian an activity for me. I had saved some money, since life in Europe was cheap in terms of U.S. dollars. A tram ride cost eight cents, a fancy meal only a dollar or two. My room, with breakfast (bread, butter, cheese, an egg, Nescafé and a vitamin pill) was only $25 a month. A brand-new red convertible TR-3 sports car cost under $2000 if purchased in Paris, which suggested a marvelous way to spend the Christmas holiday.

I stayed in Saint-Mandé, in suburban Paris, at the home of the mother of a friend from Copenhagen. When I first got the car I took Mme. Lamberget, my hostess, on a drive to Versailles. She didn't say a word, but later my friend Monique informed me that a predecessor had done exactly the same thing just a year before, with a brand-new green TR-3. At least, I hoped, I might do some original things in physics.

In the early spring I was invited to speak about my work at Imperial College, the London haunt of Abdus Salam, with whom I was to share the Nobel Prize twenty years later. Then, as now, Salam had the presence of an oriental potentate with a Cambridge education and the gift of speaking perfect English. A practicing Muslim, he neither drinks nor smokes.

I explained what progress I had made on the electroweak synthesis and why I believed that the theory was renormalizable. I was received politely but coolly, as one might expect in England. Some months later, Salam and his cronies published a paper showing that my arguments were hopelessly naïve and dead wrong. Boy, was I ever stupid! I didn't get any sympathy from my friends at the Institute: there was no reaction at all. Nobody cared about the work I was doing.

During the summer of 1959 I attended the Enrico Fermi summer school on the shore of Lake Como at the tiny town of Varenna, Italy. Among my Copenhagen colleagues, Nick, Danny and Chris Fronsdal (a Norwegian, now a professor of physics at UCLA) also came along. Summer schools were very important to the rebirth of European physics. Upon the rise of Nazism, many European physicists had come to the States. Those who remained in war-torn Europe could not train the students who would replace them. A generation of scientists and teachers had been lost, so that graduate education was generally second-rate. Many American scientists, lured by the charms of the Old World to spend their paid vacations abroad, taught the new generation of European physicists their trade. It was a kind of Marshall Plan in physics, but cheap.

Talks were given in a converted monastery. A kindred nunnery had once been conveniently sited just across the narrow lake, an easy trip by rowboat. Centuries later, talk of scandal had not yet abated. Almost all of the sixty students at the summer school were young, ambitious European men. Only four were women. Nicola Cabibbo, for example, was then, like me, just beginning his professional career. Today he is the head (*il Presidente*) of the Italian equivalent of the Department of Energy. In the interim he did some things in physics that I could shoot myself for not having thought of first. In 1963 he introduced into particle physics a key concept known as the Cabibbo angle, which is an essential ingredient of the Glashow-Iliopoulos-Maiani mechanism (1970), which itself was generalized by Kobayashi and Maskawa (1975) into today's formalism describing the interplay between the strong and weak forces. Often, in my life, I have encountered superstars before they began to shine. Remember Nicola, for we shall meet him again.

The subject of the summer school session was the weak interactions. Many famous physicists spoke, yet there was not one mention of my beloved intermediate vector boson. Everyone was pleased with the newest experimental successes of the V-A "theory." It gave a correct description of weak interaction phenomena and it was certainly something that had to be encompassed by any future theory. But the time was not yet ripe to build a real theory of the weak force.

The Italian physicist Georgio Fidecaro described to the students at Varenna an early and dramatic European triumph in high-energy physics, which, in the years immediately following World War II, had been an exclusively American preserve: they had all the old cathedrals, but we had the big particle accelerators. In February 1952, Europe had set out to build an international high-energy physics facility all its own. A meeting had been held in Geneva to establish the provisional organization that has since become known as CERN (Conseil Européen pour la Recherche Nucléaire, or, in British, the European Centre for Nuclear Research). Niels Bohr, as well as the American physicists I. I. Rabi and Viktor Weisskopf, were among CERN's staunchest supporters. Almost all of the Western European countries enthusiastically joined the new venture that, despite its name, was then and is today totally devoted to the study of elementary particle physics. By the late 1950s the two most powerful accelerators in the world were the Alternating Gradient Synchrotron (AGS) at Brookhaven, Long Island, and the CERN Proton Synchrotron (PS). The rivalry among groups of experimenters at the two labs was friendly and professional, but very intense.

Fidecaro was boasting of CERN's first victory. According to the V-A scheme, one charged pion out of every ten thousand or so was expected to decay into an electron rather than a muon. American experimenters had failed to detect this effect, and some American physicists even went so far as to claim that the data were in contradiction with the theory. An Italian group working at CERN, however, had recently succeeded in observing the elusive electron decay mode. Fidecaro explained to us how they had shown that the pion behaves exactly as the theory says it should.

(CERN's victory was actually a draw. Columbia's Jack Steinberger had done the same thing at the same time at a modest accelerator at the Nevis Laboratory in New York.)

It was clear to me that CERN was an up-and-coming institution. I decided to spend part of my European tenure there. It was to be the first of many memorable sojourns to Geneva, as an unpaid guest, a salaried visitor and, eventually, a member of the CERN Science Policy Committee. Over the succeeding decades, CERN has established itself as the primary focus of European high-energy physics.

It rivals anything we have in the States. Since the early 1970s, CERN has led the race toward the highest collision energies. CERN is the only example that I know of a completely successful international organization. True science recognizes no national boundaries. Indeed, CERN itself lies half in France and half in Switzerland.

My National Science Foundation Fellowship allowed me to spend two years abroad. I liked the life of a rootless alien. The second year, I commuted between CERN and the Bohr Institute—or between Maria, the gorgeous blonde daughter of a German bassoonist, and Bonnie, my dark Danish damsel. There were additional side trips (with or without a lady friend) to Norway, Sweden, Iceland, England, Scotland, Germany, Austria, Spain, Israel, Italy, France and to Davos, Switzerland, for a week of skiing. We call it the leisure of the theory class. You see, it's not so tough being a physicist.

On one of my frequent high-speed trips between Copenhagen and Geneva in my red TR-3 I got hopelessly lost somewhere near Hamburg. I found a cop to ask my way. He spoke some English, having picked it up as a prisoner of war in an American camp. "Go dot vay oontil de zerd red licht, den turn to *links*, den to *recht* ven de road pass oonder der britch . . ." and so on.

"*Danke schön*, officer," I said, remembering not a word of his lengthy instructions. He was not so easily put off. "Do not thank! Repeat!" said the cop. He would not let me go until I had memorized the route. That's Germany!

Murray Gell-Mann provided the dominant thrust in theoretical particle physics during much of the 1950s and 60s. While he is now my good friend, I must admit that he is sometimes difficult to talk with. He knows almost everything about almost everything, and he is not averse to letting you know that he does and you don't. He was spending the spring of 1960 on sabbatical leave from CalTech at the Collège de France in Paris. He was aware of my fumbling attempts to unify the weak and electromagnetic forces, and invited me to speak in Paris.

When a foreigner buys a car in France he has the right to tax-free plates for one year. My French tourist license plates had expired. The tax was too large to think of paying. I could still drive anywhere in Europe except in France. This was annoying, since Geneva is surrounded by France on almost every side. How could I drive to Paris to visit Gell-Mann? CERN provided me with an official document explaining exactly how to go about getting tax-free Swiss license plates. There were about a dozen steps. I had to take a driving examination to get a Swiss driving license. Appropriate automobile insurance was a prerequisite to taking the test, but I could obtain the insurance only with the plates in hand.

Somehow, I broke into the cycle. The driving test simply involved parking my little TR-3 between two concrete pillars. It was a cinch. The American who followed me was a businessman driving a Cadillac that simply wouldn't fit between the pillars. This was no excuse at all to the Swiss official. To get a Swiss driving license (valid, incidentally, for life) you must park between the pillars. In Switzerland the law tolerates no exceptions.

Finally, I drove to Paris, stopping at two three-star restaurants along the way: Père Bise in Talloires and the Hôtel de la Poste in Avallon. No restaurant in the world can compete with these utterly incredible establishments. It is no wonder the French have trouble with their livers.

After giving my seminar in Paris, I had a long talk with Murray over lunch. My ideas were closely related to what he was then doing, and he was evidently interested in what I had to say. He told me that a fish course at the restaurant we were in was superb, and he ordered it for me before I could object. But I hated fish! It brought back memories of horrid smells coming from the fish market I had to pass on my way to grammar school Nobody in my family except my father ate fish that wasn't gefilte. He often went fishing, cleaned the fish, cooked and ate them, then washed the dishes. He was otherwise a good husband, so my mother tolerated his "sole" perversion.

Against Gell-Mann, what choice did I have? I ate the fish, and it was good! Subsequently, or perhaps consequently, Murray offered me a position as a research fellow at CalTech. This gave me a second option, as I had already been offered a tenure-track assistant professorship at the new Belfer graduate school of the Yeshiva University in upper Manhattan. "Come help me build castles in the sky!" David Finkelstein, the inspired chairman of the new physics department, had written me.

In the end, I chose CalTech, which had been founded in 1891 as the Throop Polytechnic Institute. The institute had come a long way. Its Robert A. Millikan had won the first American Nobel Prize in physics for his measurement of the electric charge of the electron, and Carl David Anderson, whom he hired, won our third Physics

Nobel for his discovery of the positron. Thomas Hunt Morgan, while working at CalTech, was the first American Nobel Laureate in Physiology and Medicine, and the institute ran the world's largest telescope, the two-hundred-inch Hale reflector atop Mount Palomar. (When it comes to records, I am duty-bound to point out that the first unambiguously American Nobel Prize in science was awarded to Harvard's chemist T. W. Richards in 1914.) CalTech caught up with Linus Pauling, who copped Chemistry in 1954 and Peace in 1962. Murray and Dick Feynman were there, and *they* were the mainstream of particle theory in 1960. Yeshiva University was clearly outclassed, and my choice was no choice at all.

Curiously, I was awarded an honorary degree from Yeshiva in the summer of 1979. Later that year I would win the Nobel Prize and the physics department of Yeshiva University would be disbanded for reasons of fiscal austerity. Finkelstein's castle was to crumble.

My second paper on the electroweak theory—the good one—was based upon work I did while at CERN and Copenhagen. I put aside the question of renormalizability as a "radial problem," which needed too much hard work to solve, and focused upon what I called "partial symmetry," by which I meant that the weak and electromagnetic forces should be related as closely as possible.

I wrote in 1960:

> At first sight there may be little or no similarity between electromagnetic effects and the phenomena associated with weak interactions. Yet certain remarkable parallels emerge with the supposition that the weak interactions are mediated by unstable bosons. Both interactions are universal, for only a single coupling constant suffices to describe a wide class of phenomena: both interactions are generated by the vectorial Yukawa coupling of spin-one fields. Schwinger first suggested the existence of a triplet of vector fields whose universal couplings would generate both the weak interactions and electromagnetism. . . . The more recent accumulation of experimental evidence . . . indicates the need for at least one neutral intermediary.

There it is! That was the first argument for the necessary existence of an entirely new kind of weak force heretofore unsuspected by mortal man! It was mediated by a hypothetical neutral vector boson that I had invented! These so-called neutral current interactions would eventually become the subject of an international race for discovery between the United States and Europe. The race was won by CERN in 1973.

"The mass of the charged intermediaries must be [high], but the photon mass is zero . . . surely this is the principal stumbling block in any pursuit of the analogy between hypothetical vector bosons and photons. It is a stumbling block we must overlook." Heed Glashow's advice for success in science: Don't bite off more than you can chew. You are simply not going to solve all the problems of physics in one fell swoop. Put on blinders and do what you can do. Let smarter people sweep up the harder problems.

"The purpose of this note is to make less fanciful the unification of electromagnetism and weak interactions," I wrote. This is a much more modest goal than the creation of such a theory. I went on to prove that the parity violation seen in weak interactions could not be reconciled with the known fact that electromagnetism conserves parity in a framework involving only charged W-bosons and photons. "In order to achieve a [unified] theory of weak and electromagnetic interactions, we must go beyond the hypothesis of only a triplet of vector bosons and introduce an additional neutral vector boson."

In wrapping up the paper, I recognized the problem of strangeness. My theory seemed to predict the existence of certain kinds of strange particle decays. For example, it said that a kaon decay should often result in the production of a pion and two neutrinos. Experimenters had searched for such decays but had never found them. That's because they aren't there. An essential ingredient was missing from my theory, something magical that would exorcise the unwanted and unseen effects. It was something that I would christen "charm" in 1964, that I would identify as the solution to the problem in 1970, and that would be discovered in the laboratory in 1974.

"Our considerations seem without decisive experimental consequence. For this approach to be more than academic, a [theory] correctly describing all decay modes of all elementary particles should be sought." Amen for motherhood and apple pie!

It was a brilliant paper, but almost nobody ever read it. This was because almost nobody except Salam was interested in wild speculations about unseen intermediate vector bosons. But he, having seen how stupid I could be, had vowed never again to read one of my publications.

"It is not unfair to say that for the average physicist at that time, speculation about intermediate vector bosons was something like speculation about life on Mars—very interesting for those who like this sort of thing but without much hope of support from convincing scientific evidence and without much bearing on scientific thought and development." Substitute "atomic structure" for "intermediate vector bosons" and you have the turn-of-the-century attitude toward scientists such as Niels Bohr, according to the scholar E. C. D'Andrade. I flatter myself with the comparison.

One of my friends both at Copenhagen and CERN was the English physicist Jeffrey Goldstone, who was to play a pivotal behind-the-scenes role in the development of the electroweak theory. Today he is not only a friend but a neighbor. Forsaking his position at Trinity College, Cambridge, Jeffrey is now a professor at MIT and has moved to Brookline, Massachusetts, with his wife and little Andrew. He was the first physicist to identify the phenomenon of spontaneous symmetry breaking in quantum field theory, and he did it while I was looking over his shoulder in our cubicle at CERN in 1960. The importance of this work to the theory of the weak force was not to become clear for almost another decade.

Spontaneous symmetry breaking, believe it or not, is an everyday happening. Imagine the perfect Mexican sombrero. It looks the same from any direction: we say it has rotational symmetry. Put it on a table and roll a ping-pong ball into the brim. It rolls around rather like a roulette ball without finding a natural resting point. If there were no friction at all, the ball would roll around forever no matter how slowly it was tossed. Its energy of motion could be arbitrarily small. If there were the tiniest amount of friction, the ball would eventually come to rest. But, where? All points on the brim are just the same. Yet it does stop, and in stopping, it breaks the symmetry. The hat and the ball together no longer show a rotational symmetry: it was spontaneously broken. This Mexican hat trick seems too easy, you say? In quantum field theory, Jeffrey showed, things are more subtle and more interesting.

The sombrero shape also illustrates the mathematical function showing the potential energy of a quantum field that displays an internal symmetry. This is the kind of situation that Jeffrey imagined. The quantum version of spontaneous symmetry breaking is known as the Goldstone theorem, and the quantum-mechanical analog to the ping-pong ball is a boson that is forced to pick out one of the equivalent points along the brim of the sombrero, thereby destroying the perfect symmetry of the potential energy function. The boson sits at its chosen site, where it may gently wiggle about. It develops the ability to carry an arbitrarily small amount of energy. This means that it has become a massless particle, a particle that is now known to every student of physics as a Goldstone boson. Whenever spontaneous symmetry breaking takes place in quantum field theory, massless bosons must appear. Only a massless particle can carry an arbitrarily small amount of energy.

Nobody has ever seen a Goldstone boson, and it seems unlikely that such a thing really exists except as a mathematical possibility. The pion, however, is practically a Goldstone boson. The broken symmetry to which it corresponds is merely an approximate symmetry of nature, not an exact one. That is why the pion is very light, but not quite massless.

Quantum field theory is a science unto itself that does not necessarily deal with the things that exist in the real world. The Yang-Mills theory, for example, demands the existence of massless vector bosons. Until very recently, no such particles were known to exist. The spontaneous breaking of a conventional symmetry requires the existence of massless scalar bosons that don't seem to be there either. Here were two perfectly crazy notions. These two pathologies of quantum field theory were to be married a few years later into something quite different, something that was to explain the mystery of the W-boson mass.

The marriage could have been consummated in 1961 at Newbattle Abbey, an ancient rat-infested edifice near Edinburgh and the locale of the First Scottish Summer School in Theoretical Physics. The school had been organized by Nicolas Kemmer, the Tait Professor of Natural Philosophy at the University of Edinburgh and one of the originators of the very concept of massive vector bosons.

I had come to Scotland in my TR-3 via the ferry from Bergen to Newcastle. One thing about living in Europe is how much geography you learn. For the last time, I was a formally registered student at a formal school. (Of course, we are students all of our lives. Lifelong learning is the price of sentience.) Again, I was thrown among an ambitious group of young physicists, many of whom would make quite a splash in physics. There were fifty-three students, including just one woman. It was at this school that I met Englishman Derek Robinson and Dutchman Martinus (Tini) Veltman. Derek now teaches in Canberra, Australia, and Tini teaches in Ann Arbor, Michigan. Long before they got there, each of them was to play a crucial role in my story. Cabbibo was there, too, and so was a red-headed young Scot named Peter Higgs.

A few years later, in 1964, Peter (and some others, independently) would wed the ideas of Yang and Mills to those of Goldstone. They studied the phenomenon of spontaneous symmetry breaking in the context of gauge theory. Their result was nothing short of amazing. The massless vector boson consumed Goldstone's massless scalar boson, and from this union sprang a massive vector boson. Two massless particles, both useless as mediators of the weak force, unite to form a massive particle. It was just what the doctor ordered to give masses to the intermediate vector bosons of the weak force, except that the patient was not awake. Nobody noticed. It was not until 1967 that Steve Weinberg and Abdus Salam had the wisdom to combine the Higgs mechanism with my electroweak theory. The rest, as they say, is history.

Peter and I, at Newbattle Abbey, should have put it all together there and then. We had all the necessary ingredients. Instead we attended innumerable boring lectures, flirted with secretaries, hiked in the highlands, and drank many liters of fine Hungarian wine, courtesy of the NATO grant that supported the school.

Soon after the last lecture I drove my red Triumph to Glasgow, kissed her goodbye, and shipped her off to California. I returned to Copenhagen, where I completed the final draft of my prize-winning paper and submitted it to the European journal *Nuclear Physics*.

After a tearful farewell to Bonnie, I boarded the *Gripsholm* and sailed for home. There were no nubile girls aboard recovering from their junior year abroad. The ship was full of Swedish-American members of the Salvation Army.

7 Go West, Young Man!

I LIVED IN THE FAR WEST from September 1961 until January 1966, teaching and doing research at CalTech, at Stanford and finally at the University of California at Berkeley. My first year of exile to California was spent in sunny Pasadena under the tutelage of Murray Gell-Mann, who was at that time the dominant figure in my discipline of particle theory.

Southern California, I soon discovered, is a mighty peculiar place, somewhere between God's Country and The Land of the Fruits and the Nuts. Within hours of my arrival I was arrested and fined for the heinous crime of jaywalking. Imagine that! There was no perceptible change of climate as the year wore on: just this mad-hot sun in a smoggy sulfurous sky. No wonder that Californians, alone among humans, seem to be blissfully unaware of their own personal mortality. Not until my last few days in Pasadena could I discern the marvelous spectacle of its surrounding mountains: a rare vision like the green flash of the setting sun or a total solar eclipse.

I did not come to California for its eternal sunshine, but to work with Murray Gell-Mann, elementary-particle physics' reigning guru. During my year in Pasadena, Murray was to make several of his most spectacular contributions toward our understanding of nature's most intimate secrets: the "eightfold way" and "current algebra." Both of these developments relate most directly to the nature of the strong force, and this was to become my preoccupation for the next several years.

It was also the time of extraordinary experimental discoveries. These resulted from the deployment of large particle accelerators that could produce copious beams of energetic particles, and from the exploitation of newly developed bubble chambers that make visible the results of the collisions of these particles. Indeed, Donald Glaser, of the University of California, won the entire 1960 Nobel Prize for the discovery of this essential tool. It is said that he got the idea upon watching the trail of bubbles that would form in a glass of beer. He showed that a charged particle traversing an appropriate fluid (not beer) could, if the circumstances were just right, produce a trail of minute bubbles that would make its passage visible. In 1968, Luis Alvarez, from Berkeley, won the whole prize for his utilization of the bubble chamber for the discovery of hosts of new short-lived particles. Sometimes I feel cheated that I won only a third of a Nobel Prize. Maybe I should have become an experimenter—or maybe not. Murray was to win the whole thing, too.

The first half of the sixties were the years of the population explosion in hadrons. They were glorious years in the history of elementary-particle physics. Dozens of fascinating new particles were being found. Something very interesting and very fundamental was going on, but it was not clear what. How could there be so many different particles, each with an equal claim to being elementary? It was a little bit like the mid-nineteenth century, when scores of new chemical elements were showing up. Where was the latter-day Mendeleev who would bring order out of the chaos? He was in Pasadena, with me.

Murray and I were both fascinated by the possible application of gauge theories to physics. Quantum electrodynamics, the fundamental theory describing the interactions of electrons, positrons and photons, could be deduced from the principles of gauge invariance. It was an enviable paradigm for the construction of a more ambitious theory. I had been trying to build an extended gauge theory that could incorporate the weak force. Murray believed in the possibility of a gauge theory of the strong force. We both realized that gauge symmetry could not be exact. Except for the photon, there did not seem to exist in nature the massless vector bosons demanded by an exact gauge theory. The masses had to be inserted into our equations by hand in an ugly and undignified manner. Be that as it may, we decided to pool our resources and compile a survey article showing the potential virtues of gauge theories.

We submitted our opus, "Gauge Theories of Vector Particles," to the *Annals of Physics* in June 1961. (Many interesting things had transpired while we were writing the paper, to which I shall return.) Our paper began:

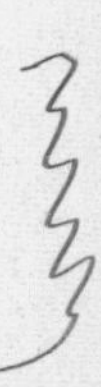

> The electromagnetic interaction of elementary particles is remarkably simple. It is of universal strength and form and is in association with a principle of gauge invariance. In fact, starting with the idea of invariance under gauge transformation with coordinate-dependent gauge functions, one can deduce the existence of a massless vector field coupled to a conserved current. If all charged fields are subjected to the same gauge transformation, then the electric charges of all particles are the same.
>
> The fact that the weak interactions are vectorial in character . . . and nearly universal in strength has suggested to many physicists [for example, me!] that they may be mediated by vector fields and that there may be a useful parallel between them and electromagnetism, perhaps even extending to the notion of gauge invariance.
>
> The strong interactions, too, seem to display some degree of universality. Moreover, the approximate conservation laws of isotopic spin and of strangeness, as well as the exact law of conservation of baryons, present an analogy with the conservation law of charge and suggest that some principles of gauge invariance may be at work. Until recently, it seemed that the strong couplings were not vectorial, but there is mounting evidence that there are objects that can be interpreted as vector mesons and that they may play a very significant role in the strong interactions.

At this point in the text there appears a reference to a recent experimental observation of a vector hadron—the rho meson—and to a seminal paper by the late John Jun Sakurai, who predicted its existence. John was one of the first serious exponents of a gauge theory of strong interactions, and one of my very close friends.

I met John as a student in Schwinger's course at Harvard. I was a freshman graduate student and he was a brash young undergraduate, a perfect candidate for early burnout—or so I thought. In early 1960 he burst upon the particle physics scene with a long diatribe modestly entitled "Theory of Strong Interactions," a paper that was generally scorned but was influential nevertheless. It began with an exceedingly apt quotation attributed to Richard Feynman.

> There is a large experimental program on production of K particles by nuclear collisions and by photons, scattering and interactions of those particles with nuclei, etc. But just between us theoretical physicists, what do we do with all these data? We can't do anything. We are facing

> a very serious problem. . . . Perhaps the results of all the experiments will produce some idiotic surprises, and some dope will be able to calculate everything. . . .

Sakurai then assumed the mantle of Feynman's dope and set forth a program so ambitious that "[t]here is one question that greatly puzzles the author: why has nobody tried this kind of approach before?"

Sakurai abandoned the pion as a truly elementary particle and the fundamental force carrier of the strong interaction, arguing instead that "it is possible to construct the pion out of a nucleon-antinucleon pair." The strong force, like electromagnetism, was to arise from a principle of gauge invariance based upon the conservation laws of isotopic spin, strangeness and baryon number. Although the pion was merely a composite system (according to Sakurai), the truly basic carriers of the strong force were the vector particles that implemented gauge invariance.

But gauge bosons are necessarily massless. Of this inconvenient fact, Sakurai said simply:

> We admit that we lack satisfactory answers to the question of the masses of the [gauge bosons]. We must assume that they are all massive lest the whole edifice of our theory should crumble down. One of the reasons why the present work is submitted for publication in spite of this mass problem is that the author hopes that the publication may prompt some clever ideas along this line.

Like me, Sakurai was perfectly happy to leave the hard problems for later and to press onward. So he assumed that his vector bosons somehow got their masses. Of course, nobody had yet seen the particles that Sakurai regarded as the roots of the strong force. Thus: "Every conceivable attempt should be made to detect experimentally direct quantum manifestations of the three kinds of vector fields introduced in our theory."

Miracle of miracles! Vector hadrons of just the sort that Sakurai prophesied were observed in the laboratory in 1961. A triplet of heavy vector mesons, rho$^+$, rho^0 and rho$^-$, each with a mass about three-quarters that of the proton, were observed as exceedingly short-lived particles, or resonances, that quickly came apart into two pions. Their average lifetime was a trillionth of a trillionth of a second, so short that these particles were evidently produced by, and decayed as a result of, the strong force.

Soon afterward the omega was found, a somewhat longer-lived vector meson that decayed into three pions, and then the phi, heavier still and decaying into a pair of kaons, among other ways. Rho, omega and phi could be regarded as the three types of gauge particles predicted by Sakurai's theory.

While Sakurai had correctly foreseen the existence of vector hadrons, in another respect he turned out to be dead wrong. For years physicists had been seeking a "higher symmetry scheme," a mathematical classification of the observed particles in terms of a symmetry more all-encompassing than isotopic spin and strangeness. There had been a number of unsuccessful attempts to create such a theory—by Schwinger, the Japanese physicist Shoichi Sakata and others. In despair, Sakurai wrote:

> . . . that for more than two years we have hunted fruitlessly for higher symmetries which do not even exist. After having spent considerable time (and energy) on various symmetry models, the present author is convinced that there are no simple patterns . . . and that all those symmetry models proposed up to now are mere mental exercises devoid of any physical significance whatsoever.

This time John's intuition had led him astray. Simple patterns do exist. A useful higher symmetry called "the eightfold way" was discovered during the year I spent at CalTech. It was come upon independently by Gell-Mann and by an Israeli military officer turned physicist (a student, in fact, of Abdus Salam) named Yuval Ne'eman. Yuval, like most other members of the international fraternity of theoretical physicists, is a friend of mine. While he has continued to do exciting and imaginative physics over the years, he is at least as well known today as the leader of a right-wing fringe political party in Israel as for his codiscovery of the eightfold way.

A first indication that something was missing in Sakurai's philosophy was the appearance of additional vector mesons beyond those corresponding to the known symmetries, vector mesons that carried the strangeness quantum number. There appeared a striking and unexpected correspondence between the observed spinless mesons and the newly discovered vector mesons:

SPINLESS MESONS VS. SPIN-ONE MESONS

Spin-0			Spin-1	
Symbol	Mass in MeV	Strangeness	Mass in MeV	Symbol
π^+	139	0	750	ρ^+
π^0	135	0	750	ρ^0
π^-	139	0	750	ρ^-
K^+	494	1	890	K^{*+}
K^0	498	1	890	K^{*0}
K^-	494	−1	890	K^{*-}
$\bar{K}^0$	498	−1	890	$\bar{K}^{*0}$
η^0	550	0	780	ω^0

In the early 1960s there were eight known mesons with spin zero and eight vector mesons with unit spin. Their electrical charges are shown as superscripts to their symbols. The central column gives the strangeness quantum number. Notice the exact correspondence between mesons with and without spin.

Let us turn again to my paper with Murray, keeping in mind that it was completed only after the discovery of the eightfold way. We examined the class of all gauge theories, from the simplest example of electromagnetism to the extended version of Yang and Mills, and on to the most general imaginable gauge theories. It was our hope that one or another of these theories might be relevant to the strong or weak force. We pointed out an intimate relationship between gauge theory and a nineteenth-century mathematical construct known as the Lie group.

Marius Sophus Lie was a Norwegian mathematician concerned principally with the study of differential equations. The symmetries of such equations could be described in terms of a continuous group of transformations—a mathematical entity that has become known as a Lie group. Later in the nineteenth century the French mathematician Elie Cartan showed that Lie groups had a significance far beyond their utility in solving complicated equations. In these mathematical systems, three great branches of classical mathematics—analysis, algebra and topology—converged to reveal the fundamental unity and beauty of mathematics.

One of Cartan's great accomplishments was the classification theorem for what are technically called simple Lie groups. He showed that every simple Lie group (except for five of them) belongs to one or another of three infinite classes of Lie groups, known as rotation groups, unitary groups and symplectic groups. The five exceptions were designated as G2, D4, E6, E7 and E8. It was a theorem analogous to the proof that only five regular, or Platonic, solids exist, but far more subtle.

One of the simple Lie groups corresponds to the set of all rotations of a rigid body that leave one point fixed. Called the three-dimensional rotation group, it has obvious relevance to physics. The same could not be said for most of the remaining simple Lie groups.

Are the simple Lie groups merely an invention of human thought, like the Polish language? Or do they possess an underlying and universal reality? Surely, intelligent aliens in other galaxies could hardly be expected to speak Polish, but I bet my boots they know about Lie groups. Mathematical entities, like integers, complex numbers, calculus and Lie groups, are real things, even if we cannot touch or smell them, or build rabbit hutches out of them. But it seems to be a reality quite apart from physical reality. Electrons exist in a very different sense than Lie groups. Despite this, the wild things that mathematicians beget can often turn out to be surprisingly relevant to physics.

Once upon a time the geometry of Euclid was generally accepted as the obvious and unambiguous description of the space we inhabit. Mathematicians, in their boundless curiosity, invented alternative (fictitious?) geometries. Euclid said that exactly one straight line could be drawn parallel to a given line and through a given point, and so it seems to an intelligent adult with paper and pencil. In the non-Euclidean geometries, there can be many such lines, or none at all. Each type of geometry is internally consistent and leads to no logical absurdities. Why did nature choose Euclid's over the others?

Today, thanks to Einstein, we know that nature didn't. Our space is not precisely Euclidean. In the vicinity of massive gravitating bodies (like the earth, which is conveniently at hand), space curves upon itself and becomes non-Euclidean. Here on earth, the area of a circle is a tiny bit smaller than πr^2 but not so much that you would notice. In the extreme case of a black hole, space curls up to such an extent

LIE GROUPS

One Lie group is familiar to all students of Euclidean geometry: Two figures are said to be *congruent* if there is a rigid displacement of the plane that takes one figure into the other. The displacement may consist of a translation, rotation or combination of the two. The set of all such transformations is known as the Euclidean group of the plane.

It is a theorem of geometry that any two congruent figures can be made to coincide by a simple rotation about an appropriate point in the plane. Here's how to find that point. Shown below are two congruent keys. The line BB′ connects corresponding points of the two keys. So does the line AA′. Their perpendicular bisectors meet at the point O. A rotation about this point takes the one key into the other. The same analysis can be performed upon any two congruent figures to find their centers of rotation.

Euclidean geometry, as most of us recall, is not a simple discipline. Moreover, the Euclidean group is not an example of a mathematically simple Lie group. Surprisingly, the group of rotations in three spatial dimensions *is* a simple Lie group, although the study of geometry in three dimensions is anything but.

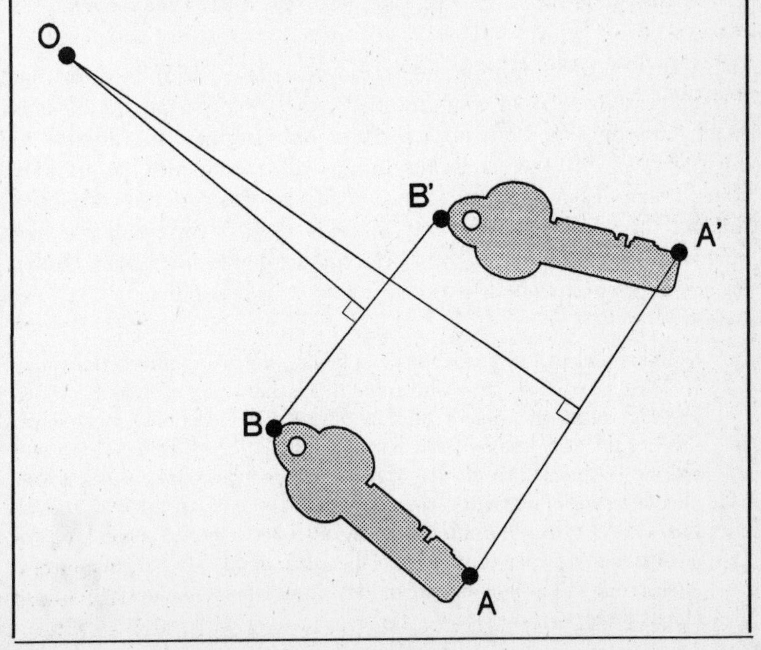

that one may enter at one's peril, but never, ever escape. Often, as in the case of geometry, the wild constructs of our mathematician brethren surprise us and have something to do with the world we live in. So it was for the Lie group.

The symmetry structure of a gauge theory is that of a Lie group, and Cartan's analysis of the simple Lie groups can be used to classify all possible types of gauge theory. Sakurai's theory, for example, did not correspond to a simple Lie group but to a combination of several. Gell-Mann and I showed three ways in which Sakurai's theory of strong interactions could be generalized so as to correspond to a simple Lie group, one of which was the underlying symmetry scheme of Murray's own eightfold way. We then went on to discuss the possible construction of a gauge theory of the weak force. The paper ended in a distinctly minor key:

> The model we have discussed is not seriously put forward as a physical theory, but it is a good illustration of the ideas involved in the gauge method. The fact that we are led to such an ugly model suggests either that partial gauge invariance does not apply to the weak interactions, or that we are missing some important ingredient of the theory.

It was both! Weinberg and Salam, years later, were to show that the gauge symmetry of weak interactions is not "partial" at all; it is exact, although spontaneously broken. Many critical ingredients to our recipe of 1961 were missing: things with cute names like quarks, color, charm, asymptotic freedom, etc. Today it is generally accepted that all the elementary-particle forces—weak, strong and electromagnetic—are generated by a principle of gauge invariance, but a lot has happened since Murray and I wrote our paper.

> We have discussed several ways in which the strong interactions may constitute a partially gauge-invariant theory, and have sketched a formal partially gauge-invariant model of the weak interactions. In general, the "weak" and "strong" will not be mutually compatible. There will also be conflicts with electromagnetic gauge symmetry, conflicts that must be resolved in favor of electromagnetism since its gauge invariance is exact. We have not attempted here to describe the three kinds of interaction together, but only to speculate about what the symmetry of each one might look like in an ideal limit where symmetry-breaking effects disappear.

My collaborative paper with Gell-Mann was written at the very end of my one-year visit to CalTech. It marked the beginning of a nine-year leave of absence from my pursuit of a gauge theory to unify the weak and electromagnetic forces. The obstacles seemed insuperable, and the time was not yet ripe for the implementation of my dream. Instead, I fell under the sway of Murray's eightfold way, which dealt with the strong force, and became one of its most ardent and dedicated disciples.

To understand my transformation from a "weakling" to a "strong man" we must return to the beginning of my crucial year at CalTech.

I arrived in Pasadena both friendless and carless. At first I was compelled to reside at the faculty club on campus. Officially known as the Athenaeum, its denizens called it the mausoleum, for it was a cold and deadly place. The ice cream served there was so deeply frozen that I had to ask the waitress to warm it in the oven. Mercifully, I was soon informed that my TR-3 had completed its ocean voyage safely and was waiting for me at the port of San Pedro, which, like Pasadena, is a part of greater Los Angeles.

With a naïve New Yorker's faith in public transportation, I set out early one morning to retrieve my car. It would have been faster on horseback. I had to spend a night in a hotel along the way! Southern Californians don't believe much in mass transport.

Anyway, I got the car and earned the right to rove the freeways. To complete my image as a playboy scientist with a sports car, I even experimented with "Vent-Air" contact lenses. (I had been four-eyed since the age of six.) The trouble was that they often popped out when I stopped the car at a red light. The ensuing search for the lenses by a suddenly legally blind driver produced incredible traffic jams. The experiment failed.

Sidney Coleman was a brilliant graduate student at CalTech who had completed all his course requirements but had yet to generate a thesis. He became my close friend in Pasadena and my scientific collaborator for years to come. He completed his thesis under my informal direction. He was my first graduate student, and my very best by far. At the end of the year, he was offered and accepted a research fellowship at Harvard, where he remains today, my colleague and still my friend. While we were waiting around for something exciting to happen in physics, we took up the habit of climbing nearby Mount Wilson on occasional weekends, to get a bit of exercise and to escape the smog.

We didn't have to wait long. Murray invented the eightfold way in January 1961.

For many years physicists had been seeking a "higher symmetry scheme" that could introduce some kind of order into the chaotic zoo of the newly discovered hadrons. Murray wrote:

> We attempt once more . . . to treat the eight known baryons as a supermultiplet, [equal in mass] in the limit of a certain symmetry but split into isotopic spin multiplets by a symmetry-breaking term. . . . The symmetry is called unitary symmetry and corresponds to the "unitary group" in three dimensions in the same way that [isotopic spin] corresponds to the "unitary group" in two dimensions.

He went on to show how the nucleons were two of the members of a family of eight baryons, how the pions were three of the members of a very similar family of eight spinless mesons, and how the vector mesons that had just been discovered constitute another family of eight. The appearance of such families of eight related particles is an essential aspect of the unitary symmetry scheme, and for this reason Murray sometimes referred to his theory as the eightfold way.

In its original version, the eightfold way was based upon a generalized principle of gauge invariance. Murray pointed out that

> . . . the utmost feature of the scheme is that it permits the description of eight vector mesons by a unified theory of the Yang-Mills type (with a mass term). Like Sakurai, we have a triplet rho of vector mesons coupled to the isotopic spin current and a singlet vector meson omega coupled to the hypercharge current. We also have a pair of doublets [of strange vector mesons].

Later, however, Murray began to back away from a gauge theory of the strong force. By June 1961 he was no longer convinced that the vector mesons were any more (or less) fundamental than any other hadrons. By 1964 he wrote of his earlier work that "a generalized Yang-Mills field theory for the vector mesons is connected with a point of view that may soon become obsolete." By this time he realized that none of the hadrons were truly fundamental particles. They were made of quarks.

There is a curious irony here. Both Murray and I began the year convinced that gauge theory was the wave of the future. He constructed a gauge theory of the strong force and I puttered with an

SUPERMULTIPLETS OF THE EIGHTFOLD WAY

According to the eightfold way, mesons and baryons should both appear in supermultiplets containing eight different particles. The quantum numbers of these particles should be such as to fill out a hexagonal array, with strangeness increasing from bottom to top, and charge increasing from lower left to upper right. Shown below are three "hexagonal octets" of hadrons: one consisting of the spinless mesons, another of the vector mesons, and a third of the eight known spin-½ baryons.

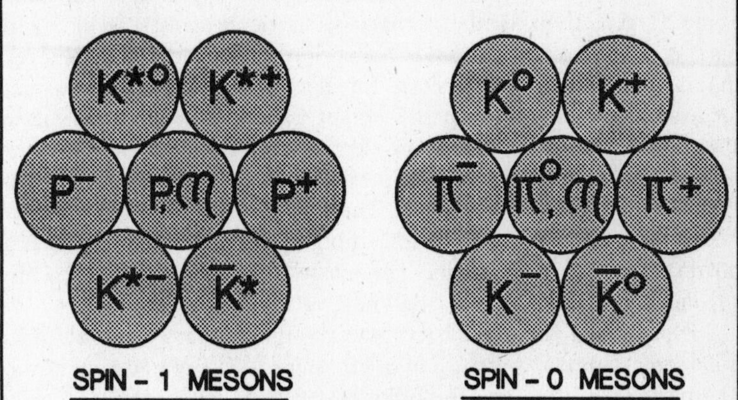

SPIN - 1 MESONS

SPIN - 0 MESONS

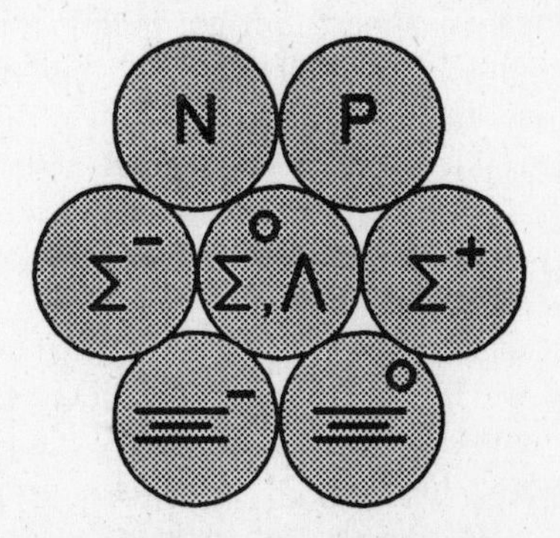

SPIN - 1/2 BARYONS

incomplete gauge theory of the weak. Each of us abandoned our pet ideas for the while. Years later, after the pivotal work of Steve Weinberg, Abdus Salam and others, the road to a gauge theory of the weak force was reopened, and I was to be honored for my pioneer contribution to the theory. Similarly, Murray and his friends, in the early 1970s, were to create today's successful (and maybe even true!) gauge theory of the strong force. In the end, gauge theory was to triumph, but it sure didn't look that way in 1961.

Let me try to explain just what the eightfold way had accomplished. At the time, eight different baryons were known. In the table below, I give their symbols, charges, strangeness assignments and masses. The measure of mass is the curious unit of equivalent energy that is preferred by the physicist, the electron volt, or eV. One gram, for example, is the equivalent of about 10^{33} eV. The electron volt is simply the amount of energy an electron gets from one volt of electricity. In other words, a penlight battery is really an electron accelerator with a power of one volt. The two-mile-long Stanford Linear Accelerator does much the same thing as the battery, only with a power of twenty billion volts. The masses of the particles below are specified in millions of electron volts, or MeV.

Notice that the eight baryons make up four isotopic multiplets (nucleons, lambda, sigma, and xi) that are well separated in mass. The particles within each multiplet have nearly equal masses, spaced apart by only a few MeV out of a thousand or so. This is because the isotopic spin symmetry is almost (but not quite) an exact symmetry. According to the eightfold way, all eight baryons belong to a single supermultiplet called the baryon octet. Clearly the mass splitting between multiplets, say between the sigmas and the nucleons, is much greater than the mass differences within multiplets. This is because the unitary symmetry is much further from being exact, much more badly broken than isotopic spin.

One of the key predictions of the theory is that all hadrons, those known and those yet to be discovered, must fill out the characteristic supermultiplet structures of the eightfold way. Only seven spinless mesons were known at the time Murray made up his theory—the three pions, and the two kaons and their antiparticles. Thus, with complete confidence, Murray claimed, "The most clear-cut prediction for the [spinless] mesons is the existence of [an eighth one] which should decay into two photons like the pi^0." Just a few months later, convincing evidence that just such a new particle exists was revealed

THE SPIN-½ BARYONS

Name	Symbol	Charge	Strangeness	Mass MeV
Nucleon	P	+1	0	938.3
	N	0	0	939.6
Lambda	Λ	0	−1	1115.6
Sigma	Σ^+	+1	−1	1189.4
	Σ^0	0	−1	1192.5
	Σ^-	−1	−1	1197.3
Xi	Ξ^0	0	−2	1314.9
	Ξ^-	−1	−2	1321.3

If the eightfold way were an *exact* symmetry of nature, then all eight baryons would have exactly the same mass. They do not. Rather, they appear as four separate submultiplets whose members *do* have about the same masses. The Gell-Mann–Okubo formula gives one relationship among these four masses:

$$\tfrac{1}{2}\,[M(N) + M(\Xi)] = \tfrac{3}{4}\,M(\Lambda) + \tfrac{1}{4}\,M(\Sigma)$$

This equation is satisfied to within a few MeV, and a similar formula holds for meson masses.

Coleman and Glashow established a relationship among the small mass differences within the submultiplets:

$$M(\Sigma^-) - M(\Sigma^+) = M(\Xi^-) - M(\Xi^0) + M(N) - M(P)$$

This equality is satisfied to within a fraction of an MeV. The empirical successes of the Gell-Mann–Okubo formula and the Coleman-Glashow formula were convincing evidence for the validity of the eightfold way.

by a group of experimenters at Johns Hopkins University. They called it the eta particle. The octet of spinless mesons was complete, just as Murray had predicted.

While Sidney and I were early converts, the rest of the physics world remained quite doubtful. For the first of several episodes in my career, I was convinced that I knew a truth that few others believed but all would soon acknowledge. It's a uniquely satisfying sensation. My old high school buddy Gary Feinberg, who had become an assistant professor of physics at Columbia, simply could not believe that the pions could be put usefully into a family of particles including kaons, which were three and a half times heavier. The stakes of our bet were inflation-proof. A year or so later Gary treated me to a magnificent dinner at a three-star restaurant. He had finally seen the light.

Gell-Mann showed that the masses of the eight baryons should satisfy an equation. This meant that he could "predict" the xi particle's mass from those of the other baryons: twice the xi mass should be equal to three times the lambda's mass, plus the sigma's mass, minus twice the nucleon's mass. I use quotation marks around the word *predict* because you cannot really predict something that has already been measured. Perhaps it is better called a retrodiction. In any case, the formula gave 1330 MeV for the mass of the xi, a result that agrees with the experimental value to better than one percent. Not a bad accomplishment!

Sidney and I were very taken with the eightfold way. It was the simplest imaginable higher symmetry scheme and it seemed to work. It displayed the elegance and beauty of truth, which is to say, it smelled right to us. We decided to look for additional predictions of Murray's theory in the realm of the electromagnetic properties of the baryons. Each of the baryons is a miniature magnet whose strength we call a magnetic moment. Sidney and I showed that all eight of the magnetic moments could be expressed in terms of those of the neutron and proton in an imaginary world where the eightfold way was exact. Thus, the values of the six remaining magnetic moments could be predicted, at least approximately. The lambda moment had already been measured, and it was in reasonable agreement with our result. In later years the other moments got measured, and all of our predictions were borne out.

We went on to explore what are called the electromagnetic mass differences among the baryons, the mass splittings within isotopic multiplets. Using the symmetry properties of the eightfold way we derived a relation among these splittings. The mass difference of the two xis should equal the mass difference of the two charged sigmas plus the mass difference of the two nucleons. Everything had already been measured except for the xi mass splitting, whose value we predicted. Our result soon became known as the Coleman-Glashow mass formula, and it turned out to be satisfied by experiment. It's the first thing that was named after me, and I'm still proud of it. It is still taught to graduate students of physics the world over.

Obviously, my year at CalTech had been one of the most significant professional experiences of my life. To a large extent, a scientist is formed by his teachers, and I feel myself to be an inferior amalgam of two of the greatest scientific minds this nation has ever produced: Schwinger and Gell-Mann. I inherited the imagination and good taste of my teachers, but by no means their incomparable technical virtuosity.

So why did I leave CalTech after one scant academic year? Why didn't I stay where the action was and where I was playing a significant role? Socially, the year had not been much of a success for me. Pasadena is full of little old ladies with sharp elbows, and I found it impossible to replace my faraway European lady friends.

Moreover, I had been a postdoc for three years. I yearned to enter the real academic world: to become an assistant professor and to begin my ascent toward a full and life-tenured university professorship. CalTech, it seemed, had no such position available for me. Out of the blue, I was surprised and delighted to receive a letter from Leonard Schiff, then chairman of the Stanford physics department, offering me an assistant professorship at a salary of $7500 for the year.

The alternative was a continuing research position (specifically not an academic job) at CalTech. When I discovered that the CalTech campus store would not cash checks for anyone (like me) lower than a genuine assistant professor, the die was cast. I accepted Stanford's offer.

With the unexpected death of my father in May 1961 I lost my deepest source of inspiration. Though he was relatively unschooled, his naïve and insatiable curiosity about how things work was always at the root of my endeavors. Pop was truly a jack-of-all-trades. He was a plumbing contractor by profession, but an expert electrician, carpenter, cabinetmaker, mason, painter, roofer, architect, landscape designer, gardener and pinochle shark as well. He was my greatest fan, and one helluva father.

Returning briefly to CalTech after the funeral, I set about to give a large cocktail party to the participants of the La Jolla Conference on Strong and Weak Interactions, the meeting at which Murray presented his updated version of the eightfold way, now staidly titled

"Symmetries of Baryons and Mesons." The party was to celebrate the marriage of John Sakurai to his demure and beautiful bride, Noriko. There is no better way to clothe one's grief than to celebrate another's joy. I served only rum drinks and cheese. It was a Zombie and Gorgonzola party.

Years later, during the Johnson administration, I returned to CalTech to deliver a seminar about my work. Sidney Coleman was also there at the time. Fred Zachariasen, a distinguished if not overly serious professor of physics at CalTech, had pinned upon his office door an envelope upon which was drawn a crude caricature of Johnson, the target for our three-sided game of darts. Later, Fred realized that the envelope contained a paper that he was supposed to referee for possible publication in *Physical Review Letters*. "This paper should not be published," advised Fred, "because its arguments are full of holes."

During our game John Sakurai appeared at the door. Fred must have enjoyed my party at La Jolla, for his immediate greeting to John was, "Hi! Where's Gorgonzola?"

The Other California

A FACULTY MEMBER OF A PHYSICS department at any major American university is expected to perform several quite separate tasks, the most important among them being teaching and research. Administration is a third aspect of academia: selecting students for admission, determining their curricula, choosing our academic successors, seeking external financial support, etc.

The two principal activities, teaching and research, overlap and complement one another. The dual-function professor is the secret to American success in science. In some countries, such as the Soviet Union, the best researchers are secreted in research institutions where they do not teach, and the university professors are not encouraged to do research. But a research scientist cannot be inspired by a teacher who does no research nor taught by a scientist who does no teaching. Our country is fortunate to have achieved a stable balance between these roles at dozens of first-rate universities.

At Stanford, I began the teaching aspect of my career. It was the fall (if such a word can be used in California) of 1961. I was assigned to teach the graduate-level course in electromagnetic theory, one of the key disciplines that must be mastered by any student of physics. As is often the case with a freshman professor, I ended up spending practically all my time preparing my lectures. It took me a long time to establish the equilibrium between teaching and research that is necessary for success in academia.

With the announcement of the award of the Nobel Prize in October, my students and I enjoyed a brief and drunken respite. Stanford broke out the champagne for its own Bob Hofstadter, who won the prize in physics "for his pioneering studies of electron scattering . . . and for his thereby achieved discoveries concerning the structure of the nucleons." As a result of Hofstadter's work, it became clear that a nucleon, quite unlike an electron, is in no sense a pointlike particle.

It is a smoothly spread-out system with a considerable radius of about 10^{-13} centimeter.

This was one of the first clues that the nucleons are not fundamental building blocks of matter, but are composite systems made up of something else. Bob and his collaborators determined the structure of the nucleon by directing a beam of high-energy electrons at the nucleus and carefully observing in what directions the electrons were scattered.

THE MUON NEUTRINO

Meanwhile, just as Hofstadter was receiving his Nobel Prize, another exciting development in particle physics was taking place. While the hadrons were occupying center stage in California, leptons were a major concern elsewhere. For the first time ever, large and ambitious neutrino experiments were being performed at Brookhaven National Laboratory and at CERN. High-energy protons, extracted from large accelerators, impinged upon stationary targets, producing a secondary beam of pions. These rapidly decayed into energetic muons and neutrinos. Passing through a barrier of earth and steel, the muons were absorbed, leaving the neutrinos alone to enter the detectors, the devices in which their effects could be studied. What would they do?

It was known that neutrinos were produced together with electrons in the process of nuclear beta decay. It was also known that pions decay into muons and neutrinos. Thus it seemed reasonable to expect that the collision of an energetic neutrino with a nucleus would sometimes produce an electron and sometimes a muon.

The American group was the first to observe neutrino-induced events, a few dozen of them. In every case, the neutrino produced a muon, never an electron. Months later, the Europeans observed thousands and thousands of neutrino interactions and confirmed the American discovery. But coming in second doesn't count for very much.

Leon Lederman (now the director of Fermilab, the largest American particle physics laboratory), Mel Schwartz (now a successful high-tech entrepreneur), and Jack Steinberger (now the leader of a large European collaboration at CERN) were then the young superstars of American experimental physics. These three physics professors at Columbia University had proven that there exist in nature at least two distinct kinds of neutrinos: muon neutrinos and electron neutrinos.

Pions decay into muons and muon neutrinos, while the beta decay of a radioactive nucleus produces electrons and electron neutrinos. Energetic muon neutrinos can interact with matter to produce muons but not electrons. The Jewish Mafia had established the existence of a fourth kind of lepton. Now there were two pairs of leptons: electrons and electron neutrinos, and muons and muon neutrinos. It was a discovery that should have won the Nobel Prize, but it did not. It confirmed Schwinger's prophecy of years before that the muon and electron neutrinos are two different kinds of particles.

The weak force can transform an electron into an electron neutrino, or it can convert a muon into a muon neutrino, and vice versa, in both cases. It cannot transform a muon into an electron neutrino or an electron into a muon neutrino. Another way to say this is in terms of conserved quantum numbers. Electron number is the number of electrons and electron neutrinos minus the number of positrons (anti-electrons) and electron antineutrinos. Muon number is the number of negative muons and muon neutrinos minus the number of positive antimuons and muon antineutrinos. No known physical process can change the values of electron number or muon number.

The Columbia experimenters began with a beam of positive pions. Each pion decays into a positive muon and a muon neutrino. The muon neutrino has a muon number of one. When it interacts in the detector, it may produce a negative muon (which also has a muon number of one), but never an electron.

When Hofstadter's party, or any of the even more staid departmental affairs ended, my students and I would often save the evening at my garden apartment in Palo Alto. My living quarters were uniquely Californian. I had one of several dozen apartments surrounding a large and delightful swimming pool. They were clean, modern, new and quite sterile. Each "unit" was inhabited by one or two transient yuppies, each with a powerful stereo system. There was no acoustical insulation, nor privacy, dogs or children. My widowed mother came to visit for a few weeks during the dead of the New York winter. She loved the sun and the swimming, and especially racing with me through the nearby hills in my open sports car, the Triumph I had bought in Paris. The occasional earthquakes, minor though they were, she found less enjoyable.

There were two other junior faculty members in my field at

Stanford: Joe Dreitlein and Marshall Baker. Marshall and I had done our doctoral work with Schwinger at the same time, and while we were at Stanford we collaborated on three papers. In one, we tried to understand the mass splitting between the muon and the electron by means of spontaneous symmetry breaking. We failed to solve what is still a leading unsolved question in physics, but we also missed the main point of spontaneous symmetry breaking: the magical appearance of a massless "Goldstone" boson.

At this point in my career I had published about twenty papers. A few played a role in the development of today's understanding of particle physics: my Copenhagen paper introducing the electroweak theory and my works with Coleman exploiting the consequences of the eightfold way. Many other papers were relevant in their time: pointing out interesting byroads that nature could have taken and making suggestions of experiments that might help us to find our way. Others were overly ambitious attempts to do today what had to be put off till tomorrow. And a few of my papers were simply wrong.

There are two distinct purposes of academic publication. On the one hand, a scholar tries to do something new that contributes to a deeper understanding of his discipline, and to do it first. The second purpose is more mundane. The scholar often must assemble a dossier of published works to assure his promotion up the academic ladder. Here is where science differs from some other disciplines: there is simply no way to achieve the second purpose without achieving the first. It doesn't help to write a dumb paper simply to increase the size of your bibliography. One brilliant idea, published or not, can win tenure at a major university. So can a dozen ideas that are interesting and marginally significant. A hundred lousy papers will get you nowhere.

"Publish or perish" is an admonition to humanists, not scientists. We face a tougher challenge. Gell-Mann puts it this way: "A theorist should be judged by the correctness of his guesses about nature; his reputation should be gauged by the number right minus the number wrong, or even the number right minus twice the number wrong."

Midyear, James Bjorken, one of the tallest of all particle physicists, joined the staff. Alone among the young particle theorists, he would remain at Stanford for many years to come, where he played a key role to convince the world that nucleons are made of quarks.

Marshall was told that he would not be promoted to tenure at Stanford, despite the fact that he was an accomplished researcher and a superb teacher. In effect, he had been fired. Joe and I saw the handwriting on the wall. Rather than giving Stanford our best years only to be curtly dismissed at the end of our contracts, we told the department that we, too, would leave at the end of the year. I bear no grudge. Stanford was simply doing its job. Any great research university must hire only the very best scientists. At Harvard, we fire assistant professors all the time. (Don't worry about them. They get tenure at Princeton or Yale or other second-string universities.) Sometimes, we too make mistakes.

Joe found a position at Washington University in St. Louis, and I solicited an assistant professorship at Berkeley, where the road to tenure was virtually assured. I would continue my tour of California.

The summer of 1962 was divided between Washington, D.C., Istanbul and Copenhagen. The Institute for Defense Analysis, a government think tank, invited me and a few dozen other young physicists to Washington to work on defense-related problems for a month. It was a recruitment effort. Those who enjoyed the work and did it well would be called upon for further consulting work. We were all carefully scrutinized by the FBI so that we could be "cleared" and have access to top-secret information.

In Washington we were offered a choice. If I might somewhat oversimplify things, we could work on activities related to war or to peace. Only three of us chose peace: Arthur Kerman (now at MIT), Charles Schwartz (now at Berkeley), and me. Our assignment was to study all the literature about spy satellites, both secret and unclassified. It was a curious challenge. We were to prepare a report, based exclusively upon correct but unclassified data, that would examine the feasibility of such systems. Our access to top-secret data allowed us to determine which, among the unclassified publications, were truthful. It was a bizarre variety of literary criticism, and to ensure its absolute pointlessness, our report itself was to be classified. It was my first and last involvement with any kind of secret or defense-related activity.

The Istanbul summer school was organized by the Turkish physicist Feza Gürsey, who is now at Yale University. It was devoted to group theoretical concepts and methods in elementary-particle physics. Sidney Coleman, as well as my old friend from Copenhagen, Nick Burgoyne, came to the meeting. Nick had become a mathematician at Berkeley, but he was still deeply interested in physics. He and I were to administer a contract between the Office of Naval Research and the University of California that supported our research activities while we were both at Berkeley. At that time, but no longer, the pursuit of nuclear and particle physics was considered a part of the Navy's mission.

Another old friend, George Sudarshan, had worked with me at Harvard when we tried, with no success, to understand why strange particles seemed to live about twenty times longer than our simplistic theory said they should. He was also one of the coinventors of the V-A theory of weak interactions. Al Tesoro, another participant in the Turkish school, had been one of my buddies at Cornell, a year behind me. He had just completed his doctorate at Columbia as the first graduate student of Gary Feinberg, my high school chum. He was about to be married to a beautiful blonde model.

Finally, there was Abdus Salam. He lectured about his recent paper with Jeffrey Goldstone and Steven Weinberg, offering several proofs of Goldstone's theorem concerning spontaneous symmetry breaking. In Istanbul, Salam said, "Here I wish to speak of the price one must pay for having this type of surreptitious breakdown of symmetry . . . a number of unwanted massless fields [i.e., particles] seem to appear in the theory."

The school was held at Roberts College, then an American-run school of exceptional quality on the shores of the Bosporus in the upstream suburb of Bebek. Istanbul, a city of extraordinary beauty, is a point of convergence for many streams in the history of Western society. It is a city to which I have returned several times, and to which I must return again. Its mosques, museums, mosaics, fortresses, nightclubs and people are unforgettable. So is its hashish. Sidney, Al and I tried to get some.

After several evenings spent fruitlessly prowling the seedier sections of town around Iki Bucuk Street (it means two and a half, at one time the price in Turkish lira of a prostitute), we succeeded in finding a criminal named Raki to assist us in our purchase. The alleged hashish resembled a thousand-year-old Hershey bar. We took it to our apartment overlooking the Bosporus and smoked, and smoked, and smoked. Nothing happened.

Another criminal, Ceri (pronounced Jerry), purveyed a superior product. This stuff really worked! I spent the night before one of my lectures hallucinating, and by morning it still hadn't worn off. I delivered one of the best lectures of my life. On another occasion,

Sidney, Salam and I visited Ceri at his home in beautiful downtown Istanbul. Salam, the faithful Muslim, watched but did not smoke. I argued so eloquently for the unification of the electromagnetic and weak interactions by a gauge theory that I almost rivaled the prophet Mohammed. So said Salam afterward, and he should know.

My lectures at the school did not, however, touch upon my subliminal preoccupation with the electroweak theory. They dealt with the eightfold way:

> As a model for the strong interactions of mesons and baryons, the validity of the unitary symmetry scheme is yet to be determined. Nonetheless, unitary symmetry is the theme of this series of lectures. . . . Perhaps the lectures may be useful—if only from an historical or methodological viewpoint—regardless of the truth of the unitary symmetry model, for they sketch many connections between the purely physical and purely mathematical. But it is our suspicion that the unitary symmetry model is physically relevant, so that what we say here may have an actual, and not just a pedagogical, significance.

Even as I was speaking, evidence was accumulating for the essential truth of the eightfold way. The high-energy physicists of the world gather together every two years at a major international conference, and in the summer of 1962, they met at CERN. I did not attend because I was too junior to be invited. I don't attend such meetings now because I'm too busy. The discovery was announced of a spin-3/2 doublet of baryons with a mass of 1530 MeV and a strangeness of minus two. A pattern was emerging among the known baryon states with this spin.

Notice that the states neatly fill up a triangular pattern involving ten particles, but with just one site unoccupied. Here is a perfect analog to the periodic table of Mendeleev. The pattern was dictated by the eightfold way, and the hole in the pattern corresponded to a particle that must exist but had not yet been found. At the conference, Murray Gell-Mann responded to the report of the new experiment with an explicit prediction:

> We should look for the last particle, called, say, Omega-minus, with strangeness minus three and isotopic spin zero. At 1685 MeV, it would be metastable and should decay by weak interactions into $K^- + \Lambda$, $\pi^- + \Xi^0$, or $\pi^0 + \Xi^-$.

PREDICTION, DISCOVERY AND SURPRISE AMONG THE HADRONS

In 1961, Murray Gell-Mann and Yuval Ne'eman introduced the symmetry scheme known as the eightfold way. What were then considered to be elementary particles were put into a kind of periodic table and formed neat patterns consisting of singlets, octets or decimets. One member of the decimet had not yet been discovered: it was the analog to a missing element in Mendeleev's table. The scheme *predicted* the existence of a new particle with specified properties. The predicted particle was first observed at Brookhaven National Laboratory in 1964. This *discovery* made all physicists believe in the empirical success, and hence physical relevance of the eightfold way.

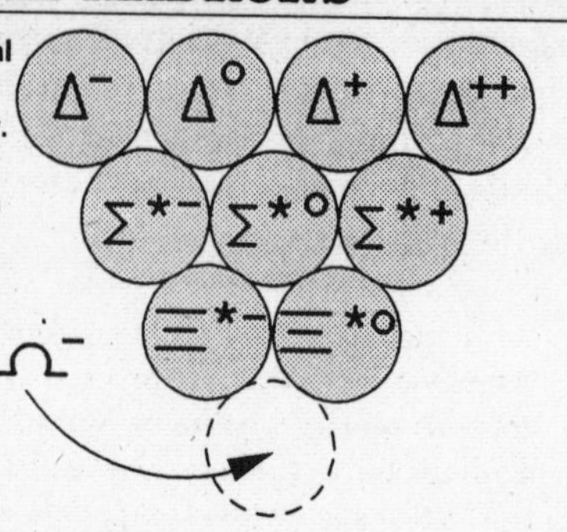

TRIANGULAR 8 - FOLD WAY DECIMET

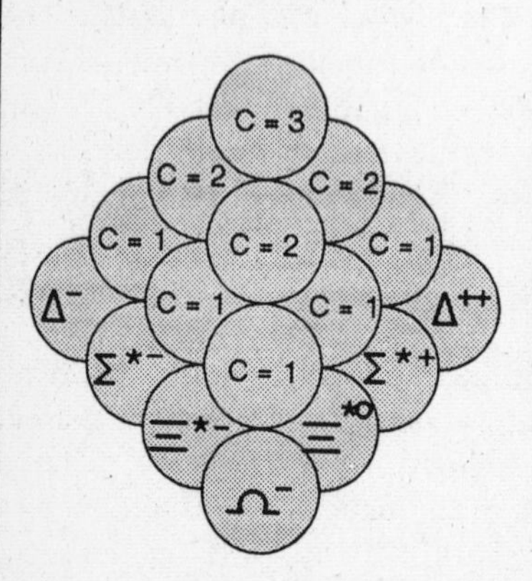

In 1974 the J/psi particle was discovered. It was the first observed particle that contained a charmed quark. These particles did not fit into the scheme of Gell-Mann and Ne'eman. The existence of the J/psi and the subsequently discovered charmed particles was a great *surprise* to most physicists. It was ultimately to comprise evidence for the quark hypothesis and for the existence of a fourth (charmed) quark species.

The new states bearing charm extend the triangular pattern into a three-dimensional tetrahedron.

Not only did Murray predict a new particle, but he was able to specify its properties in some detail, just as Mendeleev was able to specify the properties of undiscovered elements once he had invented the periodic table. The unfound particle's strangeness assignment was dictated by the symmetry scheme: the first row of the triangular pattern had zero strangeness, the next row strangeness minus-one, etc. The masses satisfy an "equal-spacing rule": the particles in the second row weigh about 145 MeV more than those in the first, those in the third row are 145 MeV heavier yet, and so the predicted omega-minus should weigh about 145 MeV more than the newly announced baryons, which had strangeness of minus two and mass of 1530 MeV. Hence the omega-minus should weigh about 1675 MeV. (Murray's original estimate was a bit different since it was based upon preliminary data.)

The pattern of masses within the spin-3/2 decimet was to become very simply understood in terms of quarks. The strangeness number of a baryon is simply the number of its constituent quarks that are strange, and the strange quark weighs somewhat more than its brethren. That does it! The omega particle is the strangest possible baryon simply because it is made of three strange quarks.

Murray's prediction of the omega-minus particle stood as a challenge to experimenters until early 1964, when a paper appeared in the European journal *Physics Letters* titled, "Observation of a Hyperon with Strangeness Minus Three." A collaboration of thirty-three experimenters working at the large proton accelerator at Brookhaven National Laboratory wrote:

> We wish to report the observation of an event which we believe to be an example of the production and decay of such a particle. The BNL 80-inch hydrogen bubble chamber was exposed to a . . . beam of 5 GeV K-mesons at the Brookhaven AGS [accelerator]. About 100,000 pictures were taken. . . . These pictures have been partially analyzed to search for the more characteristic decay modes of the Omega-minus. The event in question is shown [below].

Why was this momentous American discovery published in a European journal? At the time, our *Physics Review Letters* was universally acknowledged as the journal of record for advances in high-energy physics. *Physics Letters*, a European upstart begun in 1962, was having a difficult time getting started. As a friendly gesture, we threw them a bone. How times have changed! Our European colleagues have overtaken us at the high-energy frontier and *Physics Letters* is now the leading journal in our discipline. As a favor returned, today's European researchers submit an occasional not very important paper for publication in *Physics Review Letters*.

DISCOVERY OF THE OMEGA-MINUS PARTICLE

This is a photograph of the first discovered omega-minus particle as it was produced and decayed in the Brookhaven National Laboratory eighty-inch bubble chamber. The particles are identified in the drawing at the right. A negative kaon collides with a proton to form a neutral xi particle and a pion. The subsequent decay of the xi particle produces two characteristic V-shaped patterns. (Photo courtesy of N. P. Samios, Brookhaven National Laboratory)

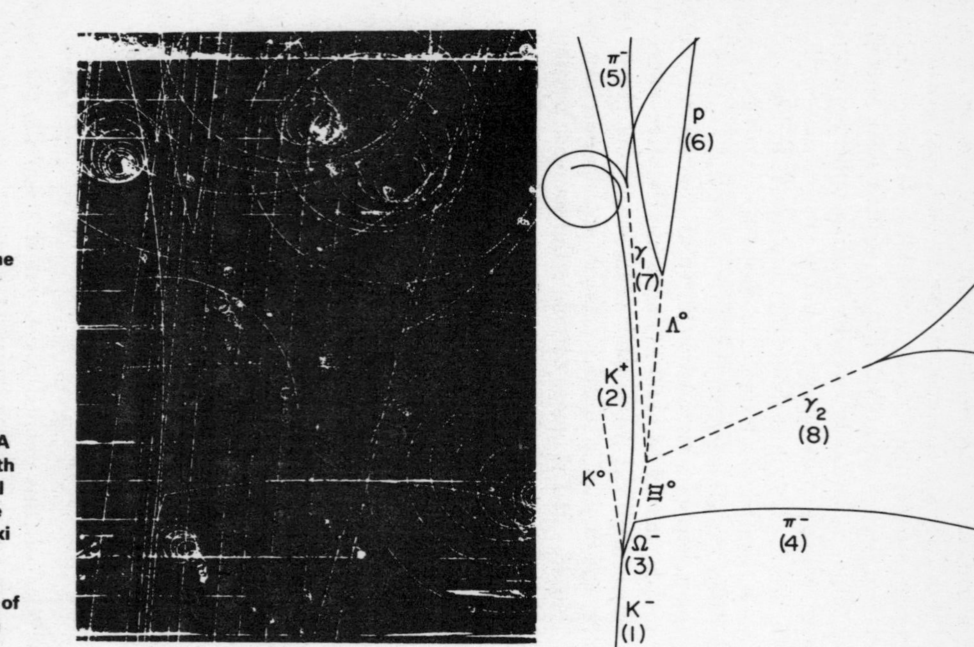

This discovery removed the last doubts of the essential relevance of the eightfold way. The mass of the particle was seen to be 1672 MeV, close to Gell-Mann's original estimate. As more and more of them were studied, their properties turned out to agree with all the theoretical predictions. The question was no longer whether the eightfold way was correct, but why.

I spent the last weeks of the summer of 1962 in Copenhagen for R&R with my Danish girlfriend Bonnie. Afterward, it was back to California to take up my teaching duties at Berkeley. I rented a small house in the Berkeley hills from a faculty member away on sabbatical. Many times I walked between the campus and my home, often being stopped by the police on my way. They were friendly but curious to find someone actually walking on the streets. Californians may jog, surf or sail, but they rarely walk to and from work. Bonnie visited me for the Christmas holidays.

Berkeley was an exciting place to be for a particle physicist in the early 1960s. Giant bubble chambers were revealing the existence of all kinds of new particles. By the end of 1962, I wrote a paper with experimenter Art Rosenfeld, pointing out how successfully the new particles could be accommodated into the supermultiplets of the eightfold way.

All known baryons could be assigned to families of one, eight or ten particles: singlets, octets and decimets. All known mesons form singlets and octets. The search for a gauge theory of the strong force appeared to have been a red herring because the spin-one particles characteristic of a Yang-Mills theory did not seem to play any special role. There were octets and singlets with spin-zero, spin-one and spin-two, all of which had an equal claim to being elementary. We showed in our paper that the properties of these many particles, in particular their masses and lifetimes, could be accounted for by the eightfold way. Our conclusion was that "with satisfaction and relief, we find that the calculated results are completely compatible with experiment."

We were a bit bothered by a question of nomenclature. The supermultiplets of the eightfold way were called octets and decimets, while the smaller isotopic multiplets were called doublets and triplets. In a footnote to our paper we wrote:

> Words signifying sets of similar particles find their origins in musical terminology. Thus, a trio, quartet . . . octet, nonet, decimet . . . is a

composition for 3, 4 . . . 8, 9, 10 . . . voices or instruments; but a triplet, quadruplet . . . octuplet, nonuplet, decuplet . . . refers to 3, 4 . . . 8, 9, 10 . . . notes played in one beat. After a triplet (of pions) and a quadruplet (of Delta particles), we use "decuplet" for the 10. Because of an unfortunate earlier misuse, "octet" has become commonplace for the 8.

Despite our etymological erudition, the family of ten particles is today still known as a decimet.

My colleagues at Berkeley included some familiar faces. Nick Burgoyne, whom I had known in Copenhagen, had become a member of the mathematics department. We ran a joint seminar in math and physics. Steven Weinberg had gotten his doctorate from Princeton and done a stint at Columbia. Having lost to Gary Feinberg in a one-on-one competition for a tenured position there, he had recently joined the Berkeley physics department. We wrote a few inconsequential papers together while we were both in California. Chuck Zemach, with whom I had shared an apartment while we were Schwinger's graduate students, became one of my closest friends. In fact, he had been instrumental in hiring me after I had decided to leave Stanford. Berkeley was certainly the place to be. At that time, its theoretical physicists, experimenters and mathematicians were working together, such as they rarely do. It was an exciting time in which the curious phenomena of particle physics were beginning to make some sense.

During my first year at Berkeley I had the good fortune to be awarded an Alfred P. Sloan Foundation Fellowship. This was a source of "free money" that I could use to escape my teaching obligations so that I could pursue my research anywhere I wished. Teaching and research do go hand in hand, but things were moving so quickly in physics that I simply had to have more time to do research so that I would know what new things to teach next year! Also, I was developing withdrawal symptoms from my yet-undetected addiction to the East Coast way of life.

I chose to spend a semester working with Sidney Coleman at Harvard and a second semester back in Copenhagen. I sold my TR-3 to a Chinese graduate student, spent another summer in Europe (where I bought a Volkswagen and shipped it back to the States), and returned to Boston just a few weeks before the assassination of John F. Kennedy.

My friend from Cornell and Harvard, Danny Kleitman, after leaving Copenhagen, had joined the physics department at Brandeis University. I moved into a spare room of his large rented house in Newton and got to work.

The known forces among elementary particles form a hierarchy with descending strengths: strong, electromagnetic and weak. During the sixties it seemed as if the strong force itself has to be divided into two distinct components: the "very strong force" and the "medium strong force."

The "very strong force," constituting the major part of the strong force, respects the unitary symmetry group of the eightfold way. So far as this force is concerned, the members of a supermultiplet, like the eight baryons, are indistinguishable from one another. If weaker forces did not exist, there would be in nature eight conserved quantum numbers associated with the symmetries of the "very strong force." But the eight baryons are anything but indistinguishable. The two xi particles, for example, are some 40 percent heavier than the nucleons. This is too large a departure from unitary symmetry to be ascribed to electromagnetic or weak forces. There had to exist, so it seemed, another component to the strong force that was weaker than the "very strong force" so that the predictions of the eightfold way could be roughly satisfied.

The hypothetical "medium strong force," the first step down our hierarchy of forces, breaks the symmetry of the "very strong force" and gives rise to the considerable mass splittings within the eightfold-way supermultiplets. This force does not conserve four of the eight quantum numbers of the eightfold way, but it does respect the conservation laws of strangeness and isotopic spin. The approximate validity of the mass formulas, which gave many relations among the various masses of observed particles, suggested that the "medium strong force" was as simple in structure as it could be. It breaks the eightfold-way symmetry in the gentlest possible way.

In truth, as it was revealed in the 1970s, there is no such confusing, contrived and unnatural bifurcation of the strong force. The empirical successes of the eightfold way simply reflect the fact that Murray's up, down and strange quarks are not so far apart in mass. The strong force displays an exact and supernal symmetry that was not yet dreamed of in the 1960s. The idea of a "medium strong force," although ultimately wrong, was an essential stepping-stone on the way toward today's more elegant and correct theory.

The electromagnetic force is considerably weaker than the "medium strong force," and at the same time it displays less symmetry. It does not conserve isotopic spin and it consequently generates the smaller mass differences of the particles within isotopic multiplets. If there were no such thing as electromagnetism, it was believed that the neutron and the proton would be indistinguishable particles, as would the three sigma particles and the two xi particles. Although the electromagnetic interaction is responsible for symmetry breaking, its structure is very simple from the point of view of the eightfold way. The precise technical statement is that the electric charge is one of the conserved quantum numbers associated with unitary symmetry. For this reason, Sidney Coleman and I were able to make many successful predictions about the measurable electromagnetic properties of the baryons.

The weak force is the weakest of all the forces of nature except gravity. It displays the least symmetry of all. It violates the conservation of strangeness, and it leads to the failure of space-reflection (mirror) symmetry. None of the symmetries of the eightfold way survive in the presence of the weak, electromagnetic and "medium strong" forces. While the weak force leads to symmetry breaking, it does so in accordance with certain simple empirical rules. For example, although its effects sometimes conserve strangeness and sometimes do not, it never leads to a change of the strangeness quantum number by more than one unit at a time. The strangeness-changing part of the weak force displayed another curious regularity. Strange baryons sometimes decayed in such a fashion as to produce leptons. For example:

$$\Sigma^- \rightarrow N + e^- + \nu^- \text{ and}$$
$$\Xi^- \rightarrow \Lambda + e^- + \nu^-$$

In these decays, both the strangeness and the electric charge of the hadron increase by one unit. The opposite type of decay, for example,

$$\Sigma^+ \rightarrow N + e^+ + \nu$$

where the change in charge is opposite to the change in strangeness, has never been seen. Could these properties be incorporated into

MASS SPECTRUM OF THE EIGHT BARYONS

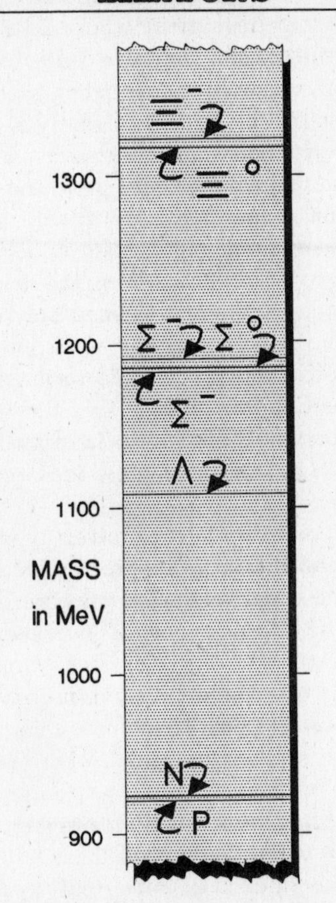

The eight spin-½ baryons constitute a single supermultiplet of the eightfold way. They are clumped into four submultiplets: nucleons (the two lightest baryons), the singlet lambda, three close-spaced sigmas and two xis. The large mass splittings *among* the submultiplets were supposed to be caused by the "medium strong force." The much smaller mass differences *within* the submutiplets were attributed to electromagnetic effects.

the V-A theory of weak interactions in a fashion that made sense from the point of view of the eightfold symmetry?

The question was answered affirmatively and decisively by Nicola Cabibbo, the young Italian physicist whom I had encountered at summer schools in Scotland and Italy. He conjectured, in the spring of 1963, that the weak analog to the electromagnetic current should be identified with one of the quantities that was conserved in the limit of exact unitary symmetry. In other words, Cabibbo suggested that the weak force should have a very simple and elegant structure from the vantage point of the eightfold way. His hypothesis immediately explained the observed regularities among the strangeness-changing decays. Moreover, it pinned down the structure of the weak current almost completely. There remained only one arbitrary parameter governing the structure of the weak interactions: it soon became known as the Cabibbo angle. The weak current that Nicola invented is called the Cabibbo current, and the theory underlying his idea is the Cabibbo theory. So my Italian friend got three things named after himself in return for just one idea—but it was a good idea.

Today Nicola is a professor at the University of Rome and president of the INFN, the Italian agency responsible for nuclear and particle research. In these days of declining budgets for pure science, it is remarkable that the INFN support has doubled under Cabibbo's stewardship.

Like fine wine, Nicola's invention has improved with time. What he accomplished was no less than a complete reinterpretation of the idea of the universality of the weak force. There are three kinds of weak processes: those involving hadrons but conserving strangeness, those changing the strangeness quantum number, and the purely leptonic processes like muon decay. Call them A, B and C, respectively. The original naïve notion of universality could be likened to an equilateral triangle in which the lengths of the sides correspond to the strengths of the three kinds of interaction. That notion was not working out. Side B, corresponding to strangeness-changing processes, appeared to be considerably shorter than the other two. That is to say, the force responsible for strangeness-changing decays was observed to be about twenty times weaker than it should have been in the equilateral-triangle picture.

Cabibbo universality replaced the equilateral triangle with a right triangle where, as we all know, the length of the hypotenuse is the

CABIBBO UNIVERSALITY

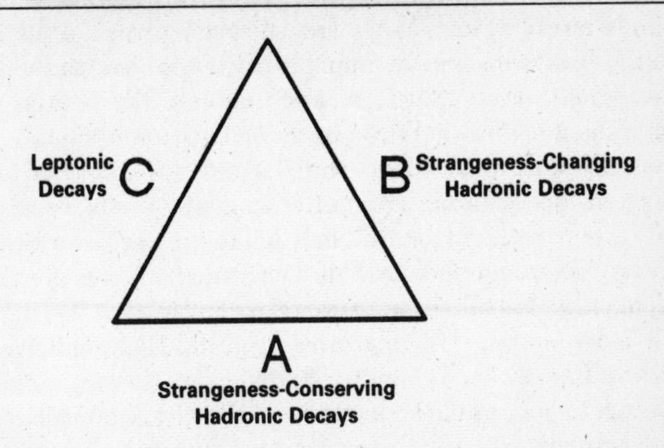

Universality prior to Cabibbo: it didn't work because Side B was observed to be too short.

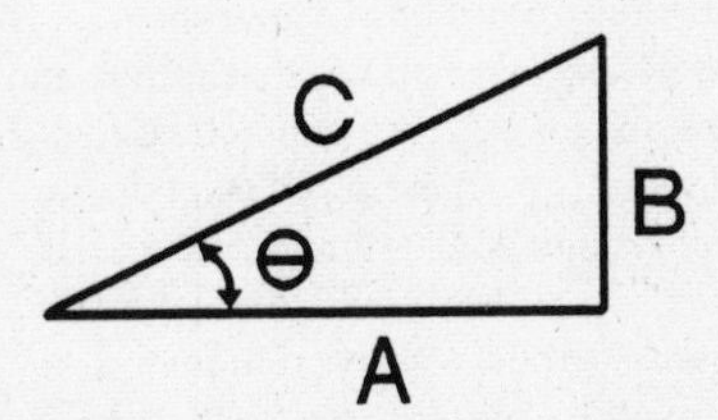

Cabibbo argued that the three effects should be described by a right triangle, in which the square of the hypotenuse is equal in length to the sum of the squares of the lengths of the two legs. This meant that strangeness-conserving decays (side A) should be slightly suppressed compared to leptonic decays (side C). Exactly this effect was observed in the laboratory.

square root of the sum of the squares of the lengths of the legs. Careful experiments soon proved that nature chose Cabibbo's interpretation of universality. In retrospect only, the validity of Cabibbo universality can be recognized to be an early clue pointing toward a gauge theory of the weak force.

Sidney Coleman and I became interested in the general question of symmetry breaking, whether due to the "medium strong force," the weak force or electromagnetism. We noticed certain striking regularities. The mass splittings within hadron supermultiplets satisfied simple formulas such as the "equal-spacing-rule" of the baryon decimet. There were similar simple relationships among the smaller electromagnetic mass splittings. For example, the neutral sigma had a mass about halfway between its charged partners. Finally, the strangeness-changing weak force, while it did not conserve isotopic spin, violated this quantum number in as gentle a way as possible.

Late one night, at Sidney's home on Lancaster Street in Cambridge, seemingly inspired by marijuana, we conceived of a simple explanation for all three of these irregularities. We began to write our magnum opus, "Departures from the Eightfold Way: Theory of Strong Interaction Symmetry Breakdown." It was to be our last significant "joint" publication. Here is the gist of our ideas, in our own words:

> These three phenomena are similar in that they all involve a mysterious dominance of unitary octets. Nevertheless, the explanations they have elicited in the literature are quite different. . . . We propose a theory of symmetry-breaking interactions that differs radically from the theories mentioned above. . . . We conjecture that there exists a unitary octet of scalar mesons. . . . If the scalar octet does exist, there is the possibility of a class of Feynman diagrams which we call "symmetry-breaking tadpole diagrams." . . . Our fundamental dynamical assumption is that symmetry-breaking processes are dominated by symmetry-breaking tadpole diagrams. For brevity, we shall refer to this assumption as "tadpole dominance." This assumption immediately explains the three instances of octet dominance we have cited above.

Theorists sometimes deal with little questions with which they have a good chance to accomplish something significant, but small. Sidney and I had become greedy. We wanted to build a general theory describing all kinds of symmetry breaking. Such grandiose ambitions are rarely fulfilled. Our drug-induced fantasy of octet dominance was interesting and perhaps even useful, but the basic idea turned out to be wrong. Nature does not use such an octet of mesons as we invented, nor are "tadpole diagrams" the universal panacea we thought they were. While I confess to being an occasional user of hashish and marijuana during the early 1960s, I cannot recommend these drugs as a secret to success in physics. Fine wines and cognacs are, however, another story.

The editors of *The Physical Review* balked at our usage of "tadpole" diagram. The word was inappropriate, they felt, to a scholarly journal.

TADPOLE DIAGRAM

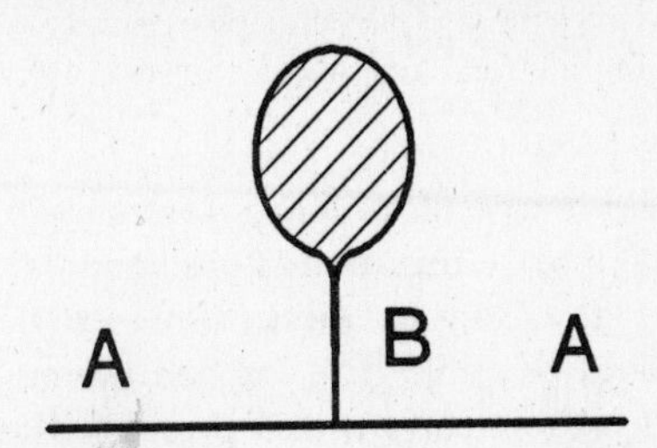

This is a diagram by which particle A can obtain a mass. Particle B doesn't go anywhere, but just ends in a blob, forming a tadpole-like design.

Yet we needed some such word and tadpole seemed to us to be properly descriptive. Nonetheless, we offered the editors *lollipop* and *sperm* as almost perfect alternatives, which led them to print our original term. The phrase *tadpole diagram* has become generally accepted for this concept throughout the world. According to my *Russian-English Phrase Book for Physicists*, even the official Soviet translation is literal, *diagramm c golovastikami*. Here is one tadpole that will never become a frog.

While we were hustling our brilliant but flawed theory, Murray Gell-Mann, back in sunny Pasadena, had a better idea. The time had come for him to invent those most fundamental particles with funny names, the quarks. George Zweig had the same idea, independently and at the same time, except that he called them aces. Quarks made it into the *Supplement to the Oxford English Dictionary*, but poor George has become a biologist.

> QUARK, *sb. Physics.* [Invented word, associated with "Three quarks for muster Mark!" (Joyce, *Finnegans Wake* (1939) II. IV).
>
> "I employed the sound 'quork' for several weeks in 1963 before noticing 'quark' in 'Finnegans Wake,' which I had perused from time to time since it appeared in 1939. . . . The allusion to three quarks

seemed perfect. . . . I needed an excuse for maintaining the pronunciation quork despite the appearance of 'Mark,' 'bark,' 'mark,' and so forth in 'Finnegans Wake.' I found that excuse by supposing that one ingredient of the line 'Three quarks for muster Mark' was a cry of 'Three quarts for Mister . . .' heard in H. C. Earwicker's pub."—M. Gell-Mann, private let. to Ed., 27 June 1978.]

Any of a group of sub-atomic particles (orig. three in number) conceived of as having a fractional electric charge and making up in different combinations the hadrons, but not detected in the free state.

Gell-Mann had decided that there were simply too many different hadrons for them to be regarded as fundamental particles in their own right. He felt that simpler things must exist in nature, out of which hadrons could be built. He was continuing a long tradition: from atoms to nuclei to neutrons and protons. But Murray argued that the buck did not stop there.

If baryons were to be made of quarks, two important questions had to be faced. First, how many quarks make a baryon? The answer to this conundrum could not be divisible by two. This odd conclusion stems from the fact that protons and other baryons are fermions, which (unlike bosons) are particles that satisfy Pauli's exclusion principle. No two identical baryons can occupy the same quantum state. This is why the atomic nucleus resembles a cluster of marbles stuck together, rather than collapsing into a still-smaller system.

If baryons have parts, the parts themselves must be fermions. This is because any composite system made of bosons is a boson. Moreover, any composite system made of fermions is a fermion or a boson depending on whether the number of constituents is odd or even. Thus, if the baryons are made up of quarks, we may conclude that quarks must be fermions and that there is an odd number of quarks in each baryon. Now, the smallest odd number is *one*, but it makes no sense to say that a composite system has only one part. Murray chose the simplest remaining possibility. He conjectured that the baryon is made up of three quarks. Each quark is a fermion with the lowest possible spin: one-half.

The second question was: How many different kinds of quarks are there? Let us address a simpler question first. How many quark "flavors" are needed to construct non-strange baryons? Experiment has demonstrated that *every* baryon has one of four values of electric charge: +2, +1, 0 or −1. Two quark flavors with different electric charges can explain this. Call them *u* for the "up" quark (the one with the larger charge) and *d* for the "down" quark. Every non-strange baryon would correspond to one of the following four quark recipes:

(*uuu*) (*uud*) (*udd*) (*ddd*).

The baryon with the highest observed electric charge, +2, must be made of three up quarks. Since the electric charge of a composite system equals the sum of the charges of its parts, it follows that the electric charge of the up quark is $Q(u) = +2/3$. The baryon with the lowest electric charge, −1, must be made of three down quarks. Thus the charge of the down quark is $Q(d) = -1/3$. Murray's simple conjecture leads inexorably to these curious fractional charges.

Already the quark model shows its power. There were many known non-strange baryons. Many had large values of spin, but none had charges larger than +2 or smaller than −1. Physicists were mystified at the experimental absence of baryons with charge +3 or −2. The quark model explained all that. Two quark flavors implied four possible electric charges, no more and no less. The proton is one example of a non-strange baryon. It is made of two up quarks and one down. Its charge is the sum of its constituents: $(+2/3)(+2/3)(-1/3) = +1$. The neutron is a non-strange baryon containing one up and two down quarks. Its charge is: $(+2/3)(-1/3)(-1/3) = 0$.

In order to include a description of strange baryons, Murray had to postulate a third species: the strange quark, *s*, whose electrical charge is the same as the down quark, $Q(s) = -1/3$. The strange quark was assumed to be somewhat heavier than the other two, and it was assigned a strangeness number of −1.

Baryons containing just one strange quark have strangeness −1 and quark structures (*uus*), (*uds*) or (*dds*), with electrical charges of +1, 0 or −1, respectively. Baryons containing two strange quarks have strangeness −2, quark content (*uss*) or (*dss*), and charges of 0 or −1. Finally, there had to be a baryon containing three strange quarks with an electrical charge of −1. This was the soon-to-be-found particle that Murray had predicted to exist in the summer of 1962. He named it the omega-minus because it was expected to be negatively charged, and because omega is the last letter of the Greek alphabet. He thought it was to be the last qualitatively different baryon that would ever be found. As it turned out, he was wrong. Physicists have since detected charmed baryons that contain a quark flavor that Murray did not foresee.

All this was precisely in accord with experiment. All the baryons then known could be understood as configurations of three quarks chosen among the three quark species, *u*, *d* and *s*.

If the first quark commandment is that a baryon is made of three quarks, the second commandment is that a meson consists of one quark and one antiquark. As with the baryons, all of the known mesons could be understood as quark configurations of this kind. Quark theory explained a long-standing mystery: why the baryons come in singlets, octets and decimets, and only those families of the eightfold way, and why the mesons appear only as octets and singlets. Moreover, as Murray noted, "We thus obtain all the features of Cabibbo's picture of the weak current." Despite the evident allure of the quark hypothesis, its inventor took a very cautious view, calling his paper merely "A Schematic Model of Baryons and Mesons" and remarking that "It is fun to speculate about the way quarks would behave if they were physical particles . . . instead of purely mathematical entities."

Why was Gell-Mann so tentative? One good reason was that nobody had ever seen a particle with a fractional electric charge. There was simply no direct evidence for the quark as an observable, physical particle. How could the proton be made up of pieces that could not be seen individually? This dilemma is still with us, since quarks remain unseen as other than the inextricable parts of hadrons. Today we have a theory suggesting that quarks are irrevocably confined within the hadrons they make up. Our theory implies the first two quark commandments, and a third: Thou cannot see an isolated quark, no matter how hard thou tryest.

Another profound difficulty with the quark theory of hadron structure soon emerged. Quarks, as we have seen, are fermions. This means that quarks have to satisfy the Pauli exclusion principle: two identical quarks cannot do exactly the same thing at the same time. However, a baryon made up out of three up quarks with a total spin of 3/2 was known to exist. Each of its three quarks is doing exactly the same thing. Impossible! This problem was solved eventually when it was realized that each of the flavors of quarks (up, down, strange) comes in three colors, and that a baryon is made up out of three differently colored quarks. I'll explain all this better later in my story, but for the moment let me only warn you of the physicist's habit of using perfectly ordinary English words in funny ways. Our definitions of color and flavor are not to be found in dictionaries.

It was fun being at Harvard among my old friends, with no

courses to teach and no students breathing down my neck. It was there, too, that I encountered Schwinger's enchanting secretary Shirley, who was partly responsible for luring me back to Harvard later on. Thank you, Alfred P. Sloan.

My memory trip continued with a return to Copenhagen in the spring of 1964. Whom did I find there but James (BJ) Bjorken, my colleague at Stanford! Together we invented the idea of a fourth fundamental hadronic constituent and gave it its name for all time, "the charmed quark." The reasons we did that were mostly wrong, but there was one buried nugget of good sense. It has to do with a historical notion that there should be some sort of parallel between the world of leptons and that of hadrons. Remember that these are the two great families of "matter particles," hadrons being those that partake of the strong force and leptons those that do not.

Once upon a time, in the 1930s, there seemed to be a kind of lepton-hadron symmetry. There were two known types of nuclear particles, protons and neutrons, and two known types of leptons, electrons and the neutrinos that Pauli had postulated. By the 1950s Robert Marshak invented a new kind of lepton-hadron symmetry involving three particles of each type: electrons, muons and neutrinos on the one hand; and neutrons, protons and lambdas on the other. With the invention of quarks, the number of fundamental strongly interacting matter particles remained three: up quarks, down quarks, and strange quarks. However, the number of different kinds of leptons had increased to four with the discovery that electron neutrinos and muon neutrinos were distinguishable particles.

BJ and I decided to reinstitute the notion of lepton-hadron symmetry. We made up a new kind of periodic table of leptons and quarks. In our table there was a very conspicuous hole, an empty space corresponding to a missing quark. The pattern of symmetry would be complete if only there existed a fourth flavor of quark, with a specific electric charge and specific weak-interaction properties. We postulated the existence of a new hadronic constituent that was not part of any then-known hadron. We assigned to our hypothetical quark a new quantum number that we called "charm." In the table-making tradition of Mendeleev and Gell-Mann, we predicted the existence of a whole family of new particles carrying the new quantum number, charmed particles. The world of four quarks and four leptons seemed very attractive to us. There appeared to be "a fundamental similarity

PERIODIC TABLE OF QUARKS AND LEPTONS

-1	-1/3	0	+2/3
electron e	down-quark d	electron-neutrino ν_e	up-quark u
muon μ '38	strange-quark s '50	muon-neutrino ν_μ '61	charm-quark c '74
tau-lepton τ '75	bottom-quark b '76	tau-neutrino ν_τ not yet seen	top-quark t not yet seen

When James Bjorken and I first devised this table in 1964 only the seven particles in the shaded portion were known. Spying the hole, we *predicted* the existence and the properties of the missing entry. We named it the charmed quark.

The *discovery* of the J/psi particle in 1974 was soon seen to confirm our prediction. It was the first of many particles containing charmed quarks to be found. The table seemed to be complete.

In 1975, Martin Perl observed a third charged lepton, the tau lepton, which was a *surprise* to us all. It seemed as if our table had an unexpected third row. The *b* (for beauty, bottom or Bicentennial) quark was established in 1976, and soon there was indirect evidence for the existence of a third neutrino. At this writing there is not yet any evidence at all for the *t* (for truth, top, but hopefully not Tricentennial) quark.

between the weak and electromagnetic interactions of the leptons and the quarks" that suggested to us "a possible intimate connection between [them]."

Our choice of the word *charm* for the designation of the fourth quark was a fortunate one. According to the *American Heritage Dictionary*, *charm* can mean "an action or formula thought to have magical power." The mysterious power that our charm had to avert evil, in the form of "strangeness-changing neutral currents," was to become clear six years later. Charm is one of the essential ingredients of today's remarkably successful theory of elementary particle physics. And those charmed particles we predicted have been found!

In the summer of 1964 I traveled to Italy, where I purchased a new sports car appropriate to my recently tenured position at Berkeley, an Alfa-Romeo 1600 Spider. I attended the Varenna summer school once again, but as a lecturer rather than a student, speaking of the successes of the eightfold way, the work I had recently completed with Sidney, and the possible existence of charmed particles. At the end of the school I drove the gorgeous blonde secretary, Silva, to her home in Milan. Holding her hair in place against the hundred-mile-per-hour wind, she explained to me her theological quandary. A certain sin of the flesh, which she committed from time to time in Italy, required upon confession the recital of a hundred *Ave Marias*. During a trip to France she did it again, but was awarded a penance of merely ten *Aves*. How could this be? I couldn't help her. There are questions that even physicists can't answer.

One of the few times I did attend the biennial International Conference on High-Energy Physics was that summer, when the meeting took place at Dubna, the site of a large but virtually useless Russian accelerator. At that time, Russian accelerators worked no better than their notorious elevators and restaurant waiters. Sidney and I were to present our latest calculations of the baryon electromagnetic mass splittings. Sidney tells the story of our trip to Dubna as "My Happiest Moment in the Soviet Union."

After an excursion in my Alfa to the beautiful city of Budapest, Sidney and I flew from Vienna to Moscow. At Sheremetyevo airport Soviet officials were gathering physicists together so that they could be bused to Dubna. A few dozen of us had to wait about angrily until it could be established that my luggage had been lost by Aeroflot and probably was on its way to Cuba. Eventually we were permitted to board a bus whose windows and doors did not close and we set off into the wet night. The road was poor. The bus leaned one way, then the other, so that we could all enjoy the rain.

Sodden, shaken and sick, we arrived at last at the one hotel in Dubna to join a jostling throng of hundreds of weary and world-famous physicists all trying to register. None were willing to form a queue. Sidney could not prevent a porter from escaping with (and promptly misplacing) his luggage. The rooms, for which we (i.e., Uncle Sam) had paid in advance, were $50 a night for Westerners but only $5 for the presumably lighter-sleeping physicists of the Eastern bloc.

Sidney and I were assigned one of two small rooms sharing a bath. Murray Gell-Mann got the other all to himself. The tiny bath contained a toilet (which broke as soon as I attempted to use it), a sink, a shower directly overhead, a radiator upon which were two pieces of toilet paper and two small towels, and a hook to hang one's clothing. Sidney took a shower. The water spewed in all directions, into the toilet and sink and onto the radiator, the toilet paper, the towels and Sidney's only remaining clothes.

Our room barely had space for two beds and a night table. There was no room for the luggage that we no longer had. Wisely, Sidney had secreted a small bottle of cognac in his briefcase. He poured two drinks so that we could toast our safe arrival in the USSR. We got into our beds, turned off the lights—and *my* bed collapsed! That was Sidney's happiest moment.

When I returned to Berkeley, the estranged and available wife of my old college friend Al Tesoro arrived in California. We lived together through the days of the student rebellion. One of its leaders, Mario Savio, was a physics major. Remember the Free Speech Movement and its less reputable successor, the Filthy Speech Movement? The government's incomprehensible policy in Vietnam was polarizing the nation. We who had supported the election of the President as "Scientists and Engineers for Johnson" felt betrayed by the carpet bombing of Vietnam and the extension of the war into Laos. We marched in opposition to what we regarded as a war America should not be fighting.

PARTICIPATION IN MARCHES EXPLAINED

THE NEW YORK TIMES SUNDAY, OCTOBER 31, 1965

Letters

to the Editor of *The Times*

To the Editor:

James Reston wrote in his Oct. 17 column that marches protesting United States actions in Vietnam are a disservice to the nation and to the cause of peace. On Oct. 18 President Johnson issued a strongly worded statement concurring with Mr. Reston's point of view. They reason that any sign of disunity in this country encourages Hanoi to avoid negotiations by giving the impression that United States resolution is weakening.

We find the implications of the above position much more alarming than the possibility that these marches may be giving false comfort to Hanoi. We are appalled to find our lawful protest answered by the demand to close ranks in the face of external enemies; it is but a short step to the identification of dissent with treason.

In the face of these attacks on free expression we participants in the Berkeley marches wish to declare the reasons which impelled us to this action.

Walter Lippmann and others have stated that decisions will soon be made in Washington concerning our conduct of the next phase of the war in Vietnam. It is argued that the American people are beginning to expect an unambiguous military victory, and that President Johnson will be under substantial political pressure to satisfy their wishes.

Capable Forces

That such a military victory could be achieved we do not doubt: our forces are capable of devastating North Vietnam by unrestricted bombing of its cities and dikes; the Vietcong could be overwhelmed by a still greater build-up of our ground forces.

We marched to make clear to the President that there is a significant element of public opinion which does not desire or expect such a victory. We marched to enter a plea now for future restraint.

Nothing can justify silence in the face of indiscriminate bombing in South Vietnam, execution of leaders of peaceful protest against the South Vietnam Government, use of weapons whose scars will remain as grisly reminders long after this war is over, torture and killing of prisoners, and large-scale bombing in North Vietnam of sites whose military importance is becoming more and more questionable. We marched to express our revulsion at the inexcusable brutality of the war.

Moral numbness has replaced an earlier apathy. The people of our country have begun to accept the Vietnam war as a natural part of the scene; a new B-52 raid seems hardly worthy of notice. The spirit of consensus has stifled debate in Congress. By marching against present United States policies we are voting with our feet for what apparently we were unable to vote for with ballots last November; we are voting for an American Government which will stand up to those who always seek military solutions.

As a first step toward a peaceful solution we urge the President to stop the bombing of North Vietnam and to make an offer of unconditional negotiations not only to Peking and Hanoi but also to the Vietcong.

Individually we have written letters and signed petitions. We are convinced that this has not been enough, that a true picture of our disenchantment and distress has not reached Washington. We are indeed, in Mr. Reston's words, "sincerely for an honorable peace, troubled by the bombing of the civil populations of both North and South Vietnam, genuinely afraid that we may be trapped into a hopeless war with China, and worried about the power of the President and the Pentagon and the pugnacious, bawling patriotism of many influential men in Congress."

If lawful public protest is a luxury which our country cannot now afford, what are the alternatives?

OWEN CHAMBERLAIN, SIDNEY COLEMAN, DANIEL FREEDMAN, SHELDON GLASHOW, ALFRED GOLDHABER, ROBERT HARRIS, MARTIN KARPLUS, JOHN RASMUSSEN, GINO SERGE, ROBERT SOCOLOW, HERBERT STRAUSS.

Berkeley, Calif., Oct. 19, 1965

The writers are scientists affiliated with various universities; Professor Chamberlain is a Nobel laureate (1959).

In the fall of 1965 my colleagues and I marched to protest the Vietnam War. Steve Weinberg did not participate in the demonstration, although he and his wife did serve coffee and cognac to the weary marchers afterward.

Time passed. I learned to sail in San Francisco Bay, to ski in the high Sierras, and to rock-climb (after a fashion) in Yosemite. I could have settled down quite comfortably in Berkeley. It was a complete surprise, in the fall of 1965, to be rung up by Bill Preston, then the chairman of the Harvard physics department. He offered me a permanent position at Harvard for a mere 30 percent cut in salary. I could be closer to my aging and ailing mother in New York, and closer still to Schwinger's beautiful red-haired secretary, Shirley.

My years of exile had come to an end. By the new year I had sold my Alfa, shed my California girlfriend, and come back home.

Return from Exile

IN JANUARY 1966 I FINALLY BECAME a full-fledged member of the physics department of Harvard University. I was among those very few alumni to achieve the ultimate Harvard dream—to do unto other students what Harvard had done to them.

Veritas!

For the next six years I was to live at 84 Prescott Street in Cambridge, in a gracious apartment house built in 1906, just a stone's throw from Harvard Yard. A trip to the used-furniture stores in Somerville yielded a motley collection of beds, tables and chairs to make my spacious four-room apartment just barely livable. My colleagues at the Harvard physics department included Sidney Coleman, who had been my unofficial graduate student at CalTech and who had become an assistant prof at Harvard, and Julian Schwinger, who had been my own research supervisor at Harvard years before. It was Schwinger who had presented me with my grand challenge in physics: to develop a unified theory of the weak force and electromagnetism.

Steve Weinberg was to join the local physics community in a few months, following me in the pilgrimage from Berkeley to Cambridge. His wife, Louise, at the ripe age of thirty-five, had finally decided upon law as a career and had been accepted to the Harvard Law School. Steve took a leave of absence from his position in California to become a Loeb Visiting Professor at Harvard. A year later he became the superstar of the MIT physics faculty, and it was there that he made his own brilliant contribution to the electroweak theory. Soon afterward, when Julian opted for sunny California, Steve was captured by Harvard as Julian's designated successor.

During my years of western exile, a number of my old Cambridge friends had forsaken physics for disciplines more to their liking. Danny Kleitman, my classmate at Cornell and Harvard and colleague at Copenhagen, was teaching physics at Brandeis when he unearthed a book of unsolved problems in mathematics. He solved many of them, and he began to realize that he would make a better mathematician than physicist.

The disciplines of mathematics and physics are closely entwined with one another. In one view, mathematics is regarded as the handmaiden of physics and as the only language in which the ultimate nature of matter can be understood. It is a mystery to me that the concepts of mathematics (things like the real and complex numbers, the calculus and group theory), which are purely inventions of the human imagination, turn out to be essential for the description of the real world of natural phenomena. But so it is.

Mathematicians, conversely, often regard the laws of physics as a mere jumping-off place for their own magnificent flights of creative fancy, which, they feel, lie far above and beyond the mundane facts of physical phenomena. Because of the intimate connections between their disciplines, mathematicians and physicists often cross the boundaries between their specialties. Remember that Isaac Newton discovered calculus!

Yet today there is another unnatural obstacle to such cross-fertilization arising from the quite different conventions for publication in the two fields. A research paper in physics is presented directly, explicitly and specifically. Subtleties are explained, analogies are drawn, and examples are given. A physics article, like a good short story, usually has a clearly stated point: "Look at things this way and you can understand why this thingamajig behaves that way one time in a zillion." Mathematics is more akin to poetry. Subtlety and brevity are essential prerequisites for the publication of a research communication in mathematics. Each logical step is concisely stated but never explained. Every word is essential as in a well-wrought poem. The reader is assumed to possess the sophisticated sensibility of the author.

Danny, like any licensed theoretical physicist, was trained to write straightforwardly, not minimally. He had one hell of a time letting professional mathematicians learn that he had solved some of their favorite puzzles. Finally, the peripatetic Hungarian mathematician Paul Erdös began to wonder who this mysterious Kleitman was who was solving his problems as quickly as he could make them up. When Erdös speaks, mathematicians listen. Danny became a much better recognized and more accomplished mathematician than he was a physicist, and he left Brandeis to become a professor at MIT, and, for a time, the popular but autocratic head of its mathematics department. His mathematical specialty is combinatorics, a classic example of which is the "traveling salesman problem": Pick a dozen random cities in the world. Starting from any one, what is the shortest round-trip passing through each city? This kind of mathematics is very, very practical. Danny has consulted for the United States government and large corporations to help them minimize their travel, telephone or shipping expenses. He even does his own income taxes, and those of Erdös as well!

Erdös is the ultimate traveling salesman whose wares are his mathematics. Although he is still a Hungarian national, he has no home of his own. He lives as a guest with his friends wherever he lectures, and subsists on his lecture fees and on the willing generosity of his mathematical fans.

Walter Gilbert had been another of my close friends. While I was a graduate student he was a junior fellow at Harvard finishing up his thesis work in theoretical physics under the direction of Abdus Salam. The Society of Fellows is a unique institution that is loosely affiliated with the university. Eight junior fellows are selected each year from among the most promising scholars in all fields of academic endeavor. One cannot apply, one is chosen. Although the salary is relatively small, the duties and obligations are nil, and the award was regarded by Uncle Sam as a tax-free gift, like the Nobel Prize. Since 1987, America has become the only nation in the world to tax the Nobel Prize, and Harvard's Junior Fellows have lost their loopholes as well.

Wally had become a member of the physics faculty by the time I returned to Harvard. He was a virtuoso theorist in the most abstract realms of mathematical physics, yet he was lured by the recent revolutionary developments in molecular biology. How could he continue to futz around with tiny particles he could never hope to see when the secrets of life were about to be revealed?

James Watson and Francis Crick had determined the structure of DNA, the key molecule of life on earth, while Wally had been at Cambridge. Watson and Wally had been close friends there, and now Jim was chief honcho of the molecular biologists at Harvard. Little by little, Wally was seduced, eventually forsaking physics to become one of America's leading biochemists. It was an amazing transition from an abstract thinker to a hands-on experimenter, and a very successful one. Among his many accomplishments, Wally discovered the "repressor" (once a holy grail of sorts among the biologists), founded the promising biomedical firm Biogen, and was awarded the Nobel Prize in chemistry in 1980.

Contemporary theoretical particle physics is often attacked, perhaps with some justification, as an irrelevant science. Who cares about our understanding of kaons, for example? We've know about these bizarre and short-lived particles for some four decades, yet they still have no conceivable practical application. They may never have one. They will not yield a cure for cancer, or even the common cold, nor will they feed the hungry in Africa or Appalachia. They do not even play an obvious role in the structure of ordinary matter. Understanding the kaon is a little bit like climbing Mount Everest. We do it because it is there. For those of you to whom this convincing argument is not entirely so, consider this: No intellectual discipline is more challenging than particle physics. Those who are trained in our rigorous science often make their marks elsewhere. Danny's combinatoric tricks save Uncle Sam more money than Reagan's deficit-laden austerity program. Wally is well on his way to a cure for cancer *and* the common cold. Remember where they began.

There are many such success stories. One of my own graduate students, Andy Yao, has switched into computer science and become a national treasure. Others do down-to-earth research for oil companies or found new and successful high-tech companies.

Harvard University operates a small cyclotron that was, decades ago, nearly a forefront research instrument. (It was a bit too small to do anything really exciting.) One of its users during the sixties was Allan Cormack, born and educated in South Africa and now a professor of physics at Tufts University. His experimental research in particle physics led him to a fascinating problem involving the analysis and interpretation of large amounts of data. He solved the problem, and his work laid the basis for the development of the CAT scanner, now a vital tool of medical diagnostics. Cormack, along with a turncoat British physicist, Sir Godfrey Hounsfield, won the Nobel Prize in medicine in my year of 1979 for pursuing a mere hobby!

While Cormack was doing his shtick in Cambridge, Andrei Sakharov, a theoretical physicist an ocean and a continent away, was making a name for himself in Russia. He became the father of the Soviet H-bomb and the coinventor of a device that may someday be the key to controlled nuclear fusion. His Tokamak offers a realistic promise for unlimited clean, safe and cheap power that depends neither on Arab goodwill nor nasty and hazardous uranium, nor the blood, sweat and tears of the coal miner. These are some of the more practical achievements of a man whose principal interest has been to unravel the deepest (and least relevant) questions about the universe: How did it all begin, and why does there exist this stuff we call matter?

Toward the end of the sixties Sakharov rebelled against the Soviet repression of its own people. He became the world's leading advocate of human rights and human survival. It was he, for example, who convinced Nikita Khrushchev to sign the test ban treaty with us. I met Sakharov in Moscow in 1972 and was deeply impressed by his unique combination of humble humanity and far-ranging imagination. It was clear at that time that he was being shunned by many of his Russian colleagues—such a man is dangerous to the Soviet government, and to those Russians who befriend him. Sakharov was awarded the Nobel Peace Prize in 1975. Characteristically, his Soviet keepers did not dare permit him to travel to Oslo to accept the honor. Indeed, for six years this Hero of the Soviet Union and of all the world was kept under virtual house arrest in the closed city of Gorky, the subject of intense and unconscionable harassment.

Gilbert, Cormack and Sakharov were all trained in the useless discipline of particle physics. Yet each has made enormous contributions to human welfare. Each of these three physicists has won the Nobel Prize, one in chemistry, one in medicine and one for peace. They are not isolated cases. The way of thinking that a physicist learns is supernally powerful, and not just in physics. There is good reason that most of the developed world requires its high school students—all of them—to understand something about the natural rules and regulations governing the universe we inhabit. This understanding is in an important part of responsible citizenship, self-fulfillment and national survival.

In the past, America was number one because of its hardworking citizens, its natural resources and its freedom of opportunity. For the future, with our resources largely spent, mere brawn will not suffice, and those who are scientifically illiterate are enslaved by their own ignorance. American education has got to get on the ball. A recent government-sponsored international survey of top high school seniors worldwide has found American students to be poorly educated in mathematics. Ours are by far the weakest students of the twelve nations surveyed. Can we afford to continue producing an infinite number of lawyers, advertisers and other scientific ignoramuses? Watch out, lest they draw and quarter you as they've done to AT&T and convince you that you like it, too. Worry a little bit that the people you entrust to make up the nation's budget probably flunked high school math, and those who built the ill-fated space shuttle *Challenger* didn't know or care what happens to rubber in cold weather.

In 1954, when I first came to Harvard as a graduate student, particle physics was in a sorry state. Physicists had no idea of what were the fundamental laws governing the behavior of atomic nuclei and their constituent nucleons. The weak and strong forces were, like some obscure diseases, of unknown origin. Upon my triumphant return twelve years later, a lot had been learned, but the big questions remained unanswered.

Thanks to the deployment of powerful accelerators and the development of ingenious particle detectors, very many new particles had been found and their properties studied. They were organized into families in terms of a purely hypothetical quark substructure.

While it was useful and predictive to recite Murray Gell-Mann's incantation that three quarks make a baryon, there was simply no systematic theory explaining how and why three quarks stuck together. Why are the forces between quarks such as to bind together three of them, but not two or five? Why, when a proton is struck a lethal blow by an oncoming energetic particle, did its quarks not come spewing out to be counted? More embarrassing yet, there was not a shred of direct evidence that quarks were inside protons, and even Father Gell-Mann sometimes regarded them as merely useful fictions of his own overly fertile imagination.

The strong nuclear force (whatever it is) holds the quarks together (if there really are such things as quarks). The weak nuclear force allows one flavor of quark to turn into another. The alert reader should remember that Murray's original recipe called for three quark flavors: two to make up ordinary matter and the strange quark to explain kaons and other strange particles. BJ Bjorken and I introduced a fourth quark flavor in 1964: *charm*. Nature surprised us all with a fifth flavor, the *bottom* or *beauty* quark, and there is certainly a sixth flavor, and perhaps even more, awaiting discovery in the Baskin and Robbins sweepstakes.

We are very fortunate, by the way, that there is such a thing as the weak force. The sun gets its energy by burning hydrogen into helium in its nulear furnace. Four hydrogen nuclei, or protons, can form a helium nucleus only if two of the protons are turned into neutrons. In terms of quarks, this means that two up quarks must be transformed into down quarks, a bit of alchemy that only the weak force can accomplish. Without it, the sun would be nothing more than a frozen ball of solid hydrogen. Three cheers for the weak interactions!

Back in 1966 we had a pretty good description of the weak force from the work of Dick Feynman, Gell-Mann, Cabibbo and others. But it was only a model, and not a genuine self-consistent theory. It described, rather than explained, and even the description was valid only up to a certain point. Any attempt to obtain precise predictions, or to describe phenomena at very high energies, led to absurd answers (like infinity) and contradictions. In the jargon of the trade, the theory—such as it was—was not renormalizable.

This deplorable situation was about to change. Beautiful theories of the weak and strong force would emerge in the seventies, and the first key experimental discoveries were to be made in California—soon after I had absconded to Harvard.

Way back when I had been nervously teaching my first classes at Stanford, Bob Hofstadter won his Nobel for the experimental work he had done at Stanford using a small electron accelerator. He showed that the electric charge of a proton was not concentrated at a mathematical point. Rather, it was spread out over a spherical volume whose radius was about 100,000 times smaller than the atom. Bob's work was an early clue that the proton had parts and was a composite system.

Stanford University set out to enlist Uncle Sam to build it a much more powerful electron accelerator so as to see more clearly what goes on within the proton. The Stanford Linear Accelerator is a two-mile-long monster, the biggest of its kind even today. Its budget compares to that of the rest of the university. This is the machine that revealed at last the existence of substructure within the proton —mysterious pointlike bits of matter that were originally called "partons" by Feynman because they were thought to be (and they are!) parts of the proton. Indeed, it turned out that the parton is a quark, after all. To the reader who may not be a particle physicist, let me try to explain by means of a fable how "we" found partons in the proton.

A young child is given a BB gun and a challenge. In the garden grows a peach tree laden with fruit. Using the gun, the child is to learn as much as she can about the nature of peaches, but under no circumstances may she touch the fruit. In fact, she cannot even see the peaches among the leaves of the tree. So she aims her little gun at the distant tree and shoots and shoots and shoots. BBs that hit a peach remain embedded within. Those that miss impinge upon the barn wall and gradually build up a shadow image of a peach. She has met her challenge by determining, through means of an indirect "scattering" experiment, the size and shape of the peaches.

But the inner structure of the peach remains a mystery. As a teenager, she continues her study of the peach with a higher-energy accelerator: a .22-caliber rifle. Most of the bullets pass right through the peachy pulp. Occasionally, a bullet strikes the peach and is deflected from its path, sometimes emerging at a large angle from its original trajectory, at right angles or even backwards. How does she explain this? She argues that there must be something small but very hard that is hidden within the peach. Her high-energy experiment has demonstrated the existence of the peach pit.

She's got promise as a physicist.

The parable of the peach has a lot to do with the history of twentieth-century elementary-particle physics. Take one:

It is 1911, and Ernest Rutherford is directing a series of experiments in which a beam of alpha particles emitted by a naturally radioactive material impinges upon a target of gold foil. Most of the time the projectiles pass straight through the foil. This is as expected in terms of the then-current "plum pudding" model of atomic structure, in which the atom is regarded as a large, soft and spongy pudding with electrons embedded in it. But, to Rutherford's amazement, the alpha particles were sometimes deflected backwards.

As in the case of the young lady and the peach, there was only one sensible explanation of the experimental data. Rutherford's discovery marked the beginning of our correct understanding of the atom as a composite structure made up of a small hard nucleus surrounded by a tenuous cloud of electrons. The definitive theory of atomic structure—quantum mechanics—emerged in the 1920s, but it left the mystery of the structure of the nucleus for the next decade. With the discovery of the neutron in 1932, it was realized that the nucleus itself is a composite system made up of protons and neutrons.

Take Two:

Hofstadter measures the size and shape of nucleons with his higher-energy electron gun in the late 1950s. But to find the quarks inside the protons and neutrons, an even larger weapon was needed. Enter SLAC, the Stanford Linear Accelerator Center.

Take Three:

The experimental procedure was much like Rutherford's, except that the projectiles were 20 GeV electrons with thousands of times more energy than Rutherford's alpha particles. This meant that the SLAC electron-scattering experiment could resolve subatomic structure thousands of times smaller. The target of the electron beam was not the atom as a whole but the proton itself, the simplest of all atomic nuclei and the prototypical hadron.

Once again, the results of the experiment showed that even the proton could not be regarded as a homogeneous sphere. Far too frequently, on the basis of such a picture, the electron was deflected by a large angle. Careful calculations done by Dick Feynman and James (BJ) Bjorken showed that any rational explanation of the data would require the existence of small charged bodies, partons, jiggling about inside the not-so-elementary proton. However, these curious partons could not be shaken out of the hadrons they were part of. They seemed to behave like freely moving particles that were somehow permanently confined within the hadron they constituted.

The partons were nothing other than Gell-Mann's hypothetical quarks. This was established by the successful quantitative description of the SLAC data by means of the quark model, and the equally successful description of the neutrino-proton scattering data that was being accumulated at laboratories elsewhere in America and abroad. The quark was no longer quite so hypothetical. As the seventies began, physicists realized that quarks had been seen at SLAC as clearly as they possibly could—as clearly, in fact, as our young lady sharpshooter could see the peach pits.

All that was missing was an understanding of the force that held quarks together so tightly that an individual quark could not escape from the hadron of which it formed a part, while paradoxically it was confined so weakly that it could move about quite freely within the hadron. The theory of such a force, now known as quantum chromodynamics, was soon to be invented.

While the quark was getting itself glimpsed in California, Steve Weinberg at MIT was preparing the way for a new and successful theory of the weak force. He, too, was pursuing my dream of a single theory of both the weak and electromagnetic forces: an electroweak unification.

Steve was familiar with my work of 1961 showing that the only possibility was a gauge theory. The job could be done if three hypothetical new particles, the weak interaction gauge bosons, existed in nature. These were the W^+ and W^- bosons, which mediate all the then-known weak processes, and the Z^0 boson, which is the mediator of a new force, one whose effects had not yet been observed in the laboratory.

It was not surprising that the W and Z particles had not been seen, because they were predicted to be very heavy, about a hundred times heavier than the proton. Not until the eighties was an accelerator built that was powerful enough to search for and find these fascinating fundamental particles.

Steve realized that my attempted electroweak synthesis was dreadfully incomplete. I had imagined no sensible mechanism that could generate large masses for W and Z while leaving the photon massless. By simply putting the masses into my theory by hand, I had destroyed all hope for a renormalizable and sensible theory. Steve was very familiar with the newly born subject of spontaneous symmetry breaking, and here is where he found a plausible source of the gauge bosons' masses.

Spontaneous symmetry breaking is the situation in which the physical system is perfectly symmetrical whereas the end result is not. A magnet is a familiar example. Most atoms are themselves small magnets. In certain materials said to be ferromagnetic, these microscopic magnets prefer to line up with their neighbors. When the material is hot, they cannot do so because they are jostled about by thermal motions. The material remains unmagnetized, with no direction in space singled out. As it cools down from a chaotic and magnetically neutral state, at a certain critical temperature all the little magnets line up with one another in a particular direction. The material becomes a magnet with a macroscopic magnetic field pointing along that direction. All directions in space are equivalent to one another, yet the ferromagnet must choose one of them along which to be magnetized.

If you lived inside the magnet, you might find it hard to believe that the laws of physics are rotationally invariant, with no particular direction favored above any other. Inside a magnet, there *is* a favored direction, and one favored direction only.

The magnetization process is spontaneous in the sense that no external agency is required: the ferromagnetic material magnetizes itself. The process is symmetry-breaking because the symmetry that says all directions are equivalent has been broken. Thus, spontaneous symmetry breaking occurs.

If that was a bit heavy, here's a simpler explanation from Abdus Salam. A dozen guests are seated about a circular table set for a formal dinner. Does one choose the napkin to one's right or one's left? It really doesn't matter as long as all the guests behave consistently. There are two equivalent solutions that are related by right-left symmetry. The first course cannot properly be served until the gutsiest guest is prepared to break the symmetry and choose a direction. It's just like the magnet: spontaneous symmetry breaking, hereafter SSB.

Jeffrey Goldstone had shown that SSB, in the context of quantum field theory, leads to the existence of massless particles called (by everyone but Jeffrey) Goldstone bosons. They have never been seen and almost certainly do not exist. SSB seemed to be simply an irrelevant pathology of the theory, having nothing at all to do with the observable world. Then, in 1964, the Scottish physicist Peter Higgs (and several others) showed that in a gauge theory the would-be Goldstone bosons are eaten by massless gauge particles that thereby acquire mass. To Higgs this seemed to be just another curious and useless property of the theory.

But Steve Weinberg and, independently, Abdus Salam in Trieste, saw that the Higgs phenomenon is just what is needed to build a sensible electroweak theory. Higgs's trick could be put to good use to give masses to the intermediaries of the weak interactions. When my old model was supplemented by a system of spinless mesons subject to SSB, everything worked out beautifully. Not only did the Higgs mechanism generate the masses of the W and Z particles, but it was also the agency through which quarks and leptons could acquire their masses. The Higgs mechanism seemed to be the key to an understanding of the origin of the masses of all the elementary particles. The laws of physics could be completely symmetric with the asymmetry and complexity of observed phenomena reflecting a spontaneous breakdown of symmetry akin to the spontaneous appearance of magnetism in a piece of iron.

The weak and electromagnetic forces appear to be so different from one another only because we happen to live at a late stage in the evolution of the universe, where it has become cold, crystalline and asymmetric. Long, long ago when our universe was hot and burning bright and unfit for mortal man, its symmetry was exquisite.

There was more good news in the Weinberg-Salam realization of my electroweak theory. The Higgs mechanism is so gentle that it leads to a consistent theory of weak interactions free from the plague of infinities. That is to say, the spontaneously broken gauge theory is renormalizable. Steve and Abdus did not know this to be true, but they suspected as much and said so in their publications. When, some time later, their conjecture was proven to be true by a young Dutch graduate student, Gerard 'tHooft, the world began to take the electroweak theory seriously.

Steve and Abdus had introduced four new players into the particle game in addition to the three heavy and yet unseen intermediate vector bosons: four spinless mesons. Three of them were potential Goldstone bosons whose fate was to combine with the massless-at-birth gauge bosons W^+, W^- and Z^0 so as to convert them into heavy particles that mediate the weak force. The fourth spinless particle must survive as an observable boson in its own right. It is known as the Higgs boson, but its existence has not yet been established in the laboratory. It remains today as the last great hurdle standing in the way of complete acceptance of the Weinberg-Salam implementation of spontaneous symmetry breaking.

Many properties of the Higgs boson are determined by the theory, but not the value of its mass. Arguments have been presented that show that it must weigh between ten and three hundred times the mass of the proton. Physicists hope that this hypothetical particle will show itself at the next generation of particle accelerators, either at CERN's almost-completed twenty-mile-around Large Electron-Positron ring (LEP) or at a large machine being built in the Soviet Union, or in America's unfunded dream machine, the Superconducting SuperCollider (SSC).

The Higgs mechanism is the simplest known way to generate spontaneous symmetry breaking and thus to produce a renormalizable theory of the weak and electromagnetic interactions together. Since I knew Goldstone and Higgs at the time they made their important discoveries, and furthermore, since I was an early champion of a unified gauge theory of weak and electromagnetic forces, I am annoyed at my own stupidity for not putting two and two together in 1964. Like all good ideas in science, the Weinberg-Salam construction is perfectly obvious in retrospect. On the other hand, I am not at all sure that the Higgs mechanism is nature's choice. It seems to me to be an ugly and contrived construction superimposed upon an otherwise elegant theory. Its virtues are simply that it works, and that no one has come up with a plausible alternative—so far.

Sometimes I compare today's very successful theory of elementary-particle physics with a gorgeous and elegantly crafted mansion. But every residence, humble or grand, must contain an object of no great beauty despite the efforts of architects, interior designers and plumbers. The flush toilet is a rather ugly thing, but it works and no one has come up with a plausible alternative. Thus, one may regard Steve Weinberg as the Thomas Crapper of elementary-particle physics. Crapper, you will recall, is the nineteenth-century London plumber who allegedly invented the flush toilet. His name is not the etymological source of the related four-letter word. On the contrary, perhaps his name influenced his choice of profession. Not a bad choice! It was good enough for my father, and Einstein once said that if he had it all to do over again he would have become a plumber. (His son did become a professor of sanitary engineering!)

Although Weinberg's paper was published in 1967, and Salam's a year later, it took more than five years for the significance of the work to sink in. Here's how Sidney Coleman put it:

> There are only two things wrong with the Weinberg-Salam theory. (i) In its original form, the theory treated leptons only. (Indeed, the title of Weinberg's paper was "A Theory of Leptons.") It was not clear how to extend the theory to strongly interacting particles. (ii) Nobody paid any attention to it. Rarely has so great an accomplishment been so widely ignored. Here is a census of the citations of Weinberg's 1967 paper as recorded in seven years of Science Citation Index: 1967, 0; 1968, 0; 1969, 0; 1970, 1; 1971, 4; 1972, 64; and 1973, 162. These include citations by Weinberg himself.
>
> Weinberg was of two minds about his theory. Friends recall that when it was published he told them it was the best thing he had ever done. At the same time, though, he worried whether the same mechanisms that hid the symmetry of the theory could also spoil its renormalizability. He struggled intermittently with the problem for years, but never obtained any results he felt were worth publishing. He spent most of his time working on other things.

> Likewise for Salam. When he presented his theory, he expressed his belief in its renormalizability, and even gave the kernel of a correct argument for his belief. Nevertheless, two years later . . . [at a relevant international conference] . . . he did not mention his theory of the weak and electromagnetic interactions. He was working on other things.
>
> It is clear from the citation census that something was happening in late 1971. What happened was Gerard 'tHooft wrote a paper on the renormalizability of gauge theories that revealed Weinberg and Salam's frog to be an enchanted prince.

I was not even aware of Steve's paper until 1970 or so. During my first years as a Harvard professor I was far from dreaming grand dreams of unified theories. My return to Harvard had been anticlimactic, and I was depressed both by the state of physics and the fact that I had broken up with Schwinger's secretary, Shirley. While we were undergraduates, soon after Steve had become engaged to Louise, Steve had advised me to get married so I could better concentrate on physics. Perhaps he was right, but then I would never have met my wife, Joan.

I wrote a few papers in the late sixties, including two collaborative ventures with Steve—nothing of any importance. Physics just didn't seem very exciting to me at the time. I played a lot of go (the oriental board game), saw a lot of movies, and finally found a new lady love, a graduate student in molecular biology at MIT. After Kay earned her doctorate, I took a sabbatical and we set off for Switzerland, where I would be employed as a visiting scientist at CERN and she would be a postdoc at the University of Geneva. I hoped that a European sojourn could inspire me to find some exciting physics to do. We found a quiet apartment in the Florissant quarter of Geneva, so quiet that you could hear a pin drop in the apartment next door. She grew her bugs at the university while I tried to have brilliant ideas at CERN. Neither of us had any great success.

We went off for a week of skiing at St. Moritz along with the university ski club. The snow was magnificent, the skies cloudless, but I broke my leg and my friend ran off with her senior professor. *Sic transit amor*. I returned to my apartment in Geneva, which was now not only quiet but lonely, wearing a cast from hip to toe. I decided to spend a week in Israel, cast and all. Upon arriving at Lod airport I discovered that El Al had performed a miracle. While my suitcase had been completely destroyed, the contents—quite intact—were presented to me on a bedsheet. So it was that I limped through the newly conquered areas of the Holy Land with my belongings in a bundle on my back. Sabbaticals are intended to refresh and invigorate. It didn't work.

One of the reasons I had come east to Harvard was to learn from and work with Schwinger. His thoughts, however, had strayed from mainline theoretical physics. He no longer believed in quantum field theory, the synthesis of quantum mechanics and relativity that forms the backbone of modern particle physics. He had invented the subject but disowned it. In its place he proposed an entirely new system of physics that he called source theory, or "sorcery." While Julian regarded his latest brainchild as something fundamentally new and qualitatively distinct, it seemed to us to be partly an ingenious reformulation of quantum field theory and partly an art form that only Schwinger could appreciate. While we often discussed physics, we had grown so far apart in our views that useful collaboration was unthinkable. Possibly, the rejection of sorcery by most of the Cambridge physics community, and not merely our notoriously poor weather, is what convinced Julian to remove himself to UCLA.

While I lived in California, Sidney Coleman and I had written about a dozen joint papers. For example, every student of particle physics today learns about the Coleman-Glashow mass formula. It's in the textbooks. I had hoped that our collaboration would flourish when we became colleagues at Harvard. Alas, it was not to be. We did write one more paper together, one that infuriated our friend T. D. Lee at Columbia. Not only was our paper wrong, but T. D. had done it first. Friendship was restored only after we retracted our paper and ceremoniously broke our pencils in his presence.

The decade of C&G papers had come to a decisive end, but we're still good buddies. He's much smarter than I am, but I've had the luckier ideas. In those days, before our marriages, we both had flats at 84 Prescott Street, at opposite ends of the building. His was cleaner and better furnished.

Theoretical physicists often summer in Europe while they have a respite from teaching. Two of my favorite watering spots were Erice and CERN. According to legend, Erice, son of Venus, founded a small settlement atop a mountain over three millennia ago. The medieval town of Erice perches on a cliff 2500 feet over the port of Trapani, in Sicily. Over the years, much of the town has been taken over by the Ettore Majorana Centre for Scientific Culture, which holds workshops, schools and conferences for fifty weeks of the year. The subjects vary from medical specialties to archeology to city planning to the consequences of nuclear war and, especially, to particle physics.

A fine old church has been desanctified and converted into a modern lecture hall, and monasteries have been recycled into luxury dormitories. It is all the creation of a particle physicist of Sicilian birth who teaches at the University of Bologna and is a staff member at CERN. I have known Antonino Zichichi for more than twenty years and have attended his International School of Subnuclear Physics on a dozen occasions. It all began as a modest little school with a director who had a grandiose dream and the drive to make it come true. The center is named in memory of the great Sicilian theoretical physicist Ettore Majorana, who mysteriously vanished on a ferry ride between Palermo and the Italian mainland in 1938.

Each summer the most renowned particle physicists (i.e., oldies but goodies) join with their younger and more creative colleagues to give a series of courses and seminars to a select group of very advanced students from all over the world. When duty doesn't call, the physicists take tortuous roads, or the occasional working cable car, down from the summit to the rocky shores of Monte Cofano, to the dazzlingly white crescent beach of St. Vito lo Capo, to the cave paintings and natural splendor of the Egadi Islands, or to the antique Greek marvels at Segesta or Selinunte. Zichichi was inspired to create a Mediterranean paradise for his colleagues to enjoy while practicing their art. Thanks, Nino!

CERN is arguably the only instance of a well-managed, creative and cooperative international organization. Its collection of atom smashers is maintained and used by a staff of two thousand who spend an annual budget of about $500 million. CERN has led the World High-Energy Sweepstakes since the early seventies. Things have just changed with the completion of America's Tevatron, which has attained collision energies three times higher than CERN can reach.

During the summers CERN becomes the country club for particle physicists from the entire world. It is located near the high Alps and the equally high cuisine of La France Gastronomique, and nearly every physicist is either a mountain climber or a gourmet. John Iliopoulos, Greek-born but very French, is both. I met him at CERN and recruited him for Harvard. John is to play a key role in this scientific autobiography.

One of the difficulties with my old electroweak model, a problem that Gell-Mann and I clearly identified in 1961, was the necessary appearance of strangeness-changing neutral-current processes. This incomprehensible exemplar of apparent gibberish, which we shall mercifully abbreviate as SCNC, refers to a certain type of decay scheme of strange particles.

STRANGENESS-CHANGING NEUTRAL CURRENTS

Strange particles earned their name because they seemed to behave paradoxically. Copiously produced in cosmic-ray collisions, they are surely subject to the strong force. Yet they live for a relatively long time, which shows that they must decay via the weak force. We know today that there is nothing very strange about them at all. A strange particle is simply a hadron containing at least one strange quark.

The strange quark inherited its name from the particle. There is nothing very strange about it either. It is just a third species of quark that happens to be a bit heavier than its fellows.

When a strange particle decays, a strange quark in it is converted into an ordinary quark through the agency of the weak force. Since the number of strange quarks changes in the process, such decay is called "strangeness changing."

Sometimes the transformation of a strange quark into an ordinary quark is accomplished by the appearance of a pair of leptons. This is the case, for example, when a negative kaon decays into a neutral pion, an electron and an antineutrino. This decay scheme can be called a strangeness-changing semileptonic charged-current process. See if you can say it three times quickly. It is semileptonic because both hadrons and leptons are involved. It is a charged-current process because the lepton pair carries a net electric charge—the negative charge of the electron. In the course of such a charged-current process the electric charge of the hadron is transferred to the leptons.

Another possible decay scheme of a strange particle could involve a pair of leptons with no net charge at all. For example, a negative kaon could decay into a negative pion, a neutrino and an antineutrino. This would be an instance of a strangeness-changing *neutral-current* process. It is called a neutral current because no electric charge is transferred from hadrons to leptons.

Strangeness-changing charged-current effects are routinely observed in the laboratory, whereas strangeness-changing neutral-current effects are simply never seen. This was a fatal problem for the electroweak theory as originally proposed, because the theory predicted the existence of SCNC, which does not exist.

The problem was neatly solved by the introduction of charm, which automatically eliminates SCNC effects and makes possible the construction of an electroweak theory that correctly describes nature.

The trouble was that any electroweak theory we could think of necessarily predicted the appearance of SCNC effects, which simply did not show up in the real world. For this reason, we gave up on the idea of a unified theory.

Steve's idea suffered from the same disease. His model of the weak interactions of leptons could not be extended into a complete theory of the weak force. Of the particles that make up ordinary matter, leptons play a very minor role in terms of mass: our world is 99.97 percent pure quarks, and Steve's theory could not handle them. Furthermore, back in 1967, most physicists didn't even believe in quarks. It was still the Dark Ages.

In the fall of 1969 Harvard imported two superlative young researchers, John Iliopoulos and Luciano Maiani, both of whom I had encountered in Europe and lured to Massachusetts. John, born and educated in Greece, now lives in Paris. Luciano is of San Marino citizenship and lives in Rome. Both had come to Harvard as postdocs in the hope of doing some significant scientific research that might further their careers, and indeed they did.

We recognized that the SCNC problem was shared by all known theories of the weak force, unified or not. More importantly, we saw a way out of the problem by extending the number of quark flavors from three to four. Six years earlier, Bjorken and I had introduced just such a fourth quark to implement an analogy between quarks and leptons, but we did not then see that this "charmed" quark would solve the SCNC problem. It all sounds perfectly crazy. Gell-Mann and I identified the SCNC problem in 1961. Bjorken and I introduced the solution, albeit unwittingly, in 1964. My problem and my solution were not wedded to one another until 1970, when John and Luciano came to Harvard and put my act together.

In our joint paper we explained the nature of the SCNC problems and showed how it could be solved if only a fourth quark flavor existed. In addition to Gell-Mann's original up, down and strange quarks, we needed a fourth and heavier kind of quark. For this, we recycled my old idea of charm. We recognized that the way was now clear to construct a unified electroweak theory that could describe hadrons as well as leptons. Our arguments for charm were far more convincing than the original appeal to aesthetics that Bjorken and I had made in 1964, and we finally had justified the choice of the word *charm* as a magical device to avert evil.

Strangeness-changing neutral currents can arise by a two-step process. First, a strange quark in a hadron becomes (via the weak force) an up quark. Then it is transformed into a down quark. The decay is strangeness-changing: a strange quark has been transmuted into a down quark. Along the way, a pair of leptons is emitted: either a neutrino and antineutrino, or an electron and positron, or a muon and antimuon. In any case, the leptons carry off no net charge. Ecch! It's an unwanted, unseen and despised SCNC effect.

There was no way to eliminate the problem with Murray's original team of three quarks. We threw the charmed quark into the game, which led to another possible play. The strange quark can become a charmed quark, which then turns into a down quark. The net result is the same: it is yet another way to produce SCNC effects. However, the rules of the game are quantum mechanical. The play involving the up quark must be added to the play involving the charmed quark to find the net probability of the SCNC process. If the charmed quark is assigned to its most natural position, the two plays turn out to be opposite in sign, so that they cancel out one another.

Charm, we found, not only restores the lost symmetry between leptons and quarks, but it also provides a natural and elegant mechanism for the suppression of strangeness-changing neutral currents. As the dictionary says, charm averts evil.

Strangeness-changing neutral currents are one example of a whole family of diseases to which theories of weak interactions are prone. All of the diseases can be cured if and only if the quarks come in matched pairs like up and down, and charmed and strange. Evidence for a fifth quark flavor, the bottom quark, appeared unexpectedly in 1977. It made its debut in a new particle called the upsilon, which was discovered at Fermilab by Leon Lederman and his friends. Because of my work with John and Luciano, *everyone* believes that there

HOW CHARM WORKS

1. This is a Feynman diagram that contributes to an unseen SCNS effect.

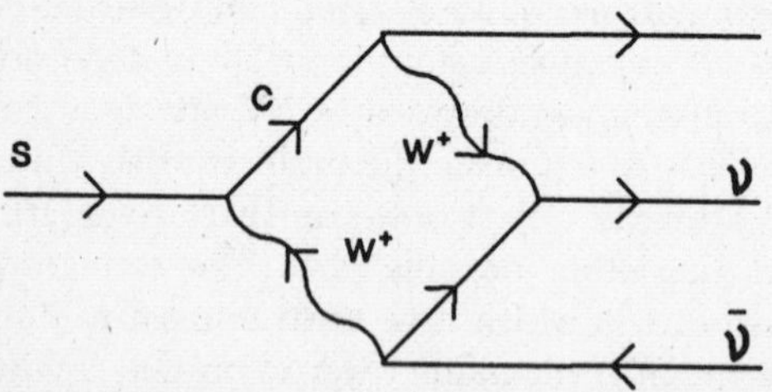

2. Here is a second diagram contributing to a SCNS effect, but involving a virtual charmed quark, C. The two contributions cancel one another.

must exist a sixth quark flavor, so as to form a third quark marriage: top and bottom.

Back in 1970, we had few believers. Murray's theory had just three quark flavors, and he was always right. No one had seen any evidence for the charmed quark—but no one had even bothered to look.

Even though our work was completed three years after the Weinberg-Salam explanation of the origin of the W and Z masses, we were completely unaware of their work. Their idea of spontaneous symmetry breaking and our charmed solution to the SCNC problem were complementary contributions that together made possible the construction of a complete and consistent theory of the weak force. We lectured on our work at MIT and, in particular, discussed it with Steve. He could have said, "Hey! That's just what I need to extend my theory of leptons to a complete theory!" But he didn't say a word.

By early 1970 he seemed to have abandoned and forgotten his most brilliant paper.

Soon after the invention of the GIM mechanism (Glashow, Iliopoulos, Maiani), as our little trick with charm came to be known, Luciano and his charming wife, Pucci (it rhymes with Gucci, and it should!), sailed back to Italy in first-class accommodations on an Italian ocean liner. Of course, we had a champagne party aboard the ship. That was the way to travel to Europe. Armed with his American triumph, Luciano went on to become one of Italy's finest theoretical physicists. John and I returned to physics and our other endeavors.

For example, together with Robert Schrader, a young German mathematical physicist, we enrolled in a scuba diving course given at the Radcliffe swimming pool. When we became certified card-carrying divers, we purchased fishing licenses, dressed from head to flippers in ¾-inch-thick rubber suits, and clambered over Boston's frozen North Shore into the deeps of the sea on midwinter hunts for lobsters. Generally, an afternoon's dive by the three of us yielded a paltry four or five beasties of barely legal size. We had to supplement our catch at a fish store in order to serve our landlubbing friends their promised lobster banquets.

On the physics front, we labored hard to remove the infinities that plagued any model of the weak interaction that involved intermediate vector bosons. By means of an incredibly ingenious argument (which we devised while diving off the Isla Mujeres in the Caribbean during the summer of 1970), we showed that almost all of the unacceptable infinities were absent in an electroweak gauge theory. The theory seemed to be trying its best to be renormalizable, but it didn't quite make it. Despite our virtuoso efforts, some infinities remained, and for a theory to have a few infinities is like a coed becoming a little pregnant—it's a serious problem.

Just before Thanksgiving 1970 I drove to my mother's apartment in Riverdale to find the turkey in the oven and my mother dead on the floor. She had been alert and active until the very end, at the age of at least eighty-five. We were never precisely certain of her age. She didn't want it known that she was older than Pop.

Even as an adult, I had remained a mamma's boy. When I visited her, or when she joined me in California, I always woke to freshly squeezed orange juice, a hearty breakfast and clean underwear. She sent me shirts from Saks Fifth Avenue, Turkish figs from Macy's, and the finest Cuban cigars and French cognacs. Nothing was too good for her brilliant and handsome baby, God's gift to some not-yet-found but certainly undeserving Jewish maiden. Though she was hardly an intellectual and had never lost her accent and Yiddish ways, she had a sharp tongue for her potential challengers. When I introduced her to Shirley, a red-haired shiksa whom Mom did not adore, she complimented her with, "What lovely hair you have. Can't you do something with it?"

The loss of my mother was a terrible blow that terminated my prolonged but comfortable adolescence. Her beloved baby was pushing forty.

I decided to spend the spring of 1971 on leave from Harvard. Derek Robinson, with whom I had toured Turkey years earlier, had become a close friend during my visits to CERN. He is a mathematical physicist, which is to say that he pursues the rigorous mathematical foundation of theories that are already in common use by less persnickety practitioners. Derek had accepted a professorship at the University of Marseille, which had a strong group in mathematical physics. He invited John Iliopoulos and me to visit for a semester. It seemed a perfect opportunity for me to do some diving in the Mediterranean, so we accepted. We arrived in the south of France at the beginning of February, when the sun was already bright and strong and the water warmer than it ever gets near Boston. The sea was much clearer than the Atlantic, though not as crystal as the waters off Maui. There were very few fish—millennia of fishermen had seen to that.

I tried delivering my lectures at the university in French, but the natives didn't like having their language mauled, so I was asked to switch to English. My student Andy Yao, who had come from Harvard to join me, and I wrote a generalization of our earlier work showing that the cancellation of most infinities was characteristic of any gauge theory, a result hardly of interest to anyone. We attended a winter school in the French Alps, where we lectured or listened all morning and skied all afternoon. I spoke on the question of neutrino identity, asking whether neutrinos produced in kaon decay are exactly the same as those produced by pions. (They are.) This delightful series of meetings had been initiated by the French Vietnamese theorist Tran Thanh Van the year before. It continues to this day. Most of the *rencontres* are held in March when the snow is best, but American teaching obligations make it difficult to attend. More recently, a series of workshops have been scheduled for the last week in January to coincide with the American semester break.

Let me say a word about all these meetings I attend at marvelous beaches in the summers and famous ski resorts in the winters. While they are important professionally, I cannot deny that they are fun vacations as well. The junkets are *not* funded by Uncle Sam or by John Harvard. For many years now, all my trips to Europe and Asia have been paid for exclusively by foreign sources, or occasionally by me.

Dick Dalitz, a theorist of Australian birth, was one of the early advocates of the quark model of hadron structure. Although there were problems with quarks, Dick simply pushed them to one side. He compared the pattern of particles that was observed in the laboratory to the pattern that would be expected if hadrons were made of quarks. The agreement was remarkable, which Dick believed to be irrefutable evidence that quarks had to exist. Dalitz had recently moved from the University of Chicago to Oxford, and he invited me to come to England to give a seminar on the GIM mechanism, whose success constituted further evidence for quarks.

Dick had become a professor at All Souls, one of the odder of the many colleges making up Oxford University. For instance, there are no students at All Souls College. There are no women, either. Once, I am told, the past master of the college was asked what would happen if a fellow of All Souls underwent a sex change operation. His response: "The rule is simply that no woman may be appointed to All Souls. What happens to a fellow afterwards is of no concern to us."

On my way back to Provence, I spent a few days sightseeing in London. In a crowded subway car at Piccadilly Station I was rudely jostled by two passengers screaming, "Let us out! Let us out!" and two others at my derriere shouting, "Let us in! Let us in!" Suddenly they vanished, the doors closed, the train began to move—and my wallet was gone. London thieves are smart. They should have become physicists. Aside from money, I had lost my scuba identification card and my driver's license.

Back in France I was reluctant to drive my leased Fiat *sans permis*. I dropped into the local town office to see what could be done. I explained that my license had been stolen by Brits. They indicated two possibilities: (1) I could obtain a French permit upon presenting my American license (not possible!); or (2) I could enroll in a twenty-four session course at a local driving school (not likely!).

I settled for an official-looking document that simply said I am who I am, and that I allege that I am the owner of a valid but stolen American driving license.

At the same time, I wrote of my plight to the Massachusetts authorities, who responded by sea mail that there was simply nothing to be done. I also wrote to the Geneva police, for I remembered that a decade earlier I had obtained, with some difficulty, a Swiss permit that I had since lost. The Swiss replied instantly that there was simply nothing to be done *unless* I sent them ten Swiss francs, three photos of myself, and an official document from the police stating that I am who I am. Within two weeks the efficient Swiss replaced my driver's license.

During the weeks that I was driving illegally, I encountered a somewhat mad American girl I had met in New Jersey the year before while visiting my brother Jules in Upper Saddle River. In the course of a drunken argument with her, late at night and far in the countryside, I had a rather nasty automobile mishap. Both cars were totaled, but mercifully no one was injured. The head-on collision took place on a tight curve in the other driver's lane. I was an unlicensed and under-the-influence driver who was obviously at fault. Fortunately for me, the perfectly innocent driver of the other car was an Arab. When the police arrived, I flashed my "I am who I am . . ." document, and my identification as a *Fonctionnaire d'Etat*. I was released, but the Arab was taken to the police station. C'est la France!

A conference on gauge theories was organized at Marseille in the summer of 1971. Sidney Coleman attended, as well as the Dutch physicist Tini Veltman, a tough old bird but a close friend. Tini, then a professor at the University of Utrecht, was one of the early champions of a unified electroweak theory. Today he teaches at the University of Michigan. At the Marseille meeting Tini told me that his very own student had produced a renormalizable gauge theory of the weak and electromagnetic interactions. It incorporated my old model of 1961, as well as the Higgs mechanism that Weinberg and Salam had appended to it. Yet the student, Gerhard 'tHooft, had invented all these things quite independently. And he had proven that the spontaneously broken gauge theory is free of unmanageable infinities. (Gerard, incidentally, has persistently refused to accept a professorship at Harvard.) The completely rigorous proof of renormalizability involved the work of 'tHooft, the late Korean physicist (and my friend and collaborator) Ben Lee, and the French physicist Jean Zinn-Justin.

Sid Coleman wrote of this moment:

> 'tHooft at this time was a graduate student at Utrecht working under the direction of Tini Veltman. A month or so before he announced his result, Veltman and I met at Marseille. Veltman said, "I have a student who has constructed a renormalizable theory of charged vector mesons." I said, "I don't believe it."

The Marseille conference marked the beginning of a decade of many profound discoveries, both theoretical and experimental. Filled with excitement, I rushed back to Harvard. My sabbatical had had its ups and downs, but it ended on a note of hysteria. The dream of electroweak unification was coming true!

10 Toward the November Revolution

SOON AFTER MY RETURN TO HARvard in the summer of 1971, Robert Schrader and I trekked to the Port of Boston to reclaim the trunk of possessions I had shipped from Marseille. I had lovingly packed it full of the latest French scuba-diving equipment: air-pressure regulators, a mask with corrective lenses, my wet suit and an impressive array of underwater armament for fishing. Together with the customs inspector, we undid the locks only to discover that everything of value had been replaced by French newspapers. There was no duty to pay. I never had the heart to replace my stolen gear. My days as a scuba diver had ended. A new era was to begin.

While I had been abroad, I had sublet my apartment to an old friend from Cornell, Laurence Mittag, a distant relation of the famous Swedish mathematician Mittag-Leffler. There is a Nobel Prize for physics, chemistry and medicine, but there is none for mathematics, the queen of the sciences. Some may say this is a result of an altercation about a woman between Alfred Nobel and Mittag-Leffler. Others maintain that they were friends, and that Mittag-Leffler had offered to endow a prize in mathematics but thereafter lost all his money. In any case, Laurence had followed in the family tradition and become a theoretical physicist. He now teaches at Boston University.

With my mother's death, my apartment had gotten rather crowded, since I had inherited a massive dining room set, an eight-foot-high breakfront, various delectable art-deco sculptures, a grand piano and several large oriental rugs. All available surfaces were covered with our books, papers and records. It was a mess.

While we were disentangling our possessions, an old Russian acquaintance dropped in: Ben Siderov, the physicist friend from Siberia who had been my housemate in Copenhagen. I rang up Danny Kleitman to suggest a reunion dinner for the four of us. That is to say, since my apartment had become practically uninhabitable, I solicited a dinner invitation for us all chez Kleitman. Danny's wife, Sharon, was accustomed to entertaining his weird friends, and she was (and is) a superb hostess, to boot.

Sharon Kleitman, née Alexander, is one of four sisters. The eldest, Lynn Margulis, is the well-known but controversial Boston University biologist who has a thing about bacteria. In Lynn's most recent popular book, *Microcosmos*, coauthored by her son, Dorian, with Carl Sagan, they wrote:

> Our own role in evolution is transient. . . . We may pollute the air and waters for our grandchildren and hasten our own demise, but this will exert no effect on the continuation of the microcosm. Our bodies are composed of ten quadrillion animal cells and another hundred quadrillion bacterial cells. . . . Beneath our superficial differences we are all of us walking communities of bacteria. . . . After we die we return to our forgotten stomping ground. The life forms that recycle the substance of our bodies are primarily bacteria. The microcosm is still evolving around us and within us. You could even say . . . the microcosm is evolving as us.

Sharon's younger sister, Joan, was newly divorced and had just moved to Boston with her two young children. Guess who was coming to dinner with Ben, Laurence and me that lovely September evening? Joan and I shucked corn and fell in love. Very soon afterward, I moved into her small apartment in Newton, with Jason (age four), Jordan (just two and a half) and Joan.

We were visited in February by Derek Robinson, my colleague at Marseille. Joan cooked a marvelous dinner of Chicken Vesuvio, after which we consumed a bottle of fine but fearsome Marc de Provence that Derek had imported from France. The five of us spent the night in Joan's apartment, enough to make any slumlord proud. Over breakfast we announced to Derek that we had become engaged. That summer, having chosen and purchased the house of our dreams in Brookline, we were married and have lived happily ever after. Since then I have had a special fondness for Marc.

I thought it would be fun to spend a month honeymooning in Europe. Joan's mother drove us to the airport. John Iliopoulos and the Martins were awaiting us at the Paris airport. As we were about to embark, the ticket agent discovered that my passport had just expired. Even my mother-in-law, who was a travel agent, couldn't sweet-talk us onto the plane. My bride began to demonstrate her manic inheritance. The three of us spent the evening getting drunk at Trader Vic's. The next day, Uncle Sam proved his mettle and saved our marriage by renewing my passport in a mere two hours. Finally we set off on our honeymoon.

We spent a week at the Hotel Select Latin in Paris. It was relatively cheap, reasonably clean, and so what if the floors tilted a bit. In the future, Joan would insist on classier accommodations, and I have gotten somewhat spoiled myself. Then we drove to the south of France to spend a few weeks in the house that Derek built amidst the wild hills of the Provençal countryside. We spent our vacation eating, drinking, sunbathing and making love.

Since that time, Derek and his family have forsaken France for Australia. He is now a professor in Canberra. He and his wife are British nationals who never felt at home in France. Derek didn't like the fact that the cleaning ladies and the graduate students were voting on tenure decisions at the University of Marseille. And they didn't want their children to grow up French. Recently, Derek told me that the greatest mistake of his life was not to have moved to Australia sooner. I myself have never been there. I doubt that I could get used to living upside down and celebrating Christmas in midsummer.

Physics was not standing still while these more personal events were transpiring. The seeds of a revolution in particle physics had been planted in the 1960s. They included the empirical success of the so-called Cabibbo theory of the weak processes, the eightfold way of Gell-Mann and Ne'eman, and the quark-parton model. The construction of the electroweak theory was essentially complete through the work of Salam, John Ward, Weinberg and me. John had collaborated with Salam in the 1960s, when they had written several papers about the electroweak theory. Today he teaches in Australia.

With the GIM mechanism added, the theory could deal with all observed weak interaction effects, as well as with some that had not yet been seen but were predicted to take place. With 'tHooft's ingenious demonstration that gauge theories subject to spontaneous symmetry breaking were renormalizable, the electroweak synthesis was promoted from a wild speculation to the first consistent theory of the weak force ever.

But first, the unique predictions of the theory had to be subjected to the ultimate test: experiment. The Z^0 particle, an integral part of the theory since its inception, generated a new form of weak interaction for which there was not yet a shred of experimental evidence: what are called neutral-current phenomena. The GIM mechanism had eliminated from the theory those neutral-current effects that caused a change of the strangeness quantum number, but it was essential that strangeness-conserving neutral-current interactions did exist.

The two laboratories in the world best equipped to search for neutral currents were Fermilab, near Chicago, and CERN, in Switzerland. They had the hottest neutrino beams, and each laboratory had an active program of neutrino physics.

The typical experimental run involves several months of machine operation, during which hundreds of thousands of neutrino-induced events are collected. The neutrinos in the beam are mostly of the muon persuasion. In the collision, a charged W particle may be exchanged between the incident lepton and a target quark. The muon-type neutrino becomes a charged muon and the quark changes its flavor. The event yields an energetic muon and a spray of hadrons from the struck quark. It is a typical "charged current" event.

The electroweak theory demands the existence of neutral Z particles as well as charged Ws. When the neutrino exchanges one of these with a quark, it remains a neutrino. As before, the struck quark spews out hadrons, but in this case no charged muon emerges. The neutrino remains a neutrino. This is the neutral-current process that was being searched for in a race for discovery between Europe and the States.

Harvard's man in Fermilab was Carlo Rubbia, our incomparable Italian import, and the leader of the American effort. Experiments of this kind are immensely difficult, requiring large teams of experimenters, lots of time, money and hard work. We theorists could only wait patiently for the vindication (or demolition) of our theories. Meanwhile we had students to train and new theoretical ideas to contemplate.

It's a sad state of affairs. Once upon a time there was no real distinction between experimental and theoretical physicists. Newton, Faraday, Maxwell, Einstein, Bohr and Fermi all did experiments and invented theories, too. Physics has gotten far more specialized. At Harvard, our theory students are not fit to do experiments, and our experimentalists can rarely understand the theory underlying their own work. Contemporary experiments can involve hundreds of Ph.D. scientists, few of whom are fully aware of the scientific purpose of their experiment. I wish there was another way, but it seems to take

THE CHARGED-CURRENT PROCESS

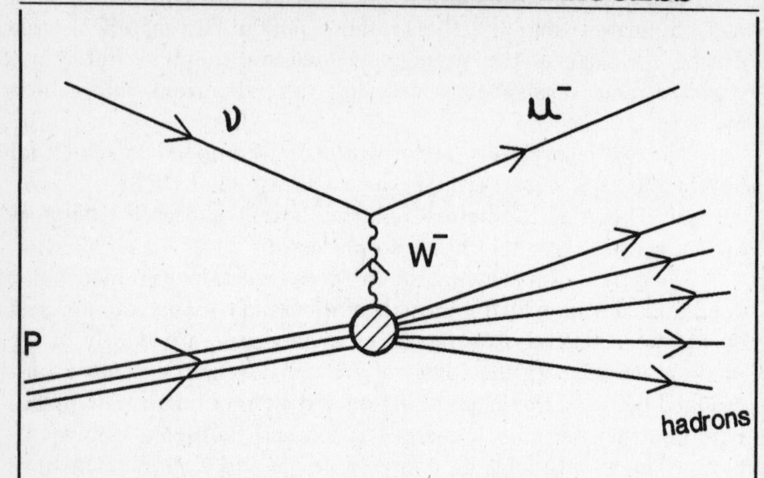

A high-energy neutrino strikes a target proton. A charged W particle is exchanged. The neutrino is transformed into a charged muon, and most of its energy is expended in the creation of new hadrons.

a lot of people—running huge, high-energy machines—to study the littlest things.

Our particle theory group at Harvard usually includes one or two postdoctoral fellows whose responsibility is to pursue their research with government contract support. We had not done so well in 1972, or so it seemed. The first six of our candidates had turned us down. Howard Georgi was number seven. He had been a Harvard undergraduate who did his doctoral research at Yale under the direction of Charles Sommerfield, one of Schwinger's students in my day. Howard's doctoral research had been of an exquisitely abstract nature and was not in a direction that any of us pursued. But we were scraping the bottom of the barrel, so we offered the job to Howard. He accepted, and Harvard is forever grateful.

Howard and I hit it off immediately. We began a collaboration that continues to this day. He was promoted to a junior fellowship and ultimately to full professor. One of our early papers, interesting in its time, proposed a rather ugly variant of the electroweak theory

THE NEUTRAL-CURRENT PROCESS

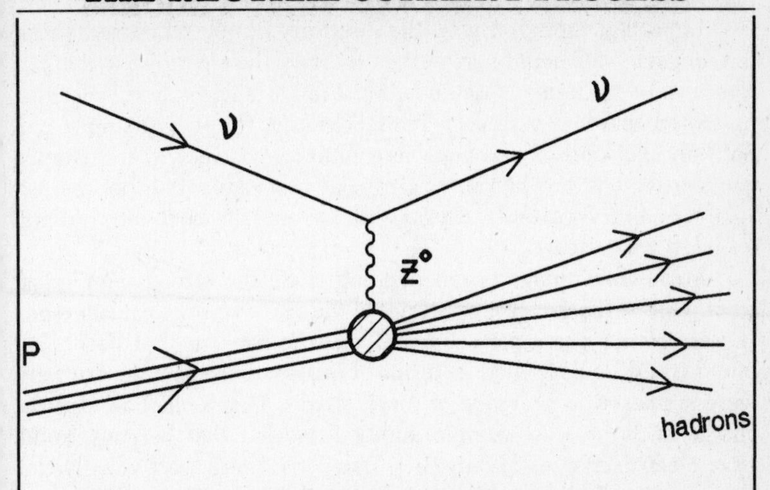

A high-energy neutrino strikes a target proton. A Z^0 particle is exchanged, carrying energy but no electric charge. The neutrino loses most of its energy, which reappears as a shower of newly created hadrons.

in which there were no neutral currents at all. We wrote: "There is considerable arbitrariness in such models, so that a detailed discussion seems premature until an appealing model is found."

You see, the experimenters had us worried because they had not yet found neutral currents. Our contrived theory was an attempt to save the concept of an electroweak synthesis from the apparent absence of neutral-current phenomena. There was no need for such artifice. As always, nature chose the simplest and most elegant theory.

In 1973 the European group, working with an enormous bubble chamber nicknamed Gargamelle, finally and conclusively observed the elusive neutral currents. The discovery paper was signed by fifty-five collaborators. Neutral currents had been there all along, and they behave exactly as the theory predicted they should. Very soon afterward, Rubbia's group at Fermilab confirmed the European discovery. Things were beginning to look good for the electroweak theory. The unification of weak and electromagnetic forces demanded the existence of the Z particle and the consequent appearance of a new form of the weak force whose effects had never before been seen. Now they were! What more could you expect of a new theory?

Important though it was, the discovery of neutral currents was not, in itself, sufficient to prove the validity of the electroweak theory. There were still some loose ends. Most importantly, where were the predicted charmed particles? Without charm, the theory simply did not work. All kinds of new hadrons containing charmed quarks simply had to exist, but they had not yet revealed themselves in experiments. Like the neutral currents, charm would eventually be found, but not for a few more years.

Meanwhile things began popping along the strong interaction front. One of the greatest mysteries of the 1970s was the considerable success of the quark-parton model. All the hadrons that had been found could be described in terms of quarks. For example, baryons were supposed to be made of three quarks. Quarks had an electric charge of ⅔ or −⅓. Simple addition showed that baryons could have electric charges of only 2, 1, 0 or −1. These were exactly the charges that all observed baryons had! At least as impressive was the extent to which the quark-parton model offered a quantitative explanation of the SLAC electron-scattering experiments, as well as the neutrino-scattering data obtained at CERN and Fermilab. The experimental results jibed with their theoretically computed values. These depended on treating the quarks in a proton as if they were free of any forces whatsoever. But that was manifestly false. There must be forces to keep the quarks confined inside the proton.

The behavior of the proton was paradoxical. When it was probed at very high energy it behaved as if it were made up of three freely moving quarks. On the other hand, at low energies things become very much more complicated. Like an atom, a nucleus or any other composite system, the proton displayed a large number of excited states, or quantum levels. Unlike any other system, the quark constituents could not be removed from the proton, nor could the quark itself be observed as a particle in its own right. What kind of theory of the strong force could account for all this?

Another obstacle to the construction of a rational theory of the strong force was the "statistics problem." The most fundamental principles of quantum mechanics and relativity have as an incontrovertible consequence the so-called spin statistics theorem. Particles whose spin is half a whole number are called fermions, and they must satisfy the Pauli exclusion principle, which states that no two fermions can do the same thing at the same time. Particles of whole-number spin, known as bosons, do not satisfy the Pauli principle. The proton is a spin-½ particle and is therefore a fermion.

Since the proton is supposed to be built out of three quarks, it follows that the quark, too, is a fermion. No two identical quarks, according to the Pauli principle, can occupy the same quantum state. However, this is just what the quarks in a baryon seemed to do. For example, several observed baryons are made up of three identical quarks all spinning in the same direction. The famous omega-minus particle is an example: it is made of three strange quarks and has spin 3/2. The quarks within baryons behave as if they were bosons. This is against the rules! Nonetheless, a few courageous physicists simply ignored the rules. Dalitz, in Oxford, and Giacomo Morpurgo, in Genoa, became dedicated champions of the quark model as early as 1965. They showed that many of the properties of so-called elementary particles could be explained if they were made of quarks. Something important was missing from the theory that let quarks pretend to be bosons, but these early disciples maintained that the reality of quarks could not be denied.

The missing ingredient was the idea of quarks in colors. To solve the statistics problem, physicists were forced to triple the number of different kinds of quarks. Each flavor of quark (up, down, whatever) was now assumed to come in three varieties, called colors. There were, for example, red, green and blue up quarks. Each of them had exactly the same charge, spin and mass. They are identical except for color. Every flavor of quark came in three colors. Once again the physicist has preempted ordinary words (color, flavor, red, etc.) for his own pursuits. None of these quark properties have anything to do with rainbows or traffic lights.

At the beginning, very few physicists took the notion of quark color seriously. It seemed like an ad hoc idea whose only purpose was to save the quark model, which only a few physicists believed in, anyway.

The notion of quarks in color was first suggested in the sixties by Wally Greenberg at the University of Maryland, but his proposal was unconvincing and unaccepted. Only in the seventies did physicists realize that the color degree of freedom was really an essential attribute of the theory of the strong force. Color is the arena in which the gauge bosons of the strong interactions act their parts. Flavor is the domain of the weak and electromagnetic gauge bosons.

If the quarks did not come in different colors, it would be simply impossible to construct gauge theories of both strong and weak interactions that do not get into one another's way so as to destroy the consistency of either theory. But we had not reached this point yet. Quantum chromodynamics, our successful theory of the strong force, had not yet been invented. The original clue to the existence of color came from the enigma that quarks in baryons behaved like bosons, though they had to be fermions.

Color solves the statistics problem in the following way. It is assumed that all hadrons contain an equal admixture of all the colors. Hadrons are essentially colorless. Thus a baryon must contain one red quark, one blue and one green. No two of the quarks in a baryon can be quite identical, and there is no conflict with the Pauli exclusion principle. But this merely exchanges the statistics contradiction for a mysterious puzzle: Why are no colored particles seen in nature? Why, for example, are no baryons seen that contain two blue quarks and one red one? Why are there no mesons made of a green quark and a red antiquark? In short, why is it that of all possible quark configurations, only the colorless states survive as observable particles? Quark color is an example of a "hidden variable." Color cannot be made manifest, say, by producing a beam of blue quarks. Golly, we cannot isolate just one quark. Color is something that simply has got to be there to make the theory work.

It sounds like mumbo-jumbo. Indeed, quarks with color were an ingenious solution to the statistics problem, but it was not in any sense a systematic theory. The success of such a kooky idea was merely a clue to the nature of the underlying mathematical explanation of the strong force.

Something more was needed: a quantum field theory of the strong interactions in which the force between quarks diminished in strength as the energy became larger. This would mean, according to quantum mechanics, that the force could be relatively weak at the very short distances probed at high energy, and yet strong enough at a long distance to ensure the absolute confinement of the quark to the interior of the hadron, about 10^{-13} centimeter in radius.

Such a force did not seem to emerge from any known quantum field theory. Quantum electrodynamics, the best-studied theory, had the opposite property: the Coulomb force is very strong at short distances and dwindles in magnitude with the inverse square of the distance between two charges. Unless someone could find a theory with the opposite property, we were in for trouble.

Any new theory of the strong force had three distinct riddles to solve: Why are quarks inescapably imprisoned within hadrons? Why does the strong force become weak when the quarks are very close together? Why are all hadrons colorless?

The key to the solution to all three problems was presented to the physics community in the spring of 1973 by a young and brilliant Harvard graduate student, Hugh David Politzer, Sidney Coleman's best student yet. (At the same time, it was discovered at Princeton by David Gross, a Princeton professor who had been a junior fellow at Harvard, and his precocious graduate student, Frank Wilczek. The basic work had been done earlier by 'tHooft, but he failed to notice its significance.)

The magic potion is known as asymptotic freedom, and it earned H. David a thesis, a junior fellowship at Harvard, and instant fame among particle physicists. It didn't hurt Gross or Wilczek, either.

In mathematics, a line that approaches a curve but never quite meets it is called an asymptote. A quantum field theory is said to be asymptotically free if the force it produces becomes weaker and weaker at shorter and shorter distances. Neither the electrical force between two charges nor the gravitational force between two masses has this property. In these cases, the force becomes weaker as the distance between the bodies grows larger, and stronger when they are closer together.

David, Frank and David exhibited a class of quantum theories that is asymptotically free; just the kind of theory that is needed to describe the interactions between quarks.

When quarks are close together, as they are within a hadron, they seem to move about freely. Thus, the instant "snapshots" of the proton provided by high-energy scattering experiments showed happily unfettered quarks. The converse is also true. At large distances, the force between quarks does not fade away; it remains large and constant. At distances the size of the proton or larger, the attractive force between two quarks corresponds to a pull of thirty tons! Quite a large force for such a tiny little thing. No wonder nuclear bombs, which depend upon the strong force among quarks, are so devastating.

We began to understand the mystery of quark confinement. Imagine that you were to grab hold of one of the quarks in a proton and try to pull it away. You might think of using a "quarkscrew" for this, but in practice a collision with an energetic electron suffices. As the quark is pulled away, more and more energy must be expended. Eventually, the string of force connecting "your" quark to the rest of the proton is broken, and the energy you have expended produces a quark-antiquark pair. The new quark falls back into the proton and restores its integrity. The antiquark combines with the liberated quark to become a meson. The two hadrons separate, and the attempt to isolate a single quark fails. All we have succeeded in doing is to produce another hadron.

QUARK CONFINEMENT

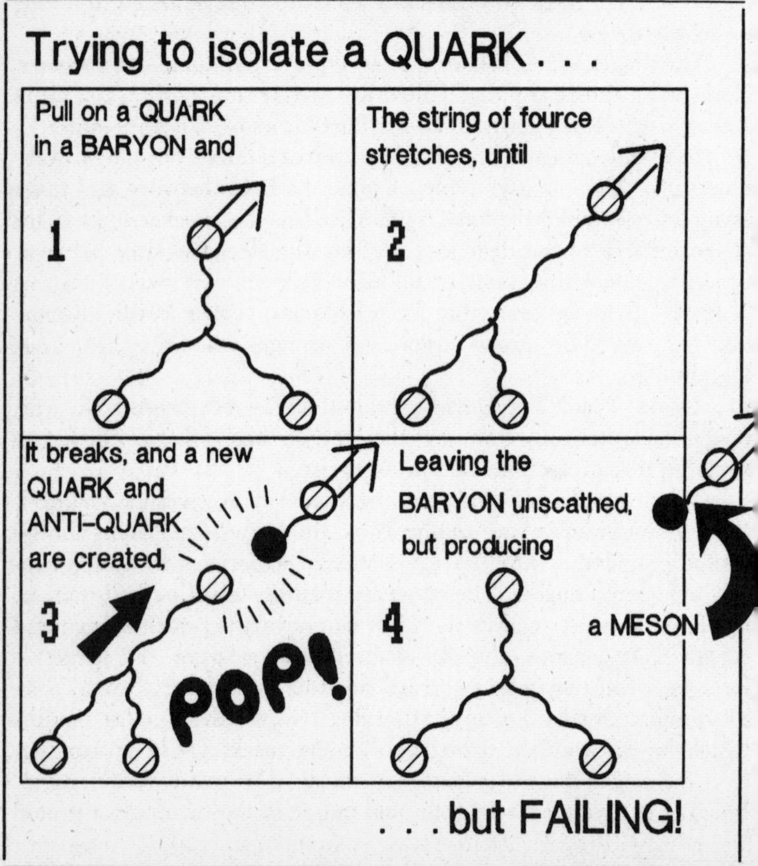

Trying to isolate a QUARK . . .

1 Pull on a QUARK in a BARYON and

2 The string of fource stretches, until

3 It breaks, and a new QUARK and ANTI-QUARK are created,

POP!

4 Leaving the BARYON unscathed, but producing

a MESON

. . . but FAILING!

The behavior of quarks is analogous to the properties of a magnet. As you know, every magnet has both a north pole and a south pole. Try to produce an isolated north pole by cutting the magnet in half. As soon as you do this, a new pair of poles is generated. Each fragment of the magnet becomes an entire magnet, displaying both a north and south pole. No one has ever seen either a magnet with only one pole (although many have searched for a "magnetic monopole") and no one has seen an isolated quark.

Actually, there was a time a decade or so ago when Bill Fairbanks and his collaborators at Stanford claimed to have evidence for the existence of isolated quarks. This result has not been confirmed by any other experiment. Few physicists today believe in the existence of free quarks. The Quark Liberation Front has joined the Filthy Speech Movement in the annals of California's glorious history.

Soon after the discovery of asymptotic freedom was announced, a full-fledged theory of the strong force was developed. Known as quantum chromodynamics, or QCD, it is a gauge theory like the electroweak theory, but it is exact and not spontaneously broken. Whearas quantum electrodynamics (QED) involves only a single gauge boson, the photon, QCD requires eight massless gauge fields called gluons. These gluons, like the quarks, are colored. Unlike photons, gluons cannot be seen as isolated particles.

The gauge theory of the strong force acts in the arena of color. That is, a quark may emit a gluon and change its color. One type of gluon changes red quarks to green, another green to blue, etc. On the other hand, the electroweak gauge particles do not sense color, but act upon the quark flavor. For example, an up quark of any color may emit a W^+ to become a strange quark or a down quark of the same color. The success of both theories depends on the associations of their gauge bosons with different quark attributes. Otherwise, they would get in one another's way and would not lead to a consistent theory. As it is, QCD and the electroweak theories are complementary and mutually consistent. Together they form what is now known as the Standard Theory of Elementary-Particle Physics. It works!

But where was charm? I remember one time sitting in my office at Harvard in the autumn of 1973. Four of my junior colleagues and I were screaming and shouting at one another in the best tradition of theoretical physicists. Howard was there, along with our new Spanish postdoc, Alvaro De Rújula. The three of us were about to embark upon an enormously fruitful and exciting collaboration. The others were our new assistant professor, Thomas Appelquist (now a senior professor at Yale), and Helen Quinn.

I had met Helen a year or two earlier at an Erice summer school in Sicily. She and her husband were both physicists postdoc-ing in Germany. Dan had found a job in the Boston area, but not Helen. She was thinking of abandoning her trade for housewifery. We encouraged her to stay with it, and to accept a temporary (and unpaid) research assistant position at Harvard. She was rapidly promoted to a faculty position, and is now a full professor and very accomplished physicist at Stanford.

In any case, there we were, five relatively mature physicists, each a devout believer in charm. But where was the evidence?

There were rumors from Rubbia's group at Fermilab of several rather peculiar phenomena. Some of the neutrino-scattering data seemed to show weird effects. There were hints of what was called "the high-y anomaly." This inexplicable observation was driving theorists up the walls. Subsequent experiments showed that the data were wrong and the anomaly nonexistent. Rubbia also reported surprising (and quite unexpected) neutrino events that produced two charged muons. In retrospect, these events were recognized as the production of charmed particles. The second muon was the decay product of charm. It seemed very likely that charm, or something like it, was about to be discovered.

Just before Christmas, Alvaro, Howard, Helen and I presented a Harvard Physics Colloquium in the form of a play entitled "Fact and Fancy in Neutrino Physics." Howard played the role of a talking computer, and I was the mad theorist. In the climactic final scene, when charm was discovered, a courier was to run onstage with a telegram to Carlo Rubbia from the Nobel Committee. As it turned out, Carlo missed the discovery of charm just as he missed the discovery of neutral currents. But Carlo didn't strike out. A decade later he would get his telegram for a home run, the discovery of all three of the weak interaction gauge bosons.

I had a hunch that 1974 would be a great year for me. After all, my first son was born in January. (Brian is clearly my clone, says Joan, sister of the great biologist.) In April, I attended a conference on Experimental Meson Spectroscopy at Northeastern University. My contribution was entitled "Charm: An Invention Awaits Discovery." Meson spectroscopists are experimental physicists who make a career of discovering new mesons. Reasonably enough, I wanted them to put aside their old-fashioned habits and search for my predicted charmed particles. I began:

> A most important question in experimental meson spectroscopy is to determine what are the hadronic quantum numbers. Charm, a conjectured strong-interaction quantum number for which the theoretical raison d'être is all but compelling, has not yet been found in the laboratory. I would bet on charm's existence and discovery, but I am not so sure it will be the hadron spectroscopist who first finds it. Not unless he puts aside for a time his fascination with such humps, resonances, and Deck effects as have been discussed at length at this meeting. Charm will not come so easily as strangeness, yet no concerted, deliberate search has been launched.

You see, all that those meson spectroscopists had been doing was collecting and comparing conventional mesons, the old-fashioned kind made of up, down and strange quarks. They couldn't get it into their heads that they should have been looking for qualitatively new particles. Charm could have been discovered years earlier than it was, had the experimenters in my audience been a bit more imaginative and daring.

I went on to speculate that charm would show up first in neutrino physics or in the study of electron-positron collisions, which is just how things actually did turn out. I concluded with a prediction of what to expect at the next meeting of the conference in 1976.

> There are just three possibilities:
> 1. Charm is not found, and I eat my hat.
> 2. Charm is found by hadron spectroscopists, and we celebrate.
> 3. Charm is found by outlanders, and you eat your hats.

Two years later, the conference was held again at Northeastern. Its organizer, Roy Weinstein (now dean of sciences at the University of Houston), passed out foul-tasting candies in the shapes of hats for the participants to consume. Charm had indeed been found, and by then it was old hat.

I was not the only advocate of charm. Every two years we particle people have a monster international meeting with a cast of thousands. Listen to John Iliopoulos, a coinventor of charm, at London in the summer of 1974: "I have already won several bottles of wine by betting for the neutral currents and I am ready to bet now a whole case that if the weak interaction sessions of this conference were dominated by the discovery of neutral currents, the entire next conference will be dominated by the discovery of charmed particles."

It was.

The discovery of charm came simultaneously from the East and West coasts on a lovely autumn day in November. Min Chen, an MIT experimenter working with Sam Ting's group, phoned me at home that morning to ask if I could come to MIT to hear about some interesting data they had obtained at Brookhaven National Laboratory. Interesting was an understatement.

Ting was following his own hunch that there should be a particle that is produced in hadronic collisions which sometimes decays into an electron-positron pair. His group had directed a beam of 7 GeV protons at a beryllium target and had studied the electron-positron pairs coming out. Such pairs are not so unusual in themselves, for they can be produced electromagnetically. By measuring the electron and positron energies and directions, the Ting group computed the mass of the system that produced the pair. In a plot of the number of events versus computed mass, there was a tremendous bump superimposed upon the background of expected events. Ting's hunch had been right. He had discovered just the sort of particle he had imagined. What he wanted from me was to know how his particle (he called it J) fit into the theory. It took me a day to realize that it had something to do with charm.

With J particles buzzing in my head, I drove to Harvard to spread the exciting news. What were these things? Could they be Z^0 particles? That didn't seem likely because they should be much heavier. Were they Steve Weinberg's latest construct, pseudo-Goldstone bosons? This was a real possibility if the J particles were spinless. Anyway, back at Harvard, things were already astir. They hadn't yet heard of Ting's work, but there had already been an independent announcement of the discovery of a new particle at SLAC, in California.

Burt Richter's group had been pursuing their studies of electron-positron annihilation at high energy in the recently constructed SPEAR machine (Stanford Positron-Electron Accelerating Ring). SPEAR is an example of an electron-positron collider, a device in which electrons and positrons are guided by magnetic fields and accelerated by electric fields. They move in circular orbits, within an evacuated ring. When the two oppositely directed beams are made to intersect, numerous collisions take place in which the electron and positron annihilate one another. When the collision energy is sufficiently high, these collisions can produce hadrons. Smaller electron-positron colliders had been in operation for years in Italy (which built the first prototype, ADA, in 1961), France and the Soviet Union. The first really big machine was the collider at the Cambridge Electron Accelerator, a joint endeavor of Harvard and MIT. The Cambridge machine gave the first indications that something peculiar was happening, but these hints were never carefully followed up. We missed the big discovery.

When the electron and positron beams at SPEAR were tuned up to just the right energy, 3.069 GeV, the number of observed interactions soared. A new particle with that mass was being produced in enormous numbers. It was immediately christened "psi."

PSI EVENT

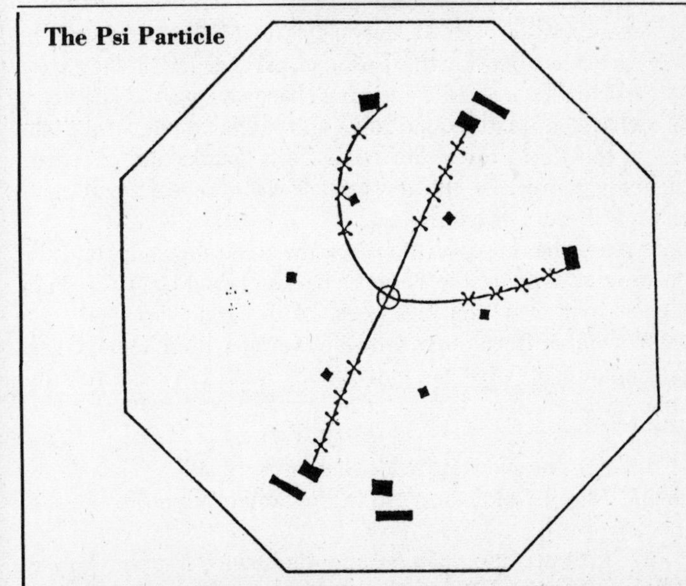

This is a computer reconstruction of one of the first psi events seen at the Stanford colliding-beam facility. A psi particle was created in an electron-positron collision. It decayed into four pions, whose tracks are shown in the figure. Notice that the event closely resembles the Greek letter psi (ψ). (Photo courtesy of R. F. Schwitters)

First they name it, then they try to figure out what it is. The psi particle was identical to Ting's J particle. Since the two discoveries were simultaneous, it goes today under the awkward name of J/psi, or, as Leon Lederman, director of Fermilab, calls it, the Gypsy particle. Ting made the J particle by banging together hadrons and seeing it decay into an electron-positron pair. Richter made the psi particle in an electron-positron collision and saw it decay into hadrons. With apologies for the pun, Burt had seen the Sam Ting.

The SPEAR discovery meant that the J/psi particle had to have spin one. That ruled out a number of possibilities. For the next few weeks everyone at physics departments throughout the world tried to figure out what this new thing was. It was a spectacularly narrow resonance with a lifetime thousands of times longer than a typical hadron. Moreover, it decayed into leptons almost as often as into hadrons. It was no ordinary hadron, that was for sure. But what was it? More precisely, what were they? Another particle was discovered by SPEAR just a few days later, the psi′, at an energy of 3.685 GeV.

By December, all of us at Harvard agreed that charm was the only rational explanation for the J/psi and psi′ particles. They were the first particles discovered containing charmed quarks. They consisted of a charmed quark bound to its antiquark. In one stroke, the discovery of the J/psi proved the existence of quarks and of charm. Far from being mysterious, the new particle was to reveal the ultimate simplicity and beauty of nature.

There were many explanations for the new phenomenon published in the first issue of the *Physical Review Letters* of 1975. They had all been written within two weeks of the great discovery, our very own November Revolution. Outside Harvard, there seemed very little agreement as to what the new particle was. Let's examine the papers one by one:

1. "Are the New Particles Baryon-Antibaryon Nuclei?" by Maurice and Alfred Goldhaber. No to the father and no to the son. They are not.

2. "Interpretation of a Narrow Resonance in e^+e^- Annihilation," by my thesis advisor, Julian Schwinger. Schwinger argued that the data confirmed his own quirky electroweak theory. Not so! Julian had only recently moved to southern California. Maybe he had sunstroke.

3. "Possible Explanation of the New Resonance in e^+e^- Annihilation," by S. Borchardt, V. S. Mathur and S. Okubo. These guys had the right idea. They argued that the new particle is made up of a charmed quark and a charmed antiquark, which indeed it is. They failed to notice that asymptotic freedom explains its long lifetime.

4. "Model with Three Charmed Quarks," by R. Michael Barnett. This is the first of three Harvard papers in that notable issue of *PRL*, and Mike is an interesting case. He had come to Harvard as an unpaid honorary research fellow and lived on his savings. He figured that it would be better for his career to starve at Harvard than to be a paid research fellow at a second-rate institution. He was right. Today he has a permanent position at Lawrence Berkeley Laboratory. His paper is an ingenious, but false, explanation of the J/psi in a theory with six quarks. Actually, it turns out that there really may be six quark flavors, and Mike is likely to have been the first person to say so. Unfortunately for his theory, the fifth and sixth quarks have nothing at all to do with the J/psi.

5. "Heavy Quarks and e^+e^- Annihilation," by Tom Appelquist and H. David Politzer. Harvard again. They "report on some theoretical work on the e^+e^- annihilation in asymptotically free colored quark-gluon models of hadronic matter. [Their] fundamental assumption is that in addition to the light quarks that make up ordinary hadrons, there is a heavy quark, such as the charmed [quark]." Tom and David bought the latest quark model festooned with color, charm and eight gluons, and they used it well. They interpreted the J/psi as a charm-anticharm bound state, and they explained its very small width in terms of QCD. Actually, they had done their work before the November denouement. Had they the courage of their convictions, they could have written and published their paper prior to the experimental discovery of the J/psi particle. They could have *predicted* the particle and not merely explained it. But the meek shall not inherit the earth.

6. "Is Bound Charm Found?" by Alvaro De Rújula and me. Needless to say, we got it all right, too. ". . . asymptotic freedom and the selection rules of color gauge theory explain the mystery of the narrow width of the J/psi," we said. We went on to argue that asymptotic freedom explains other minor mysteries concerning the decays of ordinary hadrons. Alvaro coined the word *charmonium* to describe any system that contains a charmed quark and its antiquark. Today over a dozen states of charmonium have been seen in the laboratory. The J/psi was only the first one.

7. "Possible Interpretation of the J Particle," by H. T. Nieh, T. T. Wu (Harvard again) and C. N. Yang. Had they eaten at a Chinese restaurant after submitting their paper, the fortune in their cookie should have read, "You will regret the letter you have just written." This one was well out-of-bounds.

8. "Remarks on the New Resonances at 3.1 and 3.7 GeV," by five Princetonians, including Frank Wilczek. They understand things pretty well at Princeton. If you're number two, you've got to work harder. Yet, they wrote, "Various attempts have been made to explain [the narrow J/psi width], none completely convincing." Here they refer to Appelquist, Politzer, De Rújula and me, and to our invocation of asymptotic freedom. It seemed as if Frank and his Princetonians also had a courage-of-conviction problem.

9. The last of the papers was written by my deceased friend John Sakurai. He believed that the J/psi was an intermediary of the weak interaction, that it was the Z^0 boson. That is what Luciano Maiani, my GIM coauthor, thought as well. They were soon to change their minds.

Whew! And I didn't even get to mention some of the kookier explanations that came later. It would take years for a consensus to arrive. Meanwhile, we knew for certain that charm had been discovered, but there was a big "they" that was not yet convinced. By early 1975 we were convinced of the validity of QCD and the electroweak theory. Quarks, gluons, color, charm, gauge theories, the Higgs mechanism and asymptotic freedom were revealed as threads in a tapestry that portrayed a coherent picture of the strong, weak and electromagnetic forces and the particles upon which they act. The day of the standard model had come, and it was good.

The discovery of the J/psi was undeniably the experimental thread that pulled it all together. Before that day in November, none of us could see the picture in all its glory. The new particle presented physicists with a firing range at which all of our whacky ideas could be tested. A few of the ideas survived to become established as textbook dogma. Charm, quarks, color and asymptotic freedom are among them.

It was no surprise that the 1976 Nobel Prize in physics was awarded jointly to Burton Richter and Samuel Ting for "their pioneering work in the discovery of a heavy elementary particle of a new kind." That year, America's Bicentennial, Americans made a clean sweep of all the Nobel prizes in the sciences. It had happened only once before, in 1946, and it was to happen again in 1983.

The So-Called Standard Theory

JUST BEFORE CHRISTMAS 1974, I delivered an evening lecture at Harvard intended for a popular audience. It was entitled "Something Exciting Is Happening in Particle Physics," and the lecture room was overflowing with curious humanoids, as nonscientists are known to us. I attempted to put the discovery of the J/psi into a proper historical context. I explained how the atom was once regarded as the ultimate and immutable constituent of matter, how it was shown to be made of electrons orbiting a tiny nucleus, how the nucleus was made of protons and neutrons, and how these particles, together with their sisters and brothers and cousins and aunts, were now known to be made up of quarks up, down, strange and charmed (in three colors each).

I explained the miracle of quantum chromodynamics, that the quarks had the illusion of freedom but were nonetheless irrevocably confined in the hadrons they formed. I spoke of the four forces of nature, and how they had become three, and then only two. I dreamed the dream of Einstein, in which all four forces were one, our version of the Trinity, and I discussed the big bang, our Genesis, in which the perfect symmetry of nature was burning bright and supernally clear. It was not a bad lecture. It has become the basis of a core course I teach to nonscientists, "From Alchemy to Quarks," in which I try to dispel the arrogant ignorance of your average Harvard humanoid toward things scientific.

Our understanding of the ultimate structure of matter far exceeds the dreams of those who came before us. As Isaac Newton said of his own endeavors, "If [we] have seen further than other men, it is because [we] stood on the shoulders of giants." Einstein and Bohr are the founders of the special theory of relativity and quantum mechanics—the two firm pillars on which our theory rests. Everyone knows Einstein's magic formula, $E = mc^2$, expressing the equivalence of mass and energy. Everyone has heard of Heisenberg's uncertainty principle, which is the gist of quantum mechanics. Both relationships are at the heart of "high-energy physics," the experimental search for clues to guide the theorist toward the one and true theory of everything.

The uncertainty principle says that we cannot measure both the position and the momentum of a particle to absolute precision. As a corollary, observations of smaller and smaller systems require probes of higher and higher energies. Optical microscopes were superseded by powerful electron microscopes that can "see" individual atoms. Rutherford "saw" the nucleus by using a beam of particles with energies of millions of electron volts, and the experimenters at SLAC "saw" quarks by increasing the energy of the probe to billions of electron volts.

Thus, particle accelerators can be regarded as supermicroscopes that can explore the very finest structure of matter. Quantum mechanics tells us that the higher the energy we can achieve, the smaller the structure we can study. $E = mc^2$ tells us that high-energy collisions can convert energy into new forms of matter: the higher the energy, the heavier the particle that can be made.

Cosmic rays, God's gift to high-energy physicists, produce energetic collisions in which positrons, pions, muons and strange particles were discovered. Since the 1930s we have learned to generate intense energetic beams of particles artificially in machines called cyclotrons, synchrotrons and colliders. Antiprotons were produced and detected for the first time ever by the Berkeley Bevatron in the 1950s, when collision energies of billions of electron volts were achieved. When half a trillion electron volts became available at CERN, in the 1980s, the long-sought W and Z particles were discovered.

Relativity and quantum mechanics are not only the basis of our theory, but they also form the rationale for our need for even larger accelerators. They let us *create* the heavy and short-lived particles that explain the structure of matter, and they let us *see* just what these particles do.

High-energy physics of the 1960s was dominated by particle accelerators known as proton synchrotrons, which were vastly improved versions of the original cyclotron. These large circular devices produce beams of high-energy protons—hydrogen nuclei. The protons are guided by magnetic fields and accelerated by electric fields.

They are extracted from the machine and directed upon a stationary target. The target material is often liquid hydrogen, so that proton-proton collisions may be studied most directly.

During the 1950s the really big machines were the Brookhaven Cosmotron (3 GeV) and the Berkeley Bevatron (6 GeV). It was clear to the physics community that larger accelerators were essential, since we had only scratched the surface of the proton, so to speak. But severe technical obstacles were in the way. At higher energies it became more and more difficult to keep the accelerated protons stable in their designated orbits. They had an unfortunate tendency to oscillate violently one way or another and to escape from the beam. These problems were solved, in principle, in the early 1950s with the invention of "strong focusing," a procedure involving the use of an array of magnetic "lenses" to keep the protons in their places.

Soon after the announcement of strong-focusing theory, the Brookhaven National Laboratory and CERN initiated a cooperative program leading to the construction of two similar strong-focusing proton synchrotrons, one in Geneva (the Proton Synchrotron, or PS) and one in Long Island (the Alternating Gradient Synchrotron, AGS).

These machines, operating at about 30 GeV, were responsible for many of the great achievements of the sixties, including especially the discovery of dozens of new hadrons with confusing and often unexpected properties. The omega-minus particle, the baryon predicted to exist by Gell-Mann and consisting of three strange quarks, was found in 1964 by a group of physicists led by Nick Samios, using the AGS. Nick's group went on to discover the first charmed baryon in 1975. One such discovery could be a lucky fluke, but two proves true greatness. Nick turned down the ultimate honor of a tenured position at Harvard and has become director of Brookhaven Laboratory.

The AGS has enjoyed a number of other triumphs. The fact that muon neutrinos are different from electron neutrinos, was established by work done at the AGS in 1961. This was the first key step in our realization that fundamental fermions come in well-structured "families," a vital input in today's standard and successful theory of all particle physics.

In 1964 a group of physicists from Princeton used the AGS to establish what is known in the jargon as CP violation. What this means is that the laws of microphysics are not the same if time is run backwards. Of course, in the mundane world there is certainly no such thing as time-reversal symmetry. We all age, never to become infants again who crawl into our mother's wombs to disappear forever! It is a long story, but we think we understand the arrow of time of the macroworld of everyday experience. Yet physicists believed, until 1964, that in the microworld of elementary particles, time-reversal symmetry might hold true. Val Fitch and James Cronin, the leaders of the Princeton group, were awarded the Nobel Prize for showing us that it did not. Time flows only in one direction for elementary particles, just as it does for you and me.

The most recent Great Discovery at the AGS was the finding of the J particle, by Sam Ting and his friends, in the November Revolution of 1974. The AGS has had a truly remarkable history, and it may well make further surprising discoveries in the future.

With the scientific successes of the CERN and Brookhaven proton synchrotrons so evident, the pressure was irresistible to build even larger machines. Robert R. Wilson, the cowboy physicist and sculptor from Wyoming, led the effort to build a giant accelerator in the farmlands near Chicago. The Fermilab accelerator was completed under budget and ahead of schedule in 1972. Its original target was 200 GeV, soon upgraded to 400 GeV. Meanwhile, a similar (but more expensive) machine was being built at CERN. The 300 GeV Super Proton Synchrotron (SPS) was commissioned in 1975. In two decades, physicists had pushed up the high-energy frontier by a factor of 100. Not to be left out, our Russian colleagues have begun construction of a far more powerful proton synchrotron. If and when their UNK is completed, it will achieve the remarkable energy of 3000 GeV.

However, by the early 1970s the tide unexpectedly turned. It was to be a decade in which the most dramatic discoveries were achieved with electron-positron colliders, devices in which matter and antimatter are forced into collision. What could be more elegant?

In these circular machines, beams of electrons and positrons travel in opposing directions within a single ring. At several points along the circumference of the ring the beams are arranged to hit one another practically head-on. In the resulting collisions, electrons and positron meet and annihilate one another, and in the process other particles are created.

The first successful machines of this kind were built in the 1960s in Italy, France and Siberia. They could achieve collision energies of little more than a GeV. Nonetheless, they accomplished important goals. For example, they were able to study in detail the vector mesons ρ, ω and φ, particles that play an essential role in the electromagnetic structure of the proton. Burt Richter, at Stanford, fought long and hard to build a more powerful electron-positron collider in the United States. By 1972 he succeeded, with the construction of the Stanford Positron-Electron Accelerator Ring, or SPEAR, designed to study energies from 3 to 7 GeV.

Meanwhile, we easterners had at our disposal the 5 GeV Cambridge Electron Accelerator, or CEA. It was not a collider, but simply a device for accelerating a beam of electrons. The CEA was never the powerful research instrument it was designed to be. With one important exception, it led to no great advance of science. It was more like an in-house thesis factory for Harvard and MIT. Indeed, many of today's high-energy experimenters were initiated into the discipline with this machine. In an experiment for which the machine is notorious, Harvard's Frank Pipkin alleged to discover a failure of the validity of quantum electrodynamics. I say alleged because the experiment was simply wrong. Sam Ting did it right and did it better in Germany. Quantum electrodynamics was (and still is) perfectly adequate at those relatively small energies, and Sam thereby earned himself tenure at MIT.

In the mid-sixties carelessness in handling liquid hydrogen resulted in an explosion at the CEA. Several people were injured seriously and one was killed. In terms of scientific success or safety, the CEA found itself at the opposite end of the list from the AGS.

Two brilliant accelerator physicists, Ken Robinson and Gus Voss, decided to turn the CEA into something it was never intended to be: an electron-positron collider. With an infinitesimal budget and infinite ingenuity, they succeeded. In 1972, just before the machine was turned off forever and demolished, it was used to study electron-positron encounters at collision energies around 5 GeV. The results were absolutely amazing, but were not taken very seriously because of the checkered history of the machine. To understand what was going on, we must go into a little detail about what can be learned from the study of electron-positron collisions.

What can happen when a high-energy electron meets its oppositely moving antiparticle? Sometimes, particle and antiparticle are converted into two or more photons. This is a purely electromagnetic process that is well understood. The experimental observations of this process coincided with theoretical predictions.

Sometimes the electron and positron are transformed into a single "virtual" photon, which itself is transformed into a muon and an antimuon. This, too, is a purely electromagnetic process for which theoretical calculations agree precisely with experimental results.

The most interesting annihilations are those resulting in the production of hadrons, mostly consisting of pions. This process involves an interplay between the strong nuclear force and electromagnetism. The experimenters defined a fraction called R. Its numerator is the number of events involving the production of pions. Its denominator is the number of events yielding a muon-antimuon pair. The denominator was "known," in the sense that it is precisely calculable by QED. Its measured value agreed with its calculated value. The numerator of R was more mysterious. It depended on the detailed nature of the then poorly understood strong force. A central question was how R should vary as the collision energy changes.

Many theorists expected R to become smaller and smaller as the collision energy was increased. If this were true, then electron-positron colliders would become virtually useless at high energies, since there would be very few interesting hadronic events to observe. This was the basis to arguments against the construction of such colliders, arguments that Richter fought against in order to build his dream machine. Richter developed a healthy contempt for the idle predictions of theoretical physicists.

Those theorists who found themselves on the right road, the one leading to asymptotic freedom, quantum chromodynamics and the standard model, expected something quite different. At high energies, they argued, the effects of the strong interactions could be ignored. Quarks and leptons ought to be produced in pairs, with the likelihood depending simply upon the square of their electric charges. Given the charges of the quarks (⅔ for up, −⅓ for down, and −⅓ for strange), they predicted that R should settle down to a value of 0.67.

But wait a second! Weren't the quarks supposed to come in three colors? Theorists who were really in the avant-garde took this into account and predicted a thrice larger value, R = 2. What would experiment say? Would R approach zero, 0.67 or 2? Or would some different result justify Richter's animosity toward theory?

Things came to a head at the notorious London conference in the summer of 1974, the very same meeting where Iliopoulos placed his winning bet on the imminent discovery of charm. Richter gave the summary talk for the session of electron-positron physics. At that time, SPEAR, in operation for two years or so, had accumulated considerable data in the 3 to 4 GeV range. The Italian machine ADONE was responsible for the 2 to 3 GeV domain, and the CEA collider, in its last dying gasp, had explored the 4 to 5 GeV region.

Lo and behold! It seemed that none of the theorists was right. The experimentally measured values of R seemed to increase steadily with energy. At 2 GeV, R was about 2; at 3 GeV it was 3; and at 5 GeV, as measured by the notorious CEA collider, it was 5. No theorist had suggested such a remarkable result, and Richter was deliriously happy.

For the fall semester of 1974, Richter came to Harvard as a Loeb Visiting Professor, an honorific position that we use to attract current Heroes of Modern Physics to inspire ourselves and our students. Burt explained how it could be that R is an increasing function of energy. Rejecting all of current theory, he argued that somehow the electron spent part of its time being a hadron. Burt's hypothesis sounded profound, and it did explain the experimental data, but it made no sense at all as a reasonable and rigorous theory.

Now, Richter, you will recall, was the leader of the experimental group working at SPEAR. When the cat's away, the mice will play. While Burt was gibbering away at Harvard, his colleagues at Stanford were about to discover what was really going on. As Appelquist and Politzer had prophesied, but not dared to publish weeks earlier, SPEAR had encountered the threshold for the production of charmed quarks. This was the reason that R was behaving peculiarly. It was not at all a smooth function of energy, but it was displaying a series of bumps and grinds, telling us that new particles were being produced: the psi, psi′ and a host of others.

It started to become clear on November 4, with the simultaneous announcement by Stanford and Brookhaven of the discovery of the J/psi particle. Burt had rushed back to California to meet the press and enjoy his completely unexpected triumph. A few days later he announced the discovery of a second new particle, and the fun really began.

Incidentally, the scientists working at the ADONE machine in Italy just missed the discovery of the J/psi by a nose. The maximum design energy of their machine was just a few percent shy of the mass of the new particle. With considerable ingenuity and some risk, they managed to hot-rod their accelerator up to the required energy so that they could see the new particle and confirm the SLAC/Brook-

THE STATES OF CHARMONIUM

The word *charmonium* was coined by Alvaro De Rújula to signify the composite of one charmed quark and its antiquark. Today there are ten known hadrons with this quark composition, and more are sure to be discovered. They are shown in the "energy-level diagram" below.

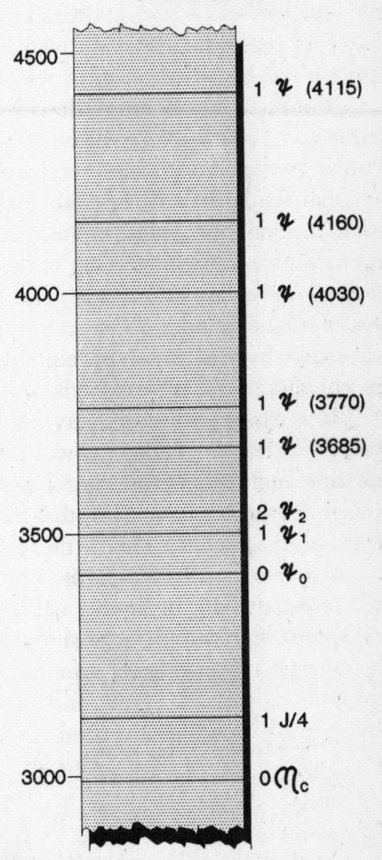

The mass scale, in MeV, is shown at the left. The spins of the various states and their names are shown at the right. All states of charmonium are unstable because the constituent charmed quark will ultimately annihilate the charmed antiquark.

haven result. Big deal! Does anyone remember who was the second person to scale Mount Everest?

After all the craziness associated with charm production was done, at collision energies beyond 5 GeV, R would be seen to settle down cozily to the value of 3.33, just as it should do in the standard model with color, asymptotic freedom and charm. These were not options on the standard model; they came with the ticket price. So what that the electron was not sometimes a hadron? So what that the theorists (some of them) had been proven to be right after all? Burt Richter had earned his share of the Nobel Prize.

Within a month after the discovery of the J/psi and its excited sister, psi′, theorists at Harvard, Cornell and Princeton recognized that the system consisting of a charmed quark bound to its antiquark, or what we called *charmonium*, should exist in a wide variety of configurations, or quantum-mechanical energy levels. It was like an atom, but on a grander scale of energy. We predicted where many of these new states should lie, and how they could be found at electron-positron colliders. Our hypothesis about the nature of the J/psi was rich in experimental consequences. In other words, the idea could be put to a decisive experimental test.

In a paper entitled "Spectroscopy of the New Mesons," Appelquist, De Rújula, Politzer and I described the expected pattern of the first eight energy states of charmonium, seven of which have subsequently been discovered at SPEAR or at a competing machine in Hamburg named DORIS. The theory worked like a charm, which is, of course, exactly what it is!

A decade later, in October 1985, hundreds of physicists gathered in Copenhagen to celebrate the centennary of Niels Bohr's birth. I was one of the speakers:

> How pleased I am—having spent twenty-seven months of the last twenty-seven years in Denmark—to return after a lamentable lapse of twenty-one years! And how honored to speak at this celebration of the genius of Niels Bohr! What a marvelous century it has been, and how wonderful Copenhagen truly is!
>
> The tantalizing discovery of the Balmer series [a mysterious regularity in the spectrum of hydrogen that Bohr was to explain] took place a year before Bohr's birth, 101 years ago. Last year, the Nobel Prize was awarded to an Italian and a Dutchman for their discovery of the intermediate bosons W and Z—the penultimate predictions of today's the-

> ory. It was a fitting climax to the century of quantum mechanics, the century we now celebrate as Niels Bohr's. . . .
>
> Bohr's modest domain of atomic sizes has been expanded and expounded by the arrogant reductionism of today's physical scientists to cover the whole shebang. Mysteries still confront us at the largest distance scales . . . but we have made remarkable progress at the smaller scales. We have what appears to be a correct, complete and consistent theory which describes all the known phenomena of the microworld: quantum chromodynamics and the electroweak theory.
>
> On a vacation upon the lovely isle of Jamaica, I discovered that Jamaicans know but one variety of cheese. It is used on pizzas, in sandwiches and in omelets. It is called standard cheese. So it is in particle physics. We have only one theory that works, and a very good theory it is. Quantum chromodynamics and the electroweak theory comprise our standard model of particle physics. It is used in cosmology, astrophysics, nuclear physics and particle physics. There is really no choice. . . .

I was reminding my audience of the heroic developments of the 1970s, when the pieces of the puzzle started fitting together and our standard theory was born. Decades of chaos had ended, and elementary-particle physics had become a mature discipline.

The year 1975 was to be another banner one for me, and not solely because of the birth of my darling daughter Rebecca. It was to be a year of furious collaboration with De Rújula, Georgi and others at Harvard. There was hardly ever a quiet moment in the halls of Lyman Laboratory. A coherent theory of elementary-particle physics was finally emerging, and everyone at Harvard wanted to get in on the show.

For the first time ever, the fundamental laws governing the microworld began to appear. Just as the 1920s was the decade in which the atom was finally understood, the 1970s was the decade in which the secrets of the subnuclear world were revealed. More importantly, they were revealed to us and not to others. Only a few other theoretical groups bought the new synthesis hook, line and sinker. Andrei Sakharov and his collaborators in Moscow took the bait; so did the Cornell group and, reluctantly, even our friends at Princeton. Most others were baffled by the discoveries and proved it with their senseless publications. The SLAC theory group even went so far as to send out a pamphlet with perforated margins, so that incorrect hypotheses could be discarded readily. Soon enough, there was nothing left but the binding, and the sounds of laughter from Harvard.

I remember sitting in my office early in the new year, chatting with Howard Georgi. It had to be early in the day, since Howard usually departs in midafternoon. He lives far enough away from the city to provide for the family horse, and must get home at a decent hour to tend the children, clean the stable, prepare the dinner, act as deacon of his church, and do whatever other chores may be necessary. His wife, Ann, is a biologist at Harvard.

Anyway, I suggested that if quantum chromodynamics is really correct, it really ought to explain the values of the masses of the observed hadrons. The baryons of lowest mass were eight in number. The nucleons contained no strange quarks, the sigma and lambda particles each contained one strange quark, and the xi particles contained two. The naïve quark model explained why the sigma and lambda were somewhat heavier than the nucleons, and the xi particles heavier yet. This was because the strange quark was heavier than the commonplace up and down quarks. But why on earth were the sigma particles about 75 MeV heavier than the lambda particle? Both kinds of particles had the same quark composition. Could QCD resolve the puzzle? We realized that it had to do with the spin dependence of the color force between the quarks, and that the nature of the spin dependence was dictated by QCD. But did the numbers work out?

Howard took the problem home and solved it after children, horse and wife were properly dealt with.

The next morning we found that we were dealing with a theory that *worked*. Qualitatively and quantitatively, the color force of QCD explains the sigma-lambda mass splitting, and much more, as well. With Alvaro De Rújula, we began to write what was to become one of my favorite papers: one of the first to point out that a systematic theory of the strong, weak and electromagnetic interaction had emerged at last. QCD and its asymptotically free color force held the quarks together; it is responsible for the strong force. The electroweak theory, with charm included so it could deal with quarks, explained the remaining forces.

We called it the standard model because it was the unique candidate for a theory of the microworld, and hence it was a standard against which any heterodox theory could be judged. Up till now, no rival has survived to take its place. The standard model has matured into the standard theory.

Our paper began:

> Once upon a time, there was a controversy in particle physics. There were physicists who denied the existence of structures more elementary than hadrons, and searched for a self-consistent interpretation wherein all hadron states, stable or resonant, were equally elementary. Others, appalled by the teeming democracy of hadrons, insisted upon the existence of a small number of fundamental constituents and a simple underlying force law. In terms of these more fundamental things, hadron spectroscopy should be qualitatively described and essentially understood just as are atomic and nuclear physics.
>
> Many recent theoretical and experimental developments seem to confirm the latter philosophy, and lead towards a unique, unified, and remarkably simple and successful view of particle physics. All hadrons are built up of a quark and antiquark or of three quarks properly chosen from among the twelve quark species and all interactions (but not yet gravity) arise from (unified?) renormalizable gauge couplings.
>
> This point of view is very tightly constrained if it is both to agree with experiment and to be theoretically consistent. In this paper, we explore the origins of the hadron mass spectrum from this standpoint. Not only do we find that the observed hadron mass splittings can be understood, but we uncover correct new relations among them. We also consider hadron states containing one or more charmed quarks in order to describe the newly observed narrow boson resonances and the necessarily (we believe) soon-to-be-discovered charmed hadrons.
>
> Let us review briefly those diverse strands of thought we now recognize to be converging. . . .

Our paper, "Hadron Masses in a Gauge Theory," contained no fundamental new principles nor daring hypotheses. All we did was to show how well the standard theory worked for particles already known and to predict (correctly!) the detailed properties of particles not yet found. Our calculations were based upon a naïve and unjustified picture of hadrons as collections of slowly moving quarks that are held together by the color force. Why this simple picture works so well is a mystery to this day. A rigorous computation of hadron masses in QCD can, in principle, be carried out with a sufficiently powerful supercomputer, or with a special purpose computer designed and built to study hadrodynamics. The first promising steps in this direction have been taken, but in large measure, computer simulation of hadron physics is a program for the future. Meanwhile, we are stuck with a theory whose naïve interpretation works better than it should. Things could be worse.

The system consisting of a charmed quark bound to a charmed antiquark, a.k.a. charmonium, is a rich laboratory in which the behavior of the quark-antiquark color force can be studied. Charmonium is a massive structure with a rest energy of 3.1 GeV. Because of asymptotic freedom, the strong (color) force between its quarks is much weaker than it is in lighter hadrons. For this reason, the charmonium system can be mathematically analyzed and its properties can be deduced from theory.

The constituents of the atom—electrons and nuclei—behave like particles in their own right, as do photons that are both the mediators of atomic forces and the particles of light. The constituents of nuclei—neutrons and protons—and the glue that binds them—pions and other mesons—also can be seen as isolated and observable particles.

This is simply not true for the quark constituents of the nucleons and mesons. Quarks cannot be isolated, nor can the gluons that bind them. The asymptotically free color force does not fall off with the square of the distance between two quarks. Unlike the force of gravity or electricity, it remains relatively strong even at large quark separation. This makes it impossible to isolate and study a single quark. Therefore quarks must be studied *in situ*, and the evidence for their reality, while compelling, is necessarily indirect. The charmonium system cannot be shaken apart into its constituent quarks, nor can it be induced to emit a gluon. Yet its behavior, as studied at electron-positron colliders, is just as is expected from its hypothetical quark substructure. The discovery of the J/psi, and the consequent development of the new discipline of charmonium spectroscopy, is what convinced the physicist on the street that quarks are not mere mathematical figments. They are real!

Although quarks and gluons cannot ever be seen as isolated particles, their spoor is evident in energetic collisions. At energies above the charmonium region, the electron-positron collisions often result in the production of two oppositely moving narrow jets of hadrons.

As interpreted by the standard theory, the electron and positron first annihilate into a virtual photon. The virtual photon then creates a pair of quarks moving off in opposing directions. As the quarks separate from one another, they are restrained by a string of color force. The string breaks, with a quark and antiquark produced at the

QUARK JETS

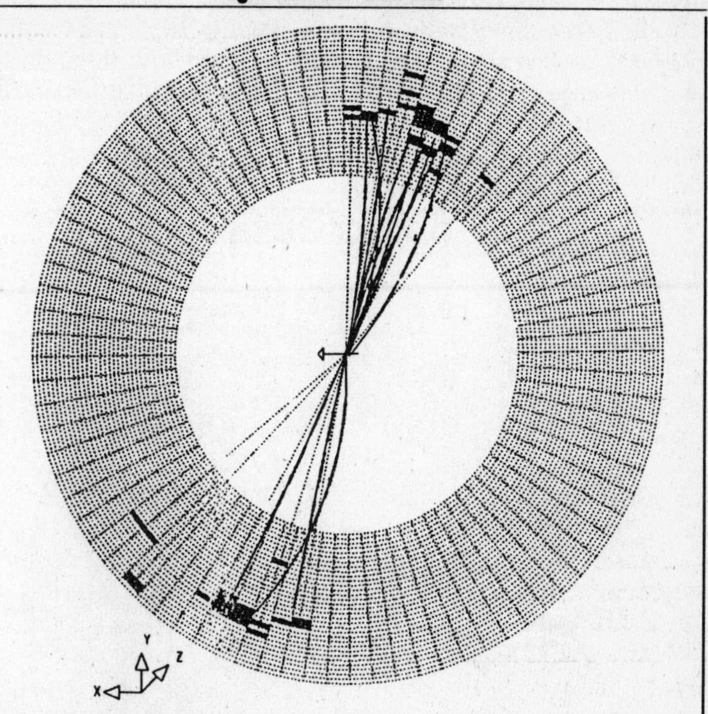

A 15-GeV positron collides with an oppositely charged electron. These particles annihilate and a quark-antiquark pair is created. In the process of hadronization, two jets of particles are produced. Solid lines are charged particles that leave visible tracks in the inner section of the JADE detector. Dotted lines are the trajectories of neutral particles that reveal themselves as they pass through the outer part of the detector. (Photo courtesy of Paul Söding, Deutsches Elektronen-Synchrotron)

loose ends. If the process ended here, a pair of mesons would emerge. In general, however, the violence of the collision leads to numerous other breaks in the string, and to the production of many quark-antiquark pairs. These aggregate into hadrons, as they must, since isolated quarks are *verboten*. The result is the production of two hadronic jets of particles, which are the observable relics of the original quarks.

Can there be relics of gluons as well? Sometimes the electron-positron collision produces a pair of quarks and a high-energy gluon as well. These three particles then engage in the "hadronization" process described above. The result is an event with three distinct jets. This manifestation of gluons was first observed at the PETRA collider in Hamburg in 1981.

GLUON JETS

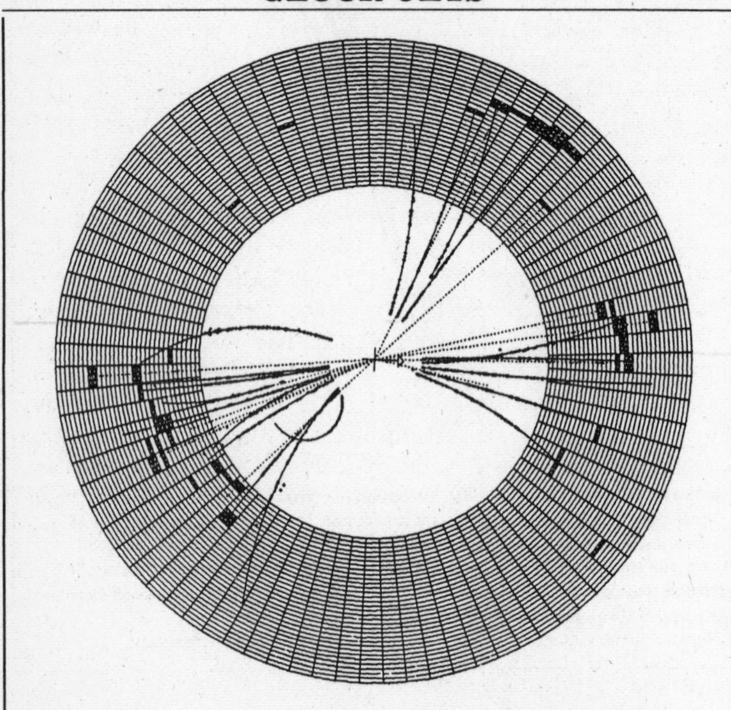

This picture shows three clearly defined hadron jets corresponding to the creation of a quark, antiquark and a gluon in electron-positron annihilation. (Photo courtesy of Paul Söding, Deutsches Elektronen-Synchrotron)

Although charmonium is made up of a charmed quark and a charmed antiquark, it is not itself a charmed particle. The charm of the quark cancels the opposite charm of the antiquark, so that charmonium has vanishing net charm. This is why it can decay rapidly, via the strong force, into ordinary particles.

Imagine, on the other hand, a particle containing one charmed quark and an ordinary antiquark. This is an example of a particle bearing "naked charm." Such a charmed meson should exist, and its discovery would represent ultimate confirmation of the idea of charm. It was called the D particle, and De Rújula and I predicted its mass would be 1860 MeV. Its strange counterpart, the F particle, a charmed quark bound to a strange antiquark, we expected to be found at 1975 MeV. (I don't want to spoil the story, but eventually, a year and a half later for the D and still later for the F, these particles were discovered. Their masses were found to be 1865 and 1971 MeV, respectively, within a fraction of a percent of our predicted value. These were just two of the many quantitative successes of the new theory.)

Over many exquisite Chinese dinners provided by Sam Ting, he became convinced that our interpretation of the J/psi particle was the likeliest explanation. He and his group set about valiantly to find our charmed mesons at Brookhaven. He failed, but only because the old AGS was just too small a machine. Its protons were just not hot enough to produce a pair of charmed particles. The AGS neutrinos were another story. Neutrinos, you see, can produce charmed particles one at a time. Nick Samios and his group, in the course of examining a million or more bubble chamber photos of AGS-produced neutrino events, found one extraordinarily interesting event. Apparently a neutrino-proton collision yielded the production of a charmed baryon and a muon. The mass of the new particle, and its decay properties, were just as we had anticipated. He knew that he had found a charmed particle and so did we, but only a few others believed it. After all, it was only one solitary event.

SPEAR, on the other hand, should have been producing hundreds of charmed particles if the theory were correct. It was almost as if the Stanford experimenters were pursuing an unconscious urge to frustrate us theorists. They even went so far as to publish papers proudly crowing of their failure to find our putative particles.

In April 1976, a year and a half after the November Revolution, charm had still evaded detection at SPEAR. I attended a conference in Madison, Wisconsin, where experimenter after experimenter related his failure to find evidence for naked charm. It was getting embarrassing—for them, not for me! I knew that the particles were

THE FIRST CHARMED BARYON

This curious event was observed in 1975 in the seven-foot bubble chamber at Brookhaven National Laboratory. The interpretation of the photograph is shown at right. A neutrino collides with a proton to form a charmed baryon and a muon. In a time too short to produce a track, the charmed baryon decays into a lambda particle and four pions. The subsequent decay of the lambda particle is clearly visible. (Photo courtesy of N. P. Samios, Brookhaven National Laboratory)

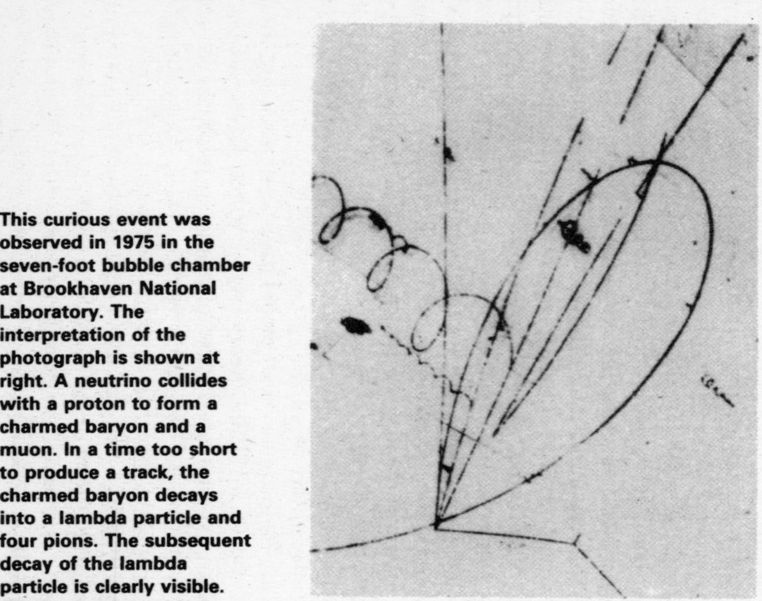

Discovery of the First Charmed Baryon

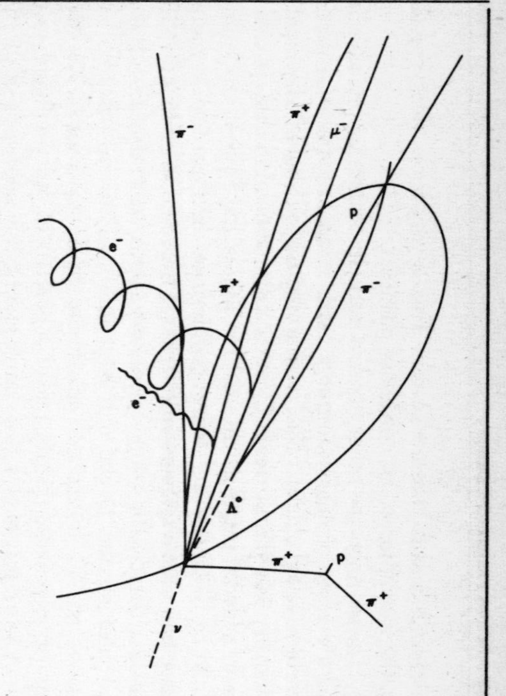

there. I collared my old friend from Berkeley, Gerson Goldhaber, on the flight to Chicago. I made it clear to him that things were getting out of hand, and that he had best find my particles. Apparently, he took my advice to heart and organized a more effective hunt. According to the standard model, charmed mesons should decay rapidly into kaons and pions. It was essential to identify the decay products as accurately as possible.

The power of positive thinking was demonstrated a few weeks later when he telephoned me to say that charmed particles at last had been found. They were there all the time and they did exactly what we had said they would. Although I had promised Gerson to restrain myself, I somehow managed to leak the story to *The New York Times*. After all, charm was my baby, and it had finally emerged after long and arduous labor.

Soon after charmonium had been discovered in 1974, Alvaro and I published a paper, "Is Bound Charm Found?" presenting our case that the J/psi particle was made of charm-anticharm. Goldhaber's announcement led to our sequel, "Is Charm Found?" The behavior of the newly discovered particle clearly indicated that it contained a new and heavy quark. More than that, the decay properties of the new particles were being measured and found to be just as qualitatively predicted in my early works in collaboration with Bjorken, Iliopoulos and Maiani, and in perfect agreement with my subsequent work with Alvaro and Howard.

For example, the D^+ was seen to decay into a negative kaon and pions, but never into a positive kaon. Thus the new quark was not just any old quark, but it was precisely my charmed quark. There was no doubt about it. Naked charm had been found, and just in time for my nation's Bicentennial. Gerson, incidentally, went on to become California's Scientist of the Year. He deserved much more.

Soon after the discovery of charmonium, and while we were all waiting for the discovery of naked charm, I discovered an amusing apparent symmetry of nature. The four known leptons and the four known quarks (including charm) could be organized into a pattern reminiscent of the periodic table of the elements.

The columns designate electric charge rather than chemical valence, as in Mendeleev's table. The rows (of which there are only two) designate replicated families of fundamental fermions. The first row I called the relevant family. These are the particles that are evidently essential to the operation of Spaceship Earth. Up and down

THE PERIODIC TABLE OF QUARKS AND LEPTONS IN 1974

Electric Charge ➧	−1	−⅓	0	+⅔
The Relevant Family ➧	electron	down quark	electron neutrino	up quark
The Irrelevant Family ➧	Muon	strange quark	muon neutrino	charm quark

The discovery of the J/psi in 1974 filled the space at the lower right corner of the table. This new particle was made up of charmed quarks, whose existence had been predicted a decade earlier.

quarks make up the atomic nuclei, which, together with electrons, are the constituents of the atoms of all earthly matter. The electron neutrinos are important to us, too. Without them primordial hydrogen could not have been transmuted into the variety of chemical elements necessary to life, nor could our ultimate energy source, the sun, burn its fuel. Without neutrinos our lovely planet would probably never have come into existence.

My periodic table had a hole in its second row corresponding to the yet-unseen charmed quark. It was again reminiscent of Mendeleev's table. His table had holes in it as well, places corresponding to the then-unknown elements scandium, gallium and germanium, whose properties he was able to predict in detail prior to their discovery.

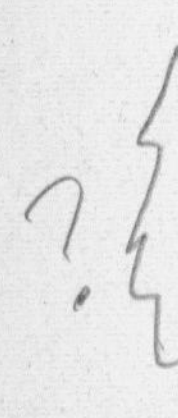

So it was with charm, all over again. The second row of my table consisted of the irrelevant particles: muons, muon neutrinos, strange and charmed quarks. None of these quarks or leptons seemed to play any essential or even useful role at all in the everyday world. They could just as well not have existed, or so it seemed—and seems still. Why there is a second family of quarks and leptons remains an Unsolved Problem of Nature. There seemed to be complete symmetry between the relevant particles and the irrelevant, between the essential and the useless, between the building blocks of the tangible world and the particles whose only purpose was to justify my salary and those of my colleagues.

It soon turned out that there was no such symmetry at all. A third family of particles was awaiting discovery. There had to be at least twice as many irrelevant particles as relevant ones in our bestiary.

Before we begin the Tale of the Tau Lepton and the Third Family, let me digress upon the meaning of irrelevance in science. What is irrelevant, exotic and eclectic today often becomes central, critical and even practical tomorrow. The phenomenon of radioactivity, when it was discovered in 1896 by Henri Becquerel, could not have seemed less relevant to earthly affairs, whereas today it is the subject of mass paranoia. A tiny amount of naturally occurring matter is intrinsically radioactive and it can, under certain circumstances, present a hazard to human health. But naturally occurring radioactivity is not all bad. Without it, Columbus would probably never have discovered America!

Natural radioactivity is a principal source of internal heating in the body of the earth. If the whole earth were a nuclear-free zone, then its molten iron core would have solidified eons ago. This would have extinguished the dynamo deep underground that generates the earth's magnetic field. Without a magnetic field, the Chinese could never have invented the magnetic compass, and without a compass to guide him it is unlikely that Columbus would have set off on his voyage of discovery.

This wouldn't have done the American Indians any good because the magnetic field of the earth plays an important role in shielding us from damaging cosmic rays. It is nature's own Strategic Defense Initiative, which renders those missiles from deep space impotent and obsolete. Without radioactivity, neither Columbus, nor the Indians, nor us, nor any other recognizable forms of life might have emerged from the primordial slime. Take that, Ms. Fonda!

The irrelevant may ultimately reveal its relevance, but it can take a long time. Radio waves were first produced and observed by Heinrich Hertz in 1887, but the first commercial radio station did not start broadcasting until 1920 (November 2, KDKA, Pittsburgh, Pennsylvania). On the other hand, it took only seven years between the discovery of nuclear fission in 1938 and its "practical" application to the destruction of Hiroshima and Nagasaki. The existence of neutrinos was first suggested in 1930, but they were not seen in the laboratory until 1956. Still, they have no practical use. However, I did write a serious paper with Alvaro and others in 1983 pointing out how an intense neutrino beam could be used to explore the interior of the earth. Such "whole earth photography" could be used to discover underground deposits of oil and gas. Perhaps, when we really do begin to run out of fossil fuels, our idea will be implemented and neutrinos will achieve technological relevance.

Meanwhile, the discovery of muons is fifty years old and we have known about kaons for forty years. Why are there such particles? They seem to be entirely without purpose, neither to humans nor, so far as we can see, to the Creator. Nonetheless, almost every particle physicist is convinced that every particle in nature must play an essential—if unknown—role. Our faith in simplicity dictates that our universe is unique, that the fundamental particles and forces simply must be as they are and cannot be otherwise. God could not have done without muons, but so far, only he or she knows why.

Martin L. Perl, one of the more brilliant and stubborn members of the SLAC experimental team, is the black sheep of his scientific family, being the student of one Nobel Prize winner, Isidor Rabi, and the thesis advisor of another, Sam Ting. Perhaps Perl was in the market for a great discovery to perfect the family tree. In any case, he was responsible for the discovery of the first known member of the Third Family of Fundamental Fermions, the tau lepton.

Among the many events seen at the SPEAR collider, a very few involved the inexplicable production of an electron, a muon and at least one unobserved neutral particle. These funny events appeared only when the collider was operated at collision energies beyond 4 GeV. Correctly, and contrary to the opinion of many of his staid colleagues, Perl recognized these events to be the signature of the production of a new kind of particle, one that was entirely unanticipated by the theorists.

In a first tentative paper written in August 1975, in which he headed a list of thirty-five coauthors (experimental teams have grown *large*), Perl wrote:

> We conclude that [these sixty-four events] cannot be explained either by the production and decay of any presently known particles or as coming from any of the well-understood interactions. . . . A possible

explanation for these events is the production and decay of a pair of new particles, each having a mass in the range [1600 to 2000 MeV].

In a brilliant series of subsequent experiments, Perl developed incontrovertible evidence that he had discovered a third charged lepton, a particle much like the electron and the heavier muon, but heavier yet. Whereas the electron weighs about half an MeV, and the muon about 100 MeV, Perl's tau lepton has now been measured to have a mass of 1784 MeV, right in the middle of the range quoted in his discovery paper. Its lifetime was measured to be a third of a millionth of a microsecond, a value that is in precise agreement with the predictions of the standard theory, if the new lepton represents the first known entry in a third row of our table. But this interpretation implied that two heavy quarks were to be found, quarks that were dubbed "top" and "bottom," or sometimes, "truth" and "beauty."

Why have we been taking the table of quarks and leptons so seriously? Why does the existence of a third charged lepton with canonical weak interaction properties compel the existence of two additional quarks with prescribed electric charges and weak properties?

It is not simply a question of completing a pretty pattern. John Iliopoulos, with two of his French colleagues, demonstrated in 1972 that quarks and leptons are linked to one another by requirements of internal consistency. Quarks and leptons simply must appear in complete families if the theory is to remain renormalizable and hence sensible. This linkage between disparate particle species suggests a more fundamental connection between weak and strong forces. It was the first clue that a synthesis of strong, weak and electromagnetic interactions was possible. It was a key step on the road to the grand unified theory, a subject to which we shall return shortly.

Leon Max Lederman is one of the superstars who made Columbia University a world center of high-energy physics. He was among the discoverers of the muon neutrino in 1961 and of parity violation in muon decay in 1957. More recently, he has become director of Fermilab and has managed the construction of its Tevatron proton-antiproton collider, the world's largest accelerator.

In his spare time he continues to pursue forefront research. In the early seventies he observed a curious "shoulder" in the yield of energetic muon pairs from proton collisions at Fermilab. In retrospect, this result is now explained in terms of the production of J/psi particles, but at the time it was just another curious effect. Because his experiment did not have sufficient precision, he had missed a great discovery. In 1976 he reported another apparent bump corresponding to a new particle with a mass of 6 GeV. He called his new particle the upsilon, but this time he had jumped the gun. His more careful experiments showed that the bump was not really there. It had been a statistical fluctuation, physics jargon for bad luck. His "particle" became known as the "Oops, Leon!"

Finally, in July 1977, Leon found the bump that was. He and his group reported on their "Observation of a Dimuon Resonance at 9.5 GeV in 400 GeV Proton-Nucleus Collisions." They had uncovered the first evidence for the existence of the fifth, bottom or beauty, quark. Too bad he didn't find it a year earlier. It could have been the Bicentennial quark and the discovery of the top quark might have had to await our Tricentennial. The bump was named, again, the upsilon particle. In a subsequent experiment, Leon was able to discern three discrete but closely spaced resonances. A rich and complex new system was discovered: bottomonium, a state consisting of a bottom quark bound to its antiquark. Like charmonium, it should display a variety of observable excitations. (Eleven states of bottomonium are now known.) This time Lederman had hit pay-dirt. It was probably the first time that a laboratory director had made such a microworld-shaking discovery. But Leon is no mere manager.

For every hit there are usually several near misses. Several months before the announced discovery of the "real" upsilon, my old friend Roy Weinstein reported another. He was involved in an experiment at Fermilab that suggested, but did not prove, the existence of a bump at 9.5 GeV. It was the upsilon, but the proof was lacking. Roy phoned me to let me know about his tantalizing, but not compelling, data. He knew that Lederman's experiment was more sensitive than his, and that if the particle were really there (as it was) the credit for its discovery would deservedly be Leon's. Roy had caught the first glimpse of the upsilon, but had no decisive proof. He just wanted me to know, even though he felt his data were too weak to publish. There are several roads to greatness. A lesser physicist would have published anyway.

Charmonium had been discovered simultaneously at a proton synchrotron and at an electron-positron collider. On the other hand, there were no rival claims to the upsilon. It was the one great discovery, among many important but less spectacular ones, made so far at the giant Fermilab synchrotron. But the best instrument to study bottomonium is an electron-positron collider. Where were the electron-positron experimentalists? SPEAR was too small an accelerator; it simply could not get to the 10 GeV energy region. California's next collider was too big: PEP was designed to operate in the 30 GeV region. The story was very similar in Germany: their big collider (PETRA) was too big and the little one (DORIS) was too small.

Goldilocks's choice, the machine that was just right to study the upsilon particle, was at that time being completed at Cornell University. The Cornell Electron Synchrotron Ring, known as CESR and pronounced Caesar, was designed to operate at collision energies of 10 to 15 GeV. The construction of the facility was never formally approved by the federal government. Nonetheless, it was being built by hook and by crook with funds that had been intended for other purposes.

Few physicists realized that such a medium-sized accelerator could do anything very interesting. However, the Cornell team had built the ideal machine. Shortly after the discovery of the upsilon particle at Fermilab, it became clear that CESR was just the instrument to unravel the mysteries of the bottom quark. Its potential dispelled any questions concerning the propriety of its funding. CESR became the pride and joy of its funding agency, the National Science Foundation.

I spoke at the dedication of my alma mater's new toy, praising the virtues of moderation. Soon after the machine was turned on, in 1980, CESR was to discover the existence of particles containing a single bottom quark: bare bottom, or naked beauty, take your choice. No longer was there any doubt about the interpretation of Lederman's upsilon particle. Two members of the third family had been found. That leaves the top quark and the tau neutrino. The latter particle has not been seen directly as yet, but indirect evidence for its existence is all but compelling. We remain, at the moment, entirely topless. All we know is that the toponium particle, if it exists at all (and it surely does, according to the standard theory) must be heavier than about 50 GeV.

The fundamental fermions comprise at least three families in our latter-day periodic table. But are there more than three? It seems that each family includes one massless, or almost massless, neutrino.

THE VIRTUE OF MODERATION

The low-energy particle accelerators have become fat and smug from their rewarding studies of charmonium. The high-energy machines are lean and starved because they can find nothing more wholesome than gluons. The medium-energy machine at Cornell is about to eat up all the goodies associated with the bottom quark. Alvaro De Rújula prepared this drawing for me to present at the dedication of the CESR collider.

THE PERIODIC TABLE OF QUARKS AND LEPTONS TODAY

electron	down quark	electron neutrino	up quark
muon	strange quark	muon neutrino	charm quark
tau lepton	bottom quark	tau neutrino	top quark
sigma lepton	high quark	sigma neutrino	low quark

Room is left for a possible fourth family of fundamental fermions. Theorists have even assigned tentative names to the undiscovered quarks and leptons. Shading shows particles that have been decisively demonstrated to exist.

All we need to do, to determine the number of fermion families, is to count the number of different kinds of neutrinos. It is an experiment that can and will be done with the next generation of electron-positron machines that are now being built. Soon, we will know.

The appearance of a third family of quarks and leptons was quite a surprise. However, it did have one redeeming virtue, noticed as early as 1972 by two young Japanese physicists, Makato Kobayashi and Toshihide Maskawa. With only two families of quarks it was difficult to accommodate the observed violation of CP invariance. Indeed, it was simply impossible to accomplish in the standard theory. CP invariance is a symmetry principle closely related to time-reversal symmetry, whose breakdown was discovered by Fitch and Cronin in 1964. The Japanese physicists realized that only with three or more fermion families could there be a natural explanation of the failure of CP symmetry. For their discovery, Kobayashi and Maskawa shared the first Sakurai Memorial Prize in 1984, an award created by John's widow upon the death of her husband.

Interestingly, Kobayashi and Maskawa may have hit upon the underlying raison d'être for the existence of three families, the teleological relevance of muons and tau leptons. It has to do with the manifest asymmetry between matter and antimatter in our universe.

Our solar system is built out of nucleons and electrons, not antinucleons and positrons. The same is true of all the hundred billion stars in our galaxy, the Milky Way. Furthermore, it is thought that all the hundred billion galaxies that constitute the observable universe are made of matter and not antimatter. Otherwise, collisions between matter galaxies and antimatter galaxies resulting in particle-antiparticle (or, star-antistar) annihilation would have produced characteristic astrophysical catastrophes of a sort that have never been seen. Thus, our universe is full of matter but virtually devoid of antimatter. Why should this be? The laws of the microworld are completely symmetrical between matter and antimatter. Why does the universe not respect this symmetry?

One possibility is that the matter dominance of the universe was just an accident of birth, but physicists don't much like accidents. Andrei Sakharov, in the Soviet Union, suggested in 1967 a scenario by which the matter asymmetry of the universe may have evolved. It depended on three properties of nature: the creation of the universe in a hot big bang, the violation of CP invariance, and the existence of a new force in nature that permitted the creation or destruction of nucleons.

The first two assertions had just been established in 1964. The observation of cosmic black body radiation at Princeton was incontrovertible evidence for the big bang. This feeble but universal radiation could be interpreted only as the last faint remnant of the titanic explosion that was the birth of the universe. In the same year, a group of particle physicists, also from Princeton, were the first to detect certain "forbidden" decay modes of kaons that could take place only if there were a miniscule violation of CP symmetry built into the laws of nature.

Sakharov's third postulate is more speculative. It requires that all matter is radioactive, although only very slightly. It turned out, in 1973, that the instability of matter is a necessary consequence of any attempt to forge the strong and weak forces into a truly unified theory. Thus, grand unification realized Sakharov's dream of a natural origin to the matter asymmetry of the universe.

There was a time, long, long ago, when the visible universe was very hot and so small that it would fit on a pinhead. The intense heat generated enormous numbers of particles and antiparticles of all species. Matter and antimatter were then on exactly the same footing. There was a great deal of matter in the universe, and there was an exactly equal amount of antimatter. For every quark there was an antiquark. Then the Sakharov mechanism came into play to generate a tiny asymmetry between matter and antimatter. Among every billion particles were one or two excess quarks. As the universe cooled, every antiquark found a quark to annihilate with, leaving only a few remaining quarks. The survivors combined in threes to form nucleons, the stuff of which our universe is mostly made.

None of this could have happened without an intrinsic violation of CP symmetry, and such a property of the theory seems to depend upon the existence of at least three fermion families. Here is an important and fundamental role for strangeness and charm, top and bottom, muons and tau leptons. Without them, there would have been an equal amount of matter and antimatter in the universe, which would have annihilated one another long ago. There would have been no earth, no sun, no stars in the heavens. Thank you, charm, and thank your friends!

In October 1976, Sam Ting and Burt Richter were delighted to learn that they had been awarded the Nobel Prize for their discovery of the J/psi particle. Although I agreed that they richly deserved this recognition, I began to suffer from the dread disease of Nobelitis. Soon after the positron was discovered, Dirac got the prize for predicting it, and Anderson, for finding it. Soon after the pion was discovered, Yukawa got the prize for predicting it, and Powell for finding it. I had predicted the existence of the charmed quark and, thereby, dozens of new particles. They were being discovered left and right. Richter and Ting had found the first of the new particles without having the foggiest notion of what they had done. Since they had gotten the prize, shouldn't I consider myself next in line? Mine was a classic case.

In the fall of 1977, Sam asked me casually if I had heard anything about my impending award. Needless to say, my symptoms were exacerbated, extending to total insomnia on the night before the prizes were to be announced. Early in the morning I learned that one of the prizes had landed close, but not close enough. My distinguished colleague at Harvard, John Van Vleck, was to share the prize for his work on magnetism. It was about time, for he was then seventy-eight. I could wait.

Our dedicated librarian keeps careful watch on her turf. Only the most brazen student "borrows" a book when Nina is on guard. Under the most special circumstances, such as when one of her professors wins a Nobel Prize, she lets us have a party in her library.

Hundreds of physicists came to honor our beloved "Van." Among them was my friend Asim Yildiz, who had been one of Schwinger's last thesis students at Harvard. Asim has earned doctorates in both physics and engineering, and now teaches at Boston University. He was one of my first converts to charm and an active participant in the November Revolution.

At Van's party, Asim presciently assured me that my turn for the prize would come soon. Since that time we and our families have become very close. Once a member of the Turkish tennis team, Asim has been our devoted tennis instructor. However, of all of Asim's tennis protégés, Julian Schwinger has proven to be the most promising pupil.

Meanwhile, the Harvard group was preoccupied by doubts about the validity of the standard theory. There were two apparent experimental discrepancies with the theory. Carlo Rubbia had been speaking for several years about what he called "the high-y anomaly." His data departed significantly from theoretical expectations and seemed to call for the existence of additional quarks with nonstandard properties. Moreover, the electroweak theory predicted that there should be a measurable violation of space-reflection symmetry (or parity) in the interaction between electrons and nuclei. Two experimental groups, one at Oxford University and one at the University of Washington, had carefully studied the properties of the bismuth atom and had found no evidence at all for the predicted effect. We wrote several long and contrived papers seeking alternative theories that could explain these contrary data. In one fifty-five-page paper, Alvaro, Howard and I wrote:

> The standard model is not only elegant and simple but has also been incredibly successful in predicting several of the important experimental discoveries of the past few years (neutral currents, charmonium, charmed particles). More recently, however, experiment has jumped ahead of theory. Candidates for the new leptons are accumulating at a fast pace. [The properties of the tau lepton had not yet been established. The upsilon particle had not yet been found. The notion of a Third Family was still very speculative.] Experiments on parity-violating effects in bismuth atoms seem to rule out the standard model. Anomalies in antineutrino scattering off nuclei hint at the existence of new quarks [with nonstandard weak interactions]. All these are indications that Nature may have chosen a richer structure than the standard model. . . .

Nonsense! Once again, we were misled by false experiments and flawed interpretations. Nature chose simplicity, not artifice. Our ambitious, ingenious and baroque "Theory of Flavor Mixing" was to join the scrap heap of discarded theories soon after its publication. Our theory was discarded because it was never needed in the first place.

Careful work at CERN, particularly by Jack Steinberger and his group, showed that there really wasn't any high-y anomaly at all. The standard theory worked just fine. Similarly for the bismuth problem. The interpretation of data turned out to be not as simple as it was first thought. Moreover, the Novosibirsk group reported positive evidence for parity violation in bismuth, in total disagreement with the English and American results. It was all up in the air again until 1978, when a decisive experiment was performed by the high-energy physicists working at the giant SLAC linear electron accelerator.

They obtained a source of polarized electrons by shining polarized light on a crystal of gallium arsenide. The polarized electrons, all spinning in the same direction, were injected into the machine, accelerated to tremendous energies, and made to collide with a simple nuclear target. Careful count was kept of the number of electrons that were deflected downward by four degrees. According to the standard theory, the number should change by about twenty-five parts per million, depending upon whether the electron spins pointed upward or down. This is exactly what they saw. The prediction of the standard theory had been confirmed.

With the demolition of the high-y anomaly, the observations of atomic parity violation at SLAC, and the discovery of the upsilon (confirming the notion of a third fermion family to accommodate the tau lepton), there were no remaining experimental difficulties with the standard theory. QCD and the electroweak theory were triumphant.

My old friend Steve Weinberg, as a key contributor to the development of the standard theory, gave the summary talk on weak interactions at the big biennial gathering of high-energy physicists, whose 1978 meeting took place in Tokyo. I did not attend, much preferring to spend the summer with my family in Aspen. I was amazed and disconcerted to see the transcript of Steve's talk. There were fully sixty references, but not one to any paper I had written. Why had my high school chum perpetrated such a revisionist history of the discipline in which I thought I had played something of a role?

In September 1978 I was invited to participate in a small meeting on quarks and leptons held by the Swedish Academy of Sciences. Steve (presumably also a victim of Nobelitis) seemed a bit distraught by my trip to the Nobel heartland. The night before I was to give my talk, as I was about to go to bed in Stockholm, Steve phoned long-distance from Harvard.

I had briefed him beforehand on what I would say in Sweden, but he decided that I should change the content of my lecture. In particular, he disapproved of my intention to describe logical alternatives to the standard theory. In a sense, he was right, since the only alternatives I could think of were as ugly as could be. Yet I felt the compulsion to remain open-minded. After all, the W and Z particles had not yet been discovered, and there was still room for surprises. Steve was adamant. I was stubborn. I was also sleepy, jet-lagged and somewhat drunk. After an hour of his interminable tirade I hung up on him. He didn't speak a word to me for the next year, even though we were then colleagues at Harvard and had been friends for three decades.

Steve and I would eventually reestablish a cordial relationship. Reconciliation depended upon two happenings: our sharing the Nobel Prize and his departure for Texas.

While I was in Stockholm, an elderly Swedish physicist famous for his theoretical analysis of X-ray scattering, Professor Ivar Waller, sought me out for an apparently innocent chat. (The reader should note my symptoms of Paranoia Nobelus.) He seemed to be more familiar with my early papers on the electroweak theory than I was. He asked me whether the angle I had introduced in 1961 was quite the same as what had become known as the Weinberg angle. I responded that my angle might have differed somewhat in its definition from Steve's. He assured me that our angles were, in fact, quite identical, mine preceding Steve's by some six years.

I began to feel that there really was a chance to win the Big One, but it wasn't entirely clear for what. Charm seemed a likely possibility. There had been many collaborators, including Bjorken, Iliopoulos, Maiani, De Rújula and Georgi, but I had been the common thread in works spanning more than a decade. For me to win the prize for my contributions to the electroweak theory seemed a more remote possibility. There had been significant contributions from Salam, John Ward, Steve, Gerhard 'tHooft and Tini Veltman as well. And Schwinger had prophesied the electroweak unification.

The Nobel Prize cannot be shared by more than three persons, and we were at least seven! Yet Waller's interrogation suggested that the electroweak theory was being considered for a prize.

Time would tell.

The night before the announcement of the 1978 Nobel Prize in physics, I took a sleeping pill. I was awakened abruptly by the telephone at some unearthly hour. I stumbled to the phone, to discover that it was only Mugy, who had dialed our number by mistake after one of her late nights.

Mugemana Musebwase, our Tutsi from Zaire, came to us, *au pair*, in February 1977 to help care for our four children. She lived with us for four years, quickly becoming one of the family. She is married today with two lovely children, to whom we are the closest things to grandparents within six thousand miles. She doesn't even sound Swedish! A few hours later I heard on the radio of the award of that year's Nobel Prize to the Russian Peter Kapitza (for the physics of exceedingly low temperatures) and to Arno Penzias and Robert W. Wilson of Princeton (for their discovery of the last whimper of the big bang).

Strike two!

During July 1979 I took Joan and the older boys to Corsica, where I gave a series of lectures about the standard theory at an annual summer school located on the beach near the town of Cargèse. We all had a wonderful time tanning, swimming and failing to learn to windsurf. Somehow the four of us and our baggage managed to fit into a rented Renault "Deux Chevaux." It was in this overgrown toy that Joan finally learned to drive a standard shift car. For the rest of the summer we all visited Brookhaven National Laboratory, on Long Island, where it rained every day. We ate a lot of mussels, played a lot of bridge, and promised ourselves never to spend another summer on Long Island, a promise that, incidentally (and happily), we did not keep. Inevitably, we returned to Brookline and that dreaded time of year came again. This time it was to be different.

12 The Nobel Prize

THE TELEPHONE RANG VERY EARLY on the morning of October 15, 1979. Joan stumbled to it first.

"It's for you, honey," she said.

Nervously, I picked up the receiver. Someone wanted to know if I would let the listeners of radio station WEEI know just how I felt about winning the Nobel Prize.

Thinking, Could this be true? I mumbled, "I feel sleepy!" and hung up, only to have it instantly ring again.

Another disembodied voice asked what I would do with all the money I had allegedly won. "Paint the house, of course!" It was starting to look like the real thing!

Meanwhile, Joan had scampered off to wake the kids, and perhaps the neighbors as well, screaming, "Pop did it! Pop did it!"

She rushed back to help me dress as I attended to the insatiable telephone. Four small children, still half asleep, came into our bedroom to find out just what their Pop had done now. The reporters would soon be at the door. Abdus Salam, Steven Weinberg and I had won the 1979 Nobel Prize in physics for our "contribution to the theory of the unified weak and electromagnetic interaction between elementary particles, including inter alia the prediction of the weak neutral current," in the language of the official citation.

A month earlier, Paul Martin, then chairman of our physics department, was alerted to the possibility of another Harvard prize in physics. A member of the Nobel committee had written to him, "I have a small practical problem and hope that you can help me. Do you think it would be possible to get photographs . . . of your two colleagues, Sheldon Glashow and Steven Weinberg? It is absolutely necessary that they get no idea that someone wants their photos. . . ."

Paul sent the pictures and kept the secret, but wisely laid in a generous supply of bubbly for the physics department.

When my family and I arrived at Harvard on the day of the Nobel announcement there was a grand celebration in the physics library, including hundreds of well-wishers, students and passersby. There were almost as many bottles of fine French champagne. After all, two years earlier only one Harvard physicist had won the prize. This year there were two. Even Harvard's dean, the great Henry Rosovsky, attended the party. Our president, Derek Bok, was too busy, but he did send a bottle of even finer champagne.

On the spur of the moment, Joan's sister Sharon Kleitman organized a party that evening at her home in Newton. Again, she proved herself to be the hostess with the mostest. Hundreds of my friends and even a few physicists were invited. The food was superb, and so was the champagne. We left the party with many gifts, all of them bottles of champagne. Our friends the Levins presented me with two magnums of vintage Mumms. What else can one give to a Nobel Prize winner? Clearly the Nobel Prize is good for the champagne business.

Almost at once, the telegrams began to arrive:

"BY THREE DID YOU DIVIDE AND CONQUER. CONGRATULATIONS. JULIAN [SCHWINGER]."

"OK. NOW I BELIEVE YOUR THEORY. THEY MADE THE PERFECT CHOICE THIS YEAR. RELAX AND HAVE A GOOD TIME IN STOCKHOLM. RICHARD FEYNMAN."

"CONGRATULATIONS. HAVE BEEN EXPECTING THIS SINCE OUR CONVERSATION IN PARIS 20 YEARS AGO. COME SEE US NEXT YEAR. MURRAY GELL-MANN."

"CONGRATULATIONS, SHELLY. NOW YOU CAN FINALLY PAY FOR THE SHIRT. LEON N. COOPER."

And many more. And telephone calls, and letters from those I had forgotten and those I never knew, and invitations to sign petitions, to speak at luncheons, to receive honors, to join societies, to testify before Congress, to support or oppose nuclear power, to consider new religions, to double my money, or to have my house painted cheap.

Abdus, Steve and I did not fail to congratulate one another. When

THE THREE PHYSICS LAUREATES OF 1979

Salam at the right and Weinberg in the middle rest uncertainly upon unverified predictions of the electroweak theory. I carry a bag of charm with one foot on the firm foundation of the J/psi particle and its kin. Our success is based upon the experimental observation of neutral currents, and we point with pride at the electroweak angle. This caricature was drawn by my colleague Alvaro De Rújula.

the media announced that Salam was proud to be the first Muslim to win the Nobel Prize, I responded by telegram: "Congratulations, Abdus. Didn't know that Sadat had converted."

A word about Harvard and its Nobel Prizes. Since World War II, America's share of the Nobels has risen from half to about two-thirds, and my university has fared remarkably well. On the occasion of Harvard's 350th birthday, we had captured fully thirty prizes, including eight awards in physics. There have been two champagne parties for laureates of our physics department since 1979. Our nearest competitor is Cambridge University, with an overall accumulation of eighteen prizes, six in physics. No other university in the world has more than about a dozen laureates, or more than six in physics.

Steve and I had to evade our Harvard responsibilities for two weeks to attend the Nobel festivities and receive our prizes. Harvard takes its teaching responsibilities seriously. It should, since parents pay about $300 for each hour their brat spends with its professors. Each of us had to write to our dean, Henry Rosovsky, asking for his permission to be absent from Cambridge during term time. Once Henry was assured that our teaching duties would be attended to in our absence, he granted our requests.

Of course, we all needed new wardrobes. Joan bought two formal gowns: a black velvet Oscar de la Renta with shoulder bows and a black and purple Calvin Klein creation. The children were outfitted more modestly at the boutiques of Harvard Square, and my rabbi kindly directed me to a shop in the shabby part of town offering custom-made clothing at half the price. Winning the Nobel Prize led us, for the first time, to exceed the credit allowances of our plastic cards. Ultimately, the Nobel Foundation would rectify the cash flow problem.

December 5: We flew to Sweden on my forty-seventh birthday, Joan and I, Mugemana Musebwase, Jason, Jordan, Brian and Rebecca. Seven people, seven checked bags and seven carryons. Twenty-one things that I would repeatedly count. The Nobel Foundation provides each winner with a first-class round-trip ticket to Stockholm, but I had traded mine in as partial payment for all our fares. We traveled steerage class via SAS from New York to Stockholm.

As an exceptional privilege, granted after we had pledged our children to silence and decorum, we were admitted to the SAS first-class lounge at Kennedy to await our flight. There we encountered Sir Anthony Lewis, Theodore W. Schultz and their families. They were to share the 1979 economics prize, which is not strictly a Nobel Prize. The five real prizes, for physics, chemistry, medicine and physiology, literature and peace, were inaugurated in 1901 under the terms of Nobel's will. The prize for economics in memory of Alfred Nobel was created and funded by the Bank of Sweden long afterward, in 1969. The Swedes still debate the wisdom of such an innovation.

Be that as it may, our flight was mercifully uneventful, the food tolerable, the movie not. SAS showed its mettle by serving a birthday cake in my honor. As we were all about to try to sleep, the delectable Danish stewardess came by with the usual request to place our dirty glasses on her tray. Brian, a trusting but myopic child, offered her his spectacles.

December 6: At the Stockholm airport we were introduced to a number of bigwigs, and to two of the most important people in our tale: our chauffeur Lars and my (female!) attaché Linda. Lasse (Lars) Juhlin and his Mercedes limo were at our service at all times during our week in Sweden, with additional cars and drivers available when needed. During the day Lars entertained our small children while we were at play, taking them to parks, museums and restaurants. Late at night he and his car were always at hand, one sober and the other freshly polished, to bring the dissolute parents home. Of all the perquisites of the rich, this is the one we envy most. We would have loved to bring Lars home with us, but alas, it was not possible. We have met him several times since, and today he is the Stockholm driver for the president of Volvo, lucky men both. These days, he drives Volvos.

Linda Steneberg, graduate of the Stockholm School of Economics and second secretary at the Ministry for Foreign Affairs, was chosen by the Nobel Foundation to be my charming, beautiful and efficient personal attendant for the Nobel Week. It was one of the first times the position was offered to a woman. She was to see to it that we were all in the right places at the right times, in the right dress and in the right mood. The seven of us, and our baggage, were set down in three luxurious rooms facing the sea at the Grand Hotel of Stockholm, the official residence for all the winners and their entourage.

We unpacked our belongings and took a brief nap before attending a champagne get-together with the other laureates and their families. This was followed by a soiree in honor of Glenn T. Seaborg, sponsored by General Motors and Science Service, held at Stockholm's famous Opera Cellar restaurant. GM, you see, invites the top winners of the International Science and Engineering Fair for high school students to the festivities with "Seaborg Nobel Prize Visit Awards." Glenn is a chemistry laureate of Swedish ancestry with a strong interest in secondary-school science education.

Also invited were the two top winners of the Westinghouse Science Talent Search and its director and driving force, Dorothy Schriver. I had first encountered Mrs. Schriver four decades earlier when I myself had been an STS finalist! Acting as hosts for the evening were Willard Cheek and his enchanting wife, Rosie. Willard was in charge of GM's educational program and had somehow leased a vintage Cadillac for the festivities. We all feasted on Swedish caviar, reindeer steak and cloudberries.

December 7: Three evenings were to be strictly formal, white tie and tails, so I began the morning by renting an ill-fitting penguin suit for a shade under $100. For the entire Nobel Week the weather was crisp and clear, with the sun so low in the sky that cars had to keep their headlights on throughout the day. Dark shadows enhanced the colors and lent an unearthly aspect to the quaint old streets and forbidding castles of Stockholm.

Even passersby seemed caught up in the passion and mystery of the Nobel pageant, the climax of the Swedish social season. After a luncheon snack I went off with my fellow laureates to a press conference at the Royal Academy of Sciences in suburban Frescati. "How will your work affect the future of mankind?" and all that.

Meanwhile, the rest of our contingent had arrived: my two older brothers, Sam the dentist and Jules the doctor, and their wives, Audrey and Judy; Joan's sister, Sharon, and her husband, Danny Kleitman; and Joan's widowed stepfather, Eddie, Harvard Class of 1925. Now we were fourteen. Sam and Jules had rented their tails in the States for a third of the price I paid. They were always smarter than me.

In the evening Joan and I returned to the Academy for a reception and dinner with the members of the committees who had been responsible for selecting this year's laureates. It was too late for them to change their minds, so we had a relaxed evening in anticipation of the festivities to come.

December 8: Each of the Nobel laureates is required to present a public lecture about his work. The written versions are collected as a book and are reprinted in many professional journals. I began my talk:

> In 1956, when I began doing theoretical physics, the study of elementary particles was like a patchwork quilt. Electrodynamics, weak interactions, and strong interactions were clearly separate disciplines, separately taught and separately studied. There was no coherent theory that described them all. Developments such as the observation of parity violation, the success of quantum electrodynamics, the discovery of hadron resonances, and the appearance of strangeness were well-defined parts of the picture, but they could not be easily fitted together.
>
> Things have changed. Today we have what has been called a "standard theory" of elementary particle physics in which strong, weak, and

electromagnetic interactions all arise from a local symmetry principle. It is, in a sense, a complete and apparently correct theory, offering a quantitative description of all particle phenomena and precise quantitative predictions in many instances. There is no experimental data that contradicts the theory. In principle, if not yet in practice, all experimental data can be expressed in terms of a small number of "fundamental" masses and coupling constants. The theory we now have is an integral work of art: the patchwork quilt has become a tapestry.

Tapestries are made by many artisans working together. The contributions of separate workers cannot be discerned in the completed work, and the loose and false threads have been covered over. So it is in our picture of particle physics. Part of the picture is the unification of weak and electromagnetic interactions and the prediction of neutral currents, now being celebrated by the award of the Nobel Prize. Another part concerns the reasoned evolution of the quark hypothesis from mere whimsy to established dogma. Yet another is the development of quantum chromodynamics into a plausible, powerful, and predictive theory of strong interactions. All is woven together in the tapestry: one part makes little sense without the other. Even the development of the electroweak theory was not as simple and straightforward as it might have been. It did not arise full blown in the mind of one physicist, or even of three. It, too, is the result of the collective endeavor of many scientists, both experimenters and theorists.

At the conclusion of the Nobel lectures, the families of the three Jewish laureates, Herbert Brown, the chemist from Purdue, Steve and I, attended a special service in our honor at the Stockholm synagogue. We were told that this first visit of Jewish Nobelists to the synagogue was the only time ever that the Orthodox rabbi of Stockholm ventured into the Conservative temple. Joan, along with the other women, was seated separately in accordance with the conservative traditions of Swedish Jewry.

At the end of the service the sexes were reunited at a reception where cookies and cakes were served. The worshippers wolfed down their goodies and rushed to us for our autographs. I must have signed hundreds of slightly soiled paper plates. It was flattering, but there was no food left for the laureates once the congregation had done with us. Jews are justly proud of the fact that a third of all the science laureates are Jewish, and 1979 was a better than average year.

In the afternoon, Joan drank *glugg* at a party for the laureates' wives. The only woman laureate that year was Mother Teresa, who probably doesn't drink, and anway, she would be receiving her prize in Oslo.

That evening, my extended family, along with my attaché Linda and her Spanish fiancé, sixteen of us, enjoyed a quiet and private dinner at an informal restaurant.

December 9: Sunday, a day of rest. The four American winners and their families were offered a turkey luncheon at the residence of the American ambassador to Sweden, His Excellency Rodney Kennedy-Minot, and treated to a recital of lovely Swedish folk music. In the evening we all attended a reception at the house of the Nobel Foundation. At the party, Salam introduced me to the Gothenburg Imam of the Ahmadiyya Movement of Islam. Salam is a member of this Muslim sect, which acknowledges the nineteenth-century prophet Ahmad.

Ahmad had written that "there is nothing more foolish than to think that the questions of heavenly or earthly bodies are limited to those which have so far been discovered through astronomy or physics." There is no better credo for a modern physicist. The Ahmadiyya Movement is open-ended and encouraging to scientific inquiry, unlike mainstream Islam, for which the Koran is the very last word, or the Creationist fringe of Christianity.

Taking our leave, we went to bed early. The big event was tomorrow: the presentation of the awards and the great banquet that traditionally takes place upon the anniversary of Alfred Nobel's death.

December 10: The ceremony is intricate and requires rehearsal. We would enter the stage of Stockholm Symphony Hall after the audience, including their royal highnesses, were seated. We had to learn where our seats were. The king would present each laureate with a surprisingly heavy leather-bound citation that included a unique original watercolor and a five-ounce gold medal in an elegant case. It would not do to drop them. Each of the ten of us was to arise on cue, approach the king, receive the award, shake the king's hand, bow to him and to the audience, then return to our place. As I would be the first to receive the prize, I learned my shtick, then played the role of king in the rehearsal. We were excused when the president of the Nobel Foundation was convinced that we would not embarrass him or spoil the fun.

To Alfred Nobel, and to me, physics is the most basic of the sciences. His will designates the immutable batting order for Nobel presentations: physics, chemistry, medicine, then literature. Economics was added later, but only to the very end of the list. Nobel gave the responsibility for the Peace Prize to the Norwegians since in Nobel's time the Swedes seemed too warlike. (Norway was Sweden's booty from its war with Denmark in 1814. It kept Norway under its thumb until 1905.)

That is why we physicists come first in the Nobel pecking order. We are the first to receive our prizes, our lectures are published at the beginning of the annual Nobel book, and we get the best seats at the formal dinners. Even within our group of three physics laureates, order is significant. Often the laureates for each of the prizes are specified in alphabetical order, although there are a dozen instances in which the physics laureates were carefully and deliberately not specified alphabetically.

Abdus, Steve and I were treated in precise alphabetical order. (That suited me: G comes first.) The wife of the first-mentioned physicist, in our case, Joan, is escorted down the grand staircase by his royal highness the king, to be the first couple seated at the central table. What happens to Nobel protocol, you may ask, when the first physics laureate is a woman? Madame Curie never found out, because she and her husband, Pierre, shared their prize with Becquerel, who came first both alphabetically and in the citation. Her second Nobel Prize, although unshared, was won for chemistry. Maria Goeppert-Mayer was alphabetically ahead of her two fellow physics laureates, but the Nobel committee had put Wigner unalphabetically in the first position. There have been no other women physics laureates, so the problem remains unresolved.

Now you will understand why, when I returned to the Grand Hotel after the rehearsal, I found my children playing tag in the corridors and Joan, shod in her highest heels, tirelessly practicing her entry down the stairway on the arm of Linda, her surrogate king. Tragedy struck as we were all dressing for the Event. Joan was stunning in her brand-new Oscar de la Renta, and the kids were more spiffy than ever before or since. But I discovered that I had neither cuff links nor shirt studs! With only minutes to spare, the situation was saved by Gösta Ekspong, who lent me his extra set. Ekspong is a leading Swedish high-energy physicist who was then a member of the select Nobel committee responsible for choosing the physics winners. He had a vested interest in seeing to it that my shirt was tucked in.

The Ceremony: Every seat was occupied in Stockholm Symphony Hall. Families and friends of the laureates had the choicest seats. Diplomats were there in force, some representing the laureates' countries (the States, Britain, Pakistan, Germany and Greece) and many others just there for the fun. Many seats were reserved for Swedish university students and for the members of the various Nobel selection committees. How the remainder of the seats were distributed is a mystery to me, but it seemed that every Swede would have been there if he or she could. The stage was bedecked by flowers contributed by the Italian city of San Remo, where Alfred Nobel had spent his last years. The members of the selection committees, the representatives of the Nobel Foundation, and any laureate of prior vintage who chose to attend were seated at the back of the stage. On stage right sat the king and his incredibly beautiful queen, and on the left were ten vacant chairs. At precisely four-thirty in the afternoon, marching to Purcell's trumpet voluntary, the ten little laureates took their places on the stage. The ceremony had begun.

The opening address was delivered by the chairman of the board of the Nobel Foundation. All the talks were in Swedish, but a booklet of English translations had been placed on each seat, even those of the laureates. It felt a bit like being in synagogue for the high holy days.

Professor Bergstrom explained the responsibility of the Swedish prize committees to select as laureates "those who shall have conferred the greatest benefit on mankind." Yet, he went on, "The Nobel Prizes in physics, chemistry and medicine have often been awarded for pioneering research achievements at the frontiers of our knowledge: enterprises, therefore, that usually can be understood and judged by small groups of specialists. This is quite in keeping with Nobel's wish that principal encouragement goes to basic and pioneering research—research whose results might subsequently lead to practical, significant developments of benefit to mankind." He concluded with an apt quotation from the works of the Greek poet Odysseus Elytis, the laureate in literature: "When people are united in collaboration, such potent and unexpected forces can arise that what had been regarded as immutable can, in fact, be changed. . . . The force of scientific development is so great that one must optimistically believe that the forces of good, too, must triumph in our problem-filled world."

The Stockholm Philharmonic then played the overture to Leonard Bernstein's *Candide*, after which our Swedish colleague Bengt

Nagel introduced the physics laureates and the significance of their accomplishments:

> . . . The importance of the [electroweak] theory is first of all intrascientific. . . . For our functioning as biological beings we rely on elements formed milliards of years ago in supernova explosions, with the new kind of weak force predicted by the theory contributing in an important way: really a fascinating connection between physics, astrophysics, and elementary particle physics.

I, then Abdus, then Steve arose on cue to accept our awards from the king. Our part in the extravaganza was done, but we still had chemistry, medicine, literature and economics to endure.

Sitting between Steve and me, Salam was a remarkable sight. He was dressed in an elaborately embroidered and gaily colored frock that he described as "Pakistani formal." Upon his head was a loosely wrapped and flowing white turban, and his feet were painfully ensnared in a pair of turned-up slippers straight out of the Arabian Nights. At the conclusion of the ceremony Abdus limped offstage to retrieve his walking shoes. I embraced my family, and we all set off in our limousine to dinner and dancing at Stockholm Town Hall.

There were 1191 dinner guests, of whom about a third were university students. The students were placed in the balcony. As the guests were searching the enormous "Blue Hall" for their assigned seats, we were following the orders of the Memorandum on Certain Procedures at the Official Nobel Events:

> The Laureates and their wives will assemble in the City Hall, where they will be introduced to Their Majesties the King and Queen and other members of the Royal Family, the Prime Minister and members of the Swedish Government. The Ambassadors of Great Britain, Greece, the Federal Republic of Germany, Norway, Pakistan, and the United States [and their wives] will be present as well as the Chairman and President of the Nobel Foundation.

Thus, drinking champagne upstairs, we missed the only untoward incident of the night, the expulsion of the Hari Krishnas who tried to crash the party. Continuing: "They will thereafter proceed to the 'Blue Hall' to the banquet with their partners at table."

The king, young and handsome, swept Joan to the head of the procession, assuring her that it would be a cinch. Abdus accompanied the wife of the Statsminister, Steve escorted Princess Desirée, and I received the arm of the Belgian wife of the Greek ambassador. Two by two, to a flourish of trumpets, and to the applause of the multitude, down the monumental marble staircase we went.

"At the beginning of the dinner the Chairman of the Board of the Nobel Foundation will propose a toast to H.M. the King. The King will respond by toasting to the memory of Alfred Nobel."

Flowers were everywhere, thanks again to the generosity of the city of San Remo. Throughout dinner, Joan seemed deeply involved in profound conversation with the king's uncle and with the king himself. Far, far away, I was seated between my Belgian companion and yet another princess. From one side I got the strong impression that being a Swedish princess is no picnic. From the other, I was told that a close acquaintance of Mme. Sekeris was the finest furrier in Stockholm, and that I simply had to seize the moment and buy my wife a new fur coat. "He'll make you a special deal," she said. Cheap at half the price, no doubt!

> When coffee has been served and when a Swedish choir, Collegium Musicum, has sung [its] songs, the Laureates are asked to address the audience [Salam carried the ball for us physicists]. . . . All speeches will be announced by the toastmaster, Dr. Wilhelm Oldelberg, after a trumpet call. . . .
>
> After dinner, university students will pay tribute to the Laureates at the balustrade of the "Blue Hall." A procession of students, preceded by their banners, will enter the hall [down the grand staircase], where the students' choir will sing some songs [like "Greensleeves" and "Oklahoma"] and the President of the Central Organizations of the Students' Union in Stockholm, Miss Agneta Bjorklund, will make a short speech to the Laureates. Sir Arthur Lewis will be asked to return the speech.
>
> The guests of honor will thereafter follow the King and the Queen to a room adjacent to the Golden Hall where refreshments will be served. The King and Queen will receive family members of the Laureates. . . . In an adjacent room, Swedish TV will make short interviews with Laureates and their family members. At the same time, dancing begins in the Golden Hall.

Except for the king and queen, who must never dance in public.

Toward one in the morning, those laureates and their wives who were still fit and willing were whisked off to the Stockholm School of Economics for the students' own Nobel Fete. The traditional Swedish

MENU

Suprêmes de Turbot Froid aux Oeufs d'Ablette
Sauce Hollandaise

Selle de Veau Rôti
Sauce aux Morilles
Pommes Sautées
Salade de Haricots Verts

Parfait Glace Nobel
Petits Fours

VINS

G. H. Mumm, Cordon Rouge, Brut
Château Trinité-Valrose 1970
Eau Minérale de Ramlose

CAFE

Baron Otard, Fine Champagne, V.S.O.P.
Liqueur Mandarine Napoléon

BUFFET

Bols Silver Top Dry Gin
Long John Whisky

"vickning" is a late-evening meal held after formal banquets. The students wore elegant caps with their tails, and we all sang Swedish student songs over schnapps, beer, herring and meatballs. Eventually, Steve and I presented a rather inept rendition of "Far Above Cayuga's Waters" and any other Cornell songs we could recall. Early in the morning, with a student cap that I had bought for Brian, Joan and I staggered home. Awaiting us in our room at the Grand Hotel were a dozen long-stemmed roses from my late friend and colleague, John Sakurai, and on our bed was a telegram from my old flame from Copenhagen: "Congratulations with love. Bonnie."

December 11: It was a short sleep. The Nobel Foundation, in an unintended act of sadism, arranged a 9:15 a.m. appointment for me at its offices so I could receive my award, a check for 266,666.65 Swedish crowns. Lars then drove me to Gotabanken, a Swedish bank, where Linda presented me to its president. In his plush office I transformed my award check into something more useful. I had most of the money sent, as dollars, to my Cambridge bank, ultimately to be put into trusts for the college education of my younger children. Some would indeed be spent to paint the house. A small amount found its way into the pocket of the friend of the Greek ambassador's wife. Sure enough, Joan had acquired a brand-new raccoon coat to replace the ancient mink she had inherited from my mother.

The Nobel Prize, amounting in my case to about $62,000, was not considered taxable income by the U.S. government. No country imposed a tax upon the award of this prize until the 1986 U.S. tax reform bill effectively closed this "loophole." Future American laureates must expect to share their winnings with Uncle Sam, unless they choose to defect. Contrariwise, Argentina offers an additional lifetime stipend to its Nobel laureates.

The physics laureates and their families spent the afternoon at the Student Union of the Royal Institute of Technology. We were asked to come for lunch and to rap with the students about physics and life. As they wrote in their letter of invitation, "We see this as an opportunity for you to meet tomorrow's generation of maybe Nobel Prize winners in your field of research." Maybe. It was, perhaps, the most scientifically rewarding day of Nobel Week. The students were almost as sharp as ours at Harvard.

It was a fairy tale night: we dined with the king and queen at their palace. It was a small dinner with only 140 guests: white tie and tails for me, Joan in her Calvin Klein. Joan's dinner partner was the Utrikesministern, whatever that might be, but mine was the queen. She is as charming and witty as she is beautiful, and with tongue in cheek she taught me the fine art of the Swedish *skol*: To drink, you see, you must either skol or be skolled. It is an intimate, transient and asymmetrical relationship between two persons. No third party may intervene. Eye contact must be maintained by both parties from initiation of the act until the sip is swallowed. And so forth. If the rules are faithfully followed, it takes a long time to get drunk, but it can be done.

MENU

Consommé Double au Fumet de Perdreau
Omble Chevalier aux Petits Légumes
Selle de Chevreuil Poivrade
Fromage Stilton
Mandarine Givrae

*

La Ina
Pommery Brut
Château Smith Haut Lafitte 1961
Quinta de Noval 1960 Vintage

Stig Ramel, an exceptionally dignified Swede who was the executive director of the Nobel Foundation, had the horrendous responsibility of organizing the festivities and avoiding incidents. It had all gone swimmingly. No laureate had been lost, mugged, kidnapped or arrested. None had dropped the medal, insulted the king, or kissed the queen. Stig had done his job well, and next year's ordeal was far away. When dinner was done and most of the guests had left, staid Stig bellowed in Nordic triumph and drunkenly slid all the way down the banister of the grand staircase of the Royal Palace.

December 12: The physics laureates were convinced to participate in a panel discussion at the Royal Institute of Technology concerning the future of our subdiscipline, elementary-particle physics. The main auditorium was overflowing with young students of physics and their not-so-young professors, seekers of our secret wisdom. We spoke of the future of particle physics, and I prophesied that the Z-boson would be discovered at the large electron-positron collider called LEP then being built at CERN. I was only half right. Four years later the Z-boson was indeed discovered at CERN, along with the W^+ and W^- bosons. However, LEP has not yet begun to work.

While I was partying my way through Nobel Week, my Harvard colleague Carlo Rubbia had been hard at work designing a quicker and cheaper way to find these particles. It worked, and that is why there was a sequel to our panel discussion five years later in which Carlo proudly participated.

Bengt Feldreich, the science editor of Swedish Television, had organized a round-table discussion for that afternoon, involving all of the science laureates. Here are the sorts of questions we were to address:

1. How do you identify creative young scientists?
2. Why are so many Nobel Prize winners Jewish?
3. Will technological innovation suffer now that research budgets are being cut?
4. What is scientific intuition?
5. What about nuclear power?
6. To Steve: Tell us about the first three minutes of the birth of the universe.
7. Should we fear the computer revolution?
8. What discoveries will win the Nobel Prize in the year 2000?
9. Can you explain the waves of antiscientific hysteria that are sweeping the world?
10. To Abdus: Does the Third World need, and can it afford, its own basic scientists?

These were tough questions, and they were seriously and thoughtfully addressed. The entire program, well over an hour in length, was telecast throughout Scandinavia. It simply couldn't happen here, where TV rarely tries to be serious. That evening Joan and I had cocktails at my attaché Linda's home. Afterward, she and her beau joined us at the opera house, where we all enjoyed a fine performance of Prokofiev's *Romeo and Juliet*, presented by the Swedish Royal Ballet Company.

December 13: Early in the morning, while we were still in bed, a bevy of lovely Swedish maidens dressed all in white with lit candles in their hair danced into our room to serenade us with traditional songs. It was the festival of Saint Lucia, commemorating (a bit prematurely) the return of light after winter's darkness. It was also the day that we would all visit the charming old town of Uppsala, a few hours' drive from Stockholm. The physics laureates would repeat their Nobel lectures for the students and faculty of the University of Uppsala.

More than half a millennium old, the university makes Harvard seem like a precocious child. The rector proudly showed us the prize possession of its world-famous library, the 1600-year-old Silver Bible, and the ancient amphitheater in which the great Carolus Linnaeus once studied and taught medicine.

We returned to Stockholm to dress for the formal Lucia Night Banquet organized by the students of Stockholm University. My dinner partner, in a formal white gown, was the president of the Student Union. She wore a many-colored sash across her chest, and her left hip was adorned by many "medals," some of them ingeniously fashioned from bottle caps and beer tabs. Her awards were not "real," in the sense that they had been presented to her by the king, but they were very real to her. They had been conferred upon her by various student groups in recognition of her outstanding academic and personal achievements.

Many of the students wore such decorations. They were simply emulating their elders. At the official functions of Nobel Week, only the lowly laureates were bare-breasted. The Swedish dignitaries proudly flaunted their plumage of genuine ribbons and medals of Royal Appointment. Medals were the subject of much of the dinner conversation. By a recent act of the socialist Parliament, the king had been stripped of his last vestige of power. No longer could he award medals at his discretion. My dinner partner had always dreamed of being married in a flowing white dress festooned with all the decorations she yearned to win. Her hopes were dashed, and she would have to go through life in her Coke caps.

After the dinner Steve and I were asked to jump about the stage in our tails in the manner of frogs. We did our best to oblige our hosts, and, panting and sweating, were duly dubbed "Knights of the Supreme Order of the Jumping and Smiling Frog," an order dating back to 1917. Of course, we were presented with the ribbon and medal appropriate to our new station. Thereupon the students performed their SPEX, which, we had been warned, was a "short, traditional, satirical and slightly confusing student farce in Swedish."

Dancing and drinking would go on till dawn, but without us. Our Nobel Week had come to an end all too soon. Early the next morning we would leave Stockholm with our wonderful memories. Joan and I had so much fun that we came back five years later when Carlo Rubbia and Simon van der Meer took the 1984 physics prize for their success in finding the W- and Z-bosons. Their discovery was the ultimate confirmation of the electroweak theory for which Abdus, Steve and I had been honored. For this reason, the three of us were invited to participate in Nobel Week 1984.

The 1979 and 1984 prizes showed a perfect symmetry: we had predicted the new particles and they had found them. We did it all over again: the ceremony, the banquet, the royal dinner and all the rest. Lars took leave from the Volvo Company to drive for us once more. Linda couldn't make it. She had married her Spanish fiancé, had a child, and was serving her country's government in Bangkok. (Her husband, who is in the Spanish Foreign Service, was also assigned to Thailand in a remarkable act of international cooperation.) Again there was a panel discussion at the Royal Academy of Sciences, where we were asked to respond to a recording of our own comments from five years before. It was even more enjoyable the second time around, as an alumnus rather than a newly anointed laureate. And I didn't have to worry that I might drop the medal.

December 14–18: Scandinavia. It was time to check out of the Grand Hotel. Mercifully, the Nobel Foundation picked up most of the tab. Lars and his colleague with the second car took us to the airport for our short flight to Gothenburg. Nobel Prize winners often travel through Scandinavia after the ceremonies to present their Nobel lectures at other universities. Besides Uppsala, which was more or less part of our official visit, I had been invited to Finland, Lund, Gothenburg, Oslo and Copenhagen.

Time was short. Christmas was coming, and a trip to CERN, the hub of European particle physics, was obligatory. I limited my Scandinavian tour to the universities of Gothenburg and Oslo.

Gothenburg is the home of the Volvo Company and Sweden's second city. I gave my lecture at the university, and we were feted once again. It was in the down-filled bed of our Gothenburg hotel room that Brian had his first attack of asthma. He has become an accomplished athlete, nonetheless, and he seems at last to be growing out of this mysterious and terrifying disease.

The next day we continued to Oslo to accept the hospitality of my friend Halstead Hogaasen and his university. Appropriately, we stayed at the Nobel Hotel. From the airport we were driven directly to Bigdø, a museum consisting of a collection of old Norwegian houses grouped together in a replica of an ancient town. It is in the spirit of Old Sturbridge Village in Massachusetts. The museum was officially closed for the winter, but it had been opened especially for our visit. With the museum director as our personal guide, and with little Rebecca riding high in the air on Halstead's powerful shoulders, we learned what life was like centuries ago as we tramped through the snow from house to church to workshop in the bitter cold of Norwegian winter.

At the end of the tour we were surprised and delighted to discover that a magnificent banquet had been prepared for us. A table for twelve had been set in the elegant dining room of the director's mansion. And what a table it was! Among the candles, flowers and Christmas decorations, our hosts had provided an incredible variety of traditional Norwegian specialties: caviars, smoked reindeer, several kinds of smoked fish, exotic vegetable breads and surely the strangest cheeses in the world. For those of us not driving, there was plenty of beer and schnapps. Driving under any perceptible influence of alcohol is forbidden, and Norwegian law is strict and sure.

During the weekend we toured by day and were graciously entertained at night. One evening all seven of us were invited to dinner along with a number of graduate students. Afterward, we danced till early in the morning and had a wonderful opportunity to do our laundry in our host's washing machine. The Vigeland Museum, on the outskirts of Oslo, is a sprawling park full of the Norwegian sculptor's pleasantly grotesque and slightly obscene images of humanity. We visited the Edvard Munch Museum and had lunch at the home of the son of Norway's second most famous painter, Hendrik Sørensen. I gave my canonical lecture at the University of Oslo, and on Tuesday morning our Scandinavian holiday came to an end.

December 18–23, Switzerland: The seven of us were picked up at the Geneva airport by a CERN van and delivered to Alvaro De Rújula's ancient rented mansion in the nearby small French town of Gex. We physics laureates three gave our lectures at CERN and Natasha, Alvaro's charming French girlfriend, organized a magnificent fete chez De Rújula. Everyone drank and everyone drove. This was France.

On Friday, as we were checking in at the airport, a party of Middle Easterners with an almost infinite amount of luggage pushed ahead of us, checked their many bags, and stubbornly refused to pay the overweight baggage charges. I got angrier and angrier as it seemed we would miss our flight, but we made it in the nick of time.

Our last stop was Zurich, where we would spend the weekend with our Swiss friends, the Osterwalders. Konrad had been a junior faculty member at Harvard until he was offered his dream position, tenure at his alma mater in Zurich. Our families are close, and we stayed at their spacious home in the countryside. We were shown the bomb shelter in the basement that every Swiss home must have, by law. Theirs doubled as a wine cellar. Vrini, Konrad's wonderful wife, gave us one last European party. As a very special surprise, another old Harvard colleague, Robert Schrader, came all the way from Berlin with his wife just to be with us at this time.

We returned to Brookline via Swissair on Sunday afternoon, just in time to attend our next-door neighbors', the Padulos, family Christmas party. We drank eggnog instead of champagne and sang Christmas carols. For once I was not the center of attention.

Enough of this celebrity life. There were still a few particle mysteries to solve, and I hungered to get back to work.

Beyond the Standard Theory

QUANTUM CHROMODYNAMICS AND the electroweak theory form what is called the standard theory. It is a gauge theory of all the elementary-particle forces: strong, weak and electromagnetic. The fundamental building blocks are of two kinds: matter particles (fermions) and force particles (bosons). All of the wondrous phenomena of the microworld may be expressed in terms of the complex interplay of these particles.

THE BIG PICTURE

All natural phenomena result from the interactions among the fundamental particles. The forces to which these particles are subject can be described by the Fundamental Act of Becoming:

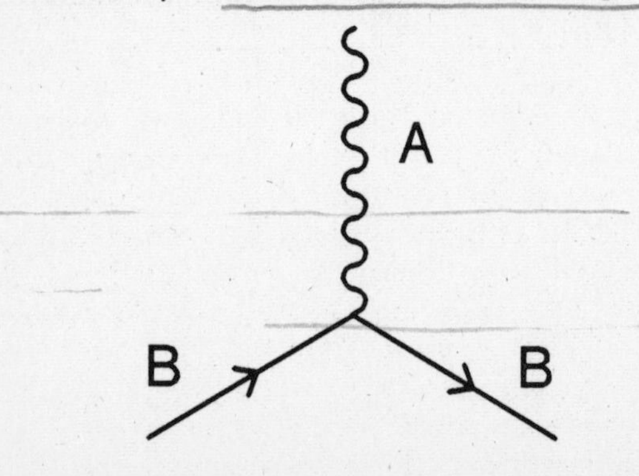

In this primitive Feynman diagram, a particle of type B (quark, lepton . . .) emits or absorbs a particle of type A (photon, gluon . . .). The identity of the emitted or absorbed particle (A) determines the nature of the force:

Photons mediate the electromagnetic force;

Gluons mediate the strong or color force;

W-bosons mediate the charged-current weak force;

Z-bosons mediate the neutral-current weak force;

Higgs bosons mediate an additional force whose direct effects have not yet been seen in the laboratory. An indirect effect of the Higgs boson is to give each particle its observed mass.

Gravitons are the quantized carriers of the familiar force of gravity. The effects of the gravitational force upon the behavior of individual particles are so small as to be entirely negligible.

The table below shows which of the elementary particles are subject to the various forces.

THE BIG PICTURE

B \ A	Photon	Gluon	$W^{\pm}$	Z^0	Higgs
Quark	E&M	Strong	Weak	Weak	Higgs
Lepton	E&M	—	Weak	Weak	Higgs
Gluon	—	Strong	—	—	—
$W^{\pm}$	E&M	—	—	Weak	Higgs
Z^0	—	—	—	—	Higgs
Higgs	—	—	—	—	Higgs

The columns of the table signify the emitted or absorbed particle A. The rows signify the particle B doing the emitting or absorbing. For example, any charged particle can emit or absorb a photon. Thus the quarks, the charged leptons and the W-bosons are active participants in the electromagnetic force. On the other hand, only particles with color are subject to the strong force. Quarks can emit or absorb gluons, as can gluons themselves.

Actual physical processes, such as take place in the laboratory, are described by more complex Feynman diagrams involving iterations of the fundamental diagram. Shown below is the mechanism underlying the decay of a neutral kaon into a pion, a positron and a neutrino:

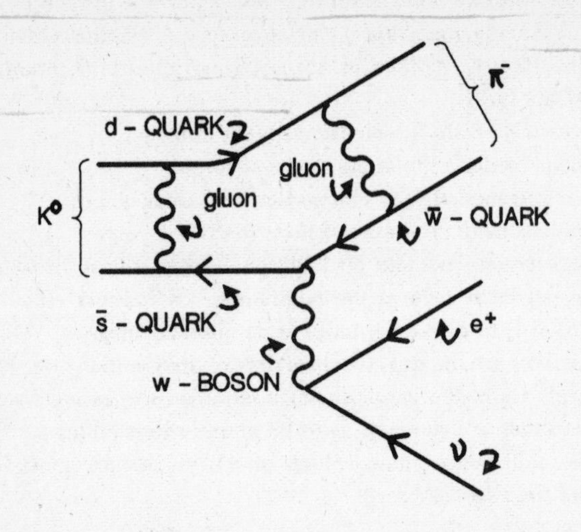

The kaon is composed of a down quark and a strange quark held together by the exchange of gluons. The strange antiquark emits a virtual W-boson and becomes an up antiquark. The W-boson, in its turn, becomes a positron and a neutrino. Notice that this diagram contains within it six fundamental acts of becoming.

The matter particles form at least three "families." Each fermion family consists of two quarks (each in three colors) and two leptons (one electrically charged, the other a neutrino). The families are identical to one another except for the masses of their members. Almost all earthly or celestial phenomena involve only the first and lightest fermion family.

The force particles consist of photons, gluons, the three recently discovered intermediate vector bosons of the weak force (Z^0, W^+ and W^-), and the Higgs boson.

Photons are responsible for binding electrons to oppositely charged nuclei to form electrically neutral atoms. A pale residue of the electromagnetic force acts between atoms to produce the chemical forces that hold together molecules and large aggregates of molecules such as you and I.

The eight colored gluons mediating the strong force are directly responsible for binding colored quarks together to form hadrons. Once again, a remnant of the strong force acts between color-neutral hadrons to produce the "nuclear force" that holds neutrons and protons together to form atomic nuclei. Gluons and photons are what hold things together: all the way from quarks to hadrons to nuclei to atoms to molecules and to mice and men.

The weak force is an agency for change. It allows neutrons to change into protons in the process of beta radioactivity. In the sun, it is the mechanism underlying the transmutation of protons into neutrons whereby more complex nuclei are built up and solar energy is generated. Because of the weak force, ordinary matter is made of up and down quarks and electrons rather than their more exotic cousins. These heavier particles decayed long ago into the mundane material that, by accident of birth, is the lightest of the matter particles. The particles that constitute the tangible universe are the runts of the litter.

The Higgs boson is the only force particle that has not yet been found by experimenters. Its role is different from the other force particles. The Higgs boson provides the mechanism by which the matter particles acquire their masses.

How many particles are there, all told, in our picture of the microworld? Let us not take color into account (so the gluons count only as one) and regard particle and antiparticle together as one. There are six quarks, six leptons, gluons, photons, Ws and Zs. Last of all, the Higgs boson brings our grand total to seventeen. (By another rational counting scheme, we could have gotten fifty-seven, as many particles as Heinz has pickles!)

By any count, the number is embarrassingly large. To be logically complete, we should also have included the mediating particle of gravity, the graviton, even though the quantum aspect of gravity cannot ever be demonstrated in the laboratory (so far as we can see) since the gravitational force is just too weak. We left it out because the standard theory does not attempt to deal with gravity. We can get by with this omission since gravity is without perceptible effect in the workings of the microworld. This is not, however, a good excuse for a theory that pretends to be complete.

The masses of some of the seventeen basic particles are prescribed in the simplest no-frills standard theory: gluons, photons and all three neutrino flavors should be massless. The values of the remaining twelve masses are not specified by the theory. They are what they are and we know not why. The top quark and the Higgs boson are the only two of the seventeen particles for which there is not yet any shred of experimental evidence or proof of existence. All we know for sure from experiment is that the top quark weighs at least twenty-five times more than the proton. All that our wonderful theory tells us is that it must weigh less than a few hundred times the proton mass.

On the other hand, the ratio of mass of the muon to that of the electron is measured to be 206.768, but so far as the theory goes it could have been just about any number at all. Experimenters can and will determine this manifestly significant number to greater and greater precision, but we have no theoretical prediction to compare it with.

In atomic physics, the situation is very different. Consider, for example, the energy levels of the hydrogen atom. They can be calculated quantum-mechanically to an extraordinary precision. All of them may be expressed in terms of the electric charge of the electron and its mass. In more complex atoms, the calculation becomes more difficult. The iron atom has twenty-six electrons whirling about it. Its energy levels are computable in principle, but in practice only a very limited precision can be attained.

For quarks and leptons, it's not that the calculation of a particle's mass is technically difficult, but that there is no calculation to do. Our theory is manifestly incomplete—it simply does not provide any mathematical procedure for determining a priori what the quark and lepton masses should be.

Including the mass ratios, the coupling constants and four numbers that describe how the weak force plays among the three fermion families, we come up with seventeen (again!) for the number of arbitrary, undetermined parameters in the theory. Each of them, like the muon-electron mass ratio, can be measured as well as you want (in principle) in the laboratory, but is not determined by the standard theory.

You may regard these seventeen arbitrary parameters as the control knobs of a Universal Television Set on which is playing The Greatest Show of the Universe, starring you, me and everything else. One knob adjusts the brightness, another the contrast, the tint, the volume, the focus, etc. Each adjustment yields a slightly different picture or sound. The question is: Why are the knobs adjusted just as they are? If the charmed quark were twice as heavy as it is, our view of the universe would be essentially unchanged. Did the Creator just set the knobs at random, or is there method in his madness? A theory more powerful than our present standard model would have fewer knobs, perhaps none at all. Ours would then be the only (and perforce, the best) of all possible worlds. There may not be such a theory, but we're sure as hell going to search for it!

Thus far, I have mentioned three of the profound questions to which the standard theory offers no answers:

1. ***Why are there so many basic particles?***
2. ***Why are there so many free parameters?***
3. ***How do we include gravity in the picture?***

There is apparently a superfluous replication of fermion families. We have evidence for three when one would seem to suffice. Furthermore, there must be a relationship between quarks and leptons implied by the similar structure of each of the families. This suggests more questions:

4. ***Why are there at least three fermion families?***
5. ***How many families are there, all told?***
6. ***Why are the families so much alike?***

The charges of the fermions are such that the hydrogen atom is neutral. That is to say, the charge of the electron seems to be *precisely* equal and opposite that of the proton. I say "seems to be" because I am describing an experimental result. The equality has been established to twenty-one decimal places and is presumably exact. It didn't have to be this way from the vantage point of the standard theory, and it simply cannot be a coincidence, so we may put one more query:

7. ***Why are the charges of the fermions the way they are?***

The lightest of our seventeen particles, not counting those that are supposed to be massless, is the electron. Among those that have been seen, the heaviest particle is the Z^0. The ratio of the masses of these two particles is a very large number, about 200,000. Even larger numbers appear in our theory if we try to take gravity into account. The gravitational attraction of an electron to a proton is *forty* powers of ten smaller than their electrical attraction. That's ten thousand billion billion billion billion times smaller!

The appearance of such very large numbers suggests that our theory is somehow incomplete. Dirac emphasized the philosophical view that all numbers that are truly fundamental to physics or mathematics, like π (3.1415 . . .) or the number of spatial dimensions (3), should be of order one; which is to say, neither very large nor very small. Thus, we may reasonably inquire:

8. ***Why do very large numbers appear to play a fundamental role in our theory?***

In the simplest realization of the standard theory, photons and neutrinos must be massless. At a slight cost of elegance, this can all be changed, yet the theory will remain consistent.

9. ***Is the photon really massless?***

Experimentally, the answer is as affirmative as it can be: if there is a photon mass, it is at least thirty-two powers of ten smaller than the electron's mass. But is there a compelling reason that the photon must be massless?

10. ***Are the neutrinos really massless?***

It is very easy to monkey with the standard theory so as to give mass to the neutrinos, but there is no solid experimental evidence for such masses. True, the Russians claimed to have proven that the electron neutrino has a mass of about 30 electron volts. More recently a Swiss group contradicted the Russians by showing that the mass is possibly zero, but certainly less than 18 electron volts. Many experiments are being done to search for neutrino masses, so that the question remains open.

If, for example, the tau neutrino weighs 100 electron volts, it could constitute the mysterious dark matter of the universe. It could even "close the universe" and lead to the eventual end of its expansion. Even the lowliest particle may have a profound effect upon how things will be billions of years from now.

Maxwell's equations display perfect symmetry between magnetism and electricity. Nature breaks this symmetry by presenting us with fundamental particles bearing electrical charge, like the electron, but none (so far) bearing individual magnetic charge. There could exist in nature particles that play the same role for magnetism as charged particles do for electricity. These conjectural carriers of magnetic charge are called magnetic monopoles, and the question of their existence was first broached by Maxwell himself in the nineteenth century. Physicists have been searching for evidence of magnetic monopoles ever since. This question, too, remains open.

11. *Are there any magnetic monopoles?*

The last of my questions concerns the stability of matter. Generally, the conservation laws of physics are linked to local symmetries. This is not the case for the conservation of baryon number. No fundamental principle guarantees that the proton, or any other atomic nucleus, is stable! It may be that all matter is ultimately radioactive, eventually to decay into leptons and photons. Diamonds may not be forever, after all. The standard theory, however, does not allow for proton decay. But the standard theory may be wrong.

12. *Does the proton live forever?*

While these questions and others have yet to be answered, it is still true that the standard theory offers a complete, consistent and correct description of all observed phenomena in the microworld. In principle, and if we had enough computer power, we could predict the results of most high-energy experiments in terms of the parameters of the theory, and we have no reason to doubt that our predictions would square with experimental data.

Searches for departures from the standard theory have yielded consistently negative results. There no longer seem to be any loose ends or false threads in the tapestry—not yet, anyway. Surely we have not exhausted nature's bag of tricks. Experimenters will simply have to look harder for the flaws that must be there, the clues to an even better theory.

Meanwhile, what should we theorists do?

As early as 1973, the quick-witted among us began to realize that particle physics had reached a plateau with the development of the standard theory. Many of the older puzzles of subnuclear physics had been resolved in terms of quantum chromodynamics and the electroweak synthesis. But surely there had to be a logical link between these two parts of the picture, a link between the strong and weak forces that could tie together the two types of matter particles: quarks and leptons.

Jogesh Pati is a meditative Indian physicist whom I met in 1960 while we were both young postdoctoral fellows at CalTech under Gell-Mann's hypnotic spell. Jogesh taught me how to recite the most powerful of all Hindu prayer words, "*Om!*" signifying the oneness of Krishna and Vishnu, and the unity of earth, sky and sun. What more appropriate mantra could there be for us seekers of the unified theory of physics? It also worked in preparation for a session with Murray, the *OM*niscient.

In more recent years, Jogesh has entered into a subcontinental alliance with Pakistani Abdus Salam to put forward a new kind of theory. Pati and Salam suggest that there may be a profound connection between the strong and weak forces by which a lepton could be regarded as simply a fourth color of quark. In this "pâté and salami" offering, the SU(3) group of quantum chromodynamics is traded for the larger group SU(4). In such a picture, new kinds of force particles must exist that can transform a quark into a lepton. While this was a fascinating idea, it was not quite a unified theory of the elementary-particle forces. It was a step toward the right goal, but in the wrong direction. Howard Georgi and I found a much better path toward the underlying relationship among the fundamental particles.

In the standard theory, the three kinds of forces are described by three different groups. (Remember, gravity, the fourth force, is not treated by the standard theory.) Strong interactions are based upon the group SU(3), whose eight gluons can "sense" the colors of quarks. Weak and electromagnetic interactions are woven into the structure of two additional groups, called SU(2) and U(1). The standard theory involves the rather complicated and ugly concatenation of three groups, SU(3) × SU(2) × U(1). Wouldn't it be neater, we felt, if we could find one single simple group that could describe all three of the particle forces together?

The possible groups upon which we could erect a unified theory were well studied by mathematicians. They are known as the simple Lie groups. Today every graduate student of physics learns all about them. All we had to do was to figure out which of these groups contained SU(3) × SU(2) × U(1) in just such a fashion as to describe the known fermion families correctly. It was the kind of purely mathematical question that is easy to answer once it is clearly stated. The hard part was to think of the question in the first place, to put the physical problem in terms of a mathematical puzzle.

Once we had done that, Howard came up with the answer overnight. There is only one group that works and accommodates the minimal standard theory. It is called SU(5). In January 1974 we submitted a paper to the *Physical Review Letters* entitled "Unity of All Elementary-Particle Forces." We began:

> We present a series of hypotheses and speculations leading inescapably to the conclusion that SU(5) is the gauge group of the world—that all elementary-particle forces (strong, weak, and electromagnetic) are different manifestations of the same fundamental interaction involving a single coupling strength. . . . Our hypothesis may be wrong and our speculations idle, but the uniqueness and simplicity of our scheme are reasons enough that it be taken seriously.

Our scheme is an elegant and compelling extrapolation of the electroweak theory. All of the elementary-particle forces, strong, weak and electromagnetic, are generated by a simple and unified gauge theory. The SU(5) symmetry cannot be reconciled with just any old collection of quarks and leptons. The fermion family must conform to one of the allowed patterns of symmetry known as representations of the group. By far the simplest allowable representation consists of two leptons and two quarks, just like the observed families.

The family structure of the matter particles, knowledge that had been dearly extracted from nature by multitudes of experimenters, is an automatic consequence of the theory. Moreover, the values of the electric charges of the quarks and leptons are prescribed. The charge of the proton simply has to be exactly equal and opposite to the charge of the electron. And the photon is automatically and necessarily a massless particle. In SU(5), or in any of its subsequent elaborations, there is no longer the option to give the photon a small mass while preserving the consistency of the theory. A photon mass would destroy the renormalizability of the theory.

Thus, at one stroke, we solved three of the outstanding puzzles confronted by the standard theory. The three elementary-particle forces were embedded within a single simple gauge group in a synthesis of supernal beauty and colossal chutzpah. We had constructed the first and the most elegant grand unified theory.

THE MAGICAL CUBE OF LEPTONS AND QUARKS

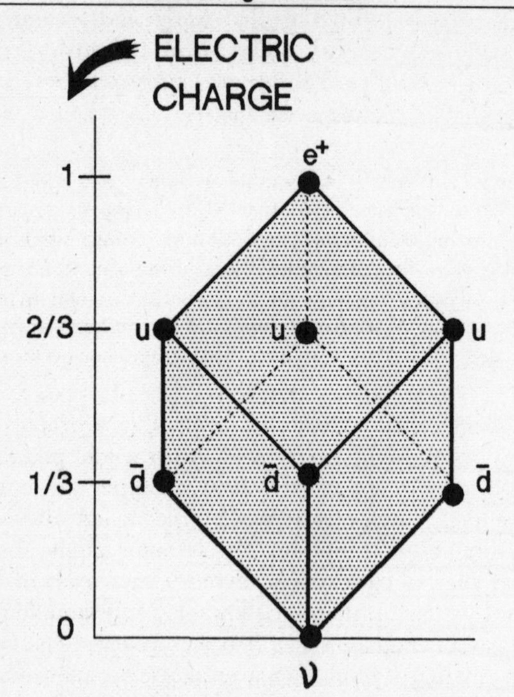

MAGICAL CUBE

Pictured above is a cube resting upon one of its corners. Each of its eight corners is associated with one of the particles in a family of fermions according to the SU(5) grand unified theory. The height of a corner above the ground, indicated on the scale to the left, determines the value of the electric charge. The positron carries unit charge. Three corners are exactly two-thirds of the way up; they correspond to the three differently "colored" up quarks with charge 2/3. Another three corners are only a third of the way up; these correspond to the three differently colored down antiquarks. The corner at the bottom corresponds to the electrically neutral neutrino. This pattern is repeated for each of the three families of fundamental fermions.

Leptons and quarks, seemingly so very different, are treated together by SU(5). The magical cube conveys only an inkling of the mathematical elegance of grand unification.

We lectured on our work at many universities. We explained how the observed quarks and leptons fitted into the "representations" of SU(5) as neatly as a hand in a glove. As if by a miracle, the "anomalies" that could spoil the theory simply canceled out of the equations. We had our audiences eating out of our hands up until the end of our talks, when we pointed out the inescapable consequences of our theory: that *all* matter is radioactive. Even Jane Fonda.

Our gauge theory involves twice as many gauge bosons as the standard theory. The extra ones mediate a new and very dangerous force of nature, a force that causes quarks to become leptons and makes matter go *poof*! We were led to the amazing conclusion that the proton is not a stable particle. It is obligated, sooner or later, to decay. All the matter on the earth and in the heavens will eventually succumb. All that will remain are a few electrons, positrons, neutrinos and photons—nothing to hold and no hand to hold it. Nothing can last forever. Forest Lawn's promise of perpetual care violates the truth-in-advertising law.

At this juncture, some of our listeners would politely point out that there is good evidence that matter is rather stable and possibly eternal. After all, it is now about fifteen billion years since the creation of the universe in the big bang. Things seem to have held up quite well. Clearly, matter lives at least that long. Maurice Goldhaber, who for years had searched in vain for evidence of proton decay, quipped that we can feel it in our bones that the proton's half-life is longer than a *billion* billion years. If it were shorter, the gradual decay of the protons within our own bodies would produce enough radioactivity to kill us.

It was clear that the new force of SU(5) had to be very feeble indeed: far, far weaker than the so-called weak force.

Let us return a moment to the standard theory to recall how very weak are the ordinary weak interactions, and what in nature has so enfeebled them. An energetic proton, with an energy of 1 GeV, can pass through only a few inches of steel before it is stopped by colliding with an atomic nucleus. This is a manifestation of the strong force. The neutrino, unimpeded by the strong force, is amazingly penetrating. A 1 GeV neutrino could pass through a hypothetical solid steel obstacle as thick as the solar system without suffering any collision at all.

The electroweak theory offers an elegant explanation for the weakness of the weak force. Although the intrinsic strengths of the weak and electromagnetic forces are the same, electromagnetism is mediated by a massless photon whereas the weak force depends on the exchange of a heavy W particle, which weighs about eighty times more than the proton. The large mass of the W particle sets the scale of the weak force. The rate of decay of an unstable particle depends upon the reciprocal fourth power of the W mass. The weakness of the weak force is simply due to the heaviness of the mediating force particle. We have traded in the question, Why is the weak force so weak? for the deeper but equally mysterious query, Why is the W-boson so heavy?

Although our SU(5) theory requires that the proton must decay eventually, it did not, at first, specify the value of the proton's lifetime. The observed stability of matter means that the new force leading to proton decay is really weak. In comparison, the ordinary weak nuclear force is a Jolly Green Giant.

The proton lifetime depends on the fourth power of our new gauge boson, X. If it were just as massive as W and Z, the new force would be just as strong as the weak force, and the proton would live for only 10^{-11} second. If the X were ten times heavier than W and Z, the proton lifetime would be stretched by ten thousand to 10^{-7} second—not nearly enough. But if the X were heavier yet, the proton could live as long as it is known to do.

Once again I was faced with a problem involving the masses of gauge bosons. Was there any way to compute what the X-boson mass must be? If there were, we could predict a value for the proton's lifetime. If this prediction conflicted with the data, we could reject the theory. Otherwise, our prediction could be a goal to which experimenters could strive.

As things were, all we could say was that to be compatible with the known level of proton stability, the X-boson has to weigh at least a billionth of a gram. This is about the mass of a bacterium: small on the scale of everyday life, but fourteen powers of ten heavier than the proton. As an elementary particle, the X-boson must be a real whopper!

There was another problem with SU(5). It was supposed to unify the strong and electroweak forces, but the strong force is really quite strong. How could it be put together with the other forces in a unified theory? In the SU(5) theory, QCD and QED are related to one another and should be equally strong. This is simply not true in nature. The QCD force between quarks is hundreds of times stronger than the electrical force between them.

Howard, along with Helen Quinn and Steve Weinberg, showed how asymptotic freedom solves both of these problems. At energies above the X-boson mass, the SU(5) symmetry is revealed, and the strong, weak and electromagnetic forces are equal in strength, as the symmetry dictates. At the X-boson mass, the grand unified symmetry is spontaneously broken down to the standard model symmetry pattern of SU(3) × SU(2) × U(1). Once again the Higgs mechanism is invoked to break the symmetry. At lower energies yet, the various forces evolve in strength according to known equations. In particular, the strong force becomes stronger at low energies because it is asymptotically free. The theory automatically explains why the strong force is so much stronger than its siblings!

The observed disparity in the strengths of the various forces allowed Howard and his friends to compute the energy scale at which the grand unified symmetry is broken, and consequently the mass of the X-boson. They attained a value of 10^{15} GeV, which corresponds to a *predicted* proton half-life of 10^{30} years.

Could experimenters test this result? Certainly not by looking at one proton and waiting for it to decay. The predicted lifetime exceeds the age of the universe by almost twenty powers of ten.

The trick is to study a large number of nucleons at the same time. A thousand tons of matter contain about 10^{33} nucleons (protons and neutrons). If the SU(5) prediction is true, about three of these nucleons ought to decay each day. Larry Sulak, our neighbor in Brookline and a junior faculty member at the Harvard physics department in the 1970s, was the first physicist to suggest an ambitious proton-decay experiment that might do the job.

His experiment employs an enormous underground tank of highly purified water. The decay of one of the protons or neutrons making up the water would generate a visible and detectable signal. After transferring to a tenured position at the University of Michigan, Larry assembled a consortium of researchers from Michigan, the University of California at Irvine and Brookhaven National Laboratory, now known as the IMB collaboration. Together they constructed the world's largest underground swimming pool deep within the Morton-Thiokol salt mine underneath suburban Cleveland.

It is a sixty-foot cube, about the size of a six-story building, located a thousand feet underground so as to be shielded from cosmic radiation that might mimic signals from the decaying nucleons. The water tank is surrounded by thousands of electronic eyes that are programmed to catch the occasional decay of a proton or neutron. In a typical decay process, a proton could transform itself into a neutral pion and an energetic positron. The pion, virtually instantaneously, becomes two gamma rays. In their passage through the water, the decay products generate a characteristic pattern of visible light, which is picked up by the electronic eyes. This is the way the dying proton could reveal itself to the hopeful IMB experimenters.

By the early 1980s, smaller and less ambitious proton decay experiments had already been launched elsewhere. In particular, a Japanese-Indian collaboration, operating out of a very deep Indian gold mine, had reported positive indications for the existence of proton decay. But their results were far from being convincing.

On my fiftieth birthday, in December 1982, Joan organized a big celebration at our home to help me through the trying passage. Two gifts still stand out in my memory. One is a professional-quality Italian espresso machine, which we have used every day since. The other is a photograph Larry Sulak presented to me under an oath of secrecy, depicting the first plausible proton-decay candidate from the IMB experiment. Maurice Goldhaber, who is both a "clean old man" and one of Larry's more distinguished collaborators, warned me that not every candidate is elected.

This one was not. Years have passed and this one event remains the only candidate for a proton decay, and one event is just not enough to convince anyone of anything. The experiment worked very well, and it proved beyond doubt that the proton does *not* decay as my theory says it should. It works still, but Maurice still does not know for sure whether the proton lives forever. If it doesn't, he still prays that it will die in his arms.

Larry's experiment has already proven that the proton, if it decays at all, is hundreds of times more stable and long-lived than the theory predicts. SU(5), in its simplest form, is a bust. So much for Howard's dream of a Nobel Prize, and for my second shot at the Swedish gold.

Wait one second!

Our brainchild is too gorgeous to relegate to the scrap heap just because of one failed prediction. The calculated value of the proton lifetime depends upon an incredible act of hubris. To get a definite answer, we theorists had been assuming that nothing very interesting happens in physics between the energies of 100 GeV (where the W- and Z-bosons play) and 10^{15} GeV (the land of the forever inaccessible X-boson).

That is to say, we casually suggested that there is a thirteen power-of-ten desert in energy that is entirely uninhabited by unexpected beasties: no new particles to be discovered, no new forces to be seen, and no surprises to be revealed by experiment. It is as if Niels Bohr, upon unraveling the mysteries of the atom, would have argued that there should be no more surprises in physics from the size of the atom up thirteen powers-of-ten to a kilometer: no DNA, no transistors, no microchips, no Teflon, no two-headed turtles, nor space shuttles nor particle accelerators. He would have been wrong, of course, and so, most likely, were we.

Perhaps there are many more subnuclear surprises awaiting us. Perhaps that is why the proton is longer-lived than we thought. The prophecy of a desert at the high-energy frontier could be self-fulfilling, however. If we don't build larger accelerators to look for new physics at the frontier, we certainly won't find any.

The grand unified theory not only predicts the eventual death and decay of all matter, but it presents a possible scenario for the origin of matter in the universe.

The effects of the X-boson are negligible in the cool and crystalline universe we now inhabit. Once upon a time, things were very different and very, very hot. The X-bosons were everywhere. Sometimes they would decay into matter particles, and a tiny bit less often, into antimatter particles. The reason for the small difference was first pointed out by Sakharov. It results from baryon number violation and the failure of time-reversal symmetry. As the universe expanded and cooled, the matter and antimatter annihilated as best it could, leaving only the tiny excess of matter. Today there are about a billion times more photons in the universe than nucleons. Nonetheless, these few remaining particles (about 10^{79} in the whole universe) are sufficient to compose all the stars of the heavens as well as the cool green hills of earth. Another mystery of the universe seems to have been solved, but not to worry, there remain quite a few more.

The Dutch physicist 'tHooft and the Russian Sasha Polyakov explained how grand unified theories impact upon the possible existence of magnetic monopoles. They showed that magnetic monopoles exist, in unified theories, as consistent mathematical constructs.

Whether they exist in nature as observable particles is a more difficult question. The 'tHooft-Polyakov monopole is necessarily very heavy. It is far too massive to be produced by any accelerator now existing or being dreamed of. You couldn't even keep one in a bottle. It would fall through the bottle to the center of the earth!

The only hope for experimenters is to discover primordial magnetic monopoles as they impinge upon the earth during their intergalactic journeys. Their magnetic charge would produce a unique and unmistakable signal as they pass through an appropriate detector. So far, none have been found, although, a few years ago, Cabrera thought he might have seen one. But the search for a magnetic analog to electrical charge has been a holy quest for decades. With new impetus from grand unified theories, scientists are organizing ever more elaborate monopole hunts.

Larry failed in his first effort to find proton decay. It was not his fault. After all, the IMB experiment performed a whole lot better than its competition, a Harvard-Purdue-Wisconsin collaboration led by Larry's former mentor and now rival, Carlo Rubbia. The HPW experiment cost big bucks but never worked very well. IMB worked just fine, but the trouble was that nature was not being very cooperative. The protons were not decaying as rapidly as we thought they should.

Things at Michigan were not very cooperative, either. Irreconcilable tensions developed among the members of the proton-decay collaboration. Furthermore, Larry's wife, Beth, and their two young children had remained in their lovely Brookline home, catty-corner from us. With her Harvard MBA, Beth was very gainfully employed nearby, and she saw no reason to forsake their friends in Brookline to build a new life among the Ann Arbor anthropoids. Larry was commuting several times a week between Massachusetts, Michigan and the Cleveland salt mine.

At parties at my home Beth's lament was often heard. Could anyone find her husband a job in Boston? While spending a sabbatical year at Boston University, I found that they were seeking a new head for their physics department. Apparently, I found a good ear to put a bug in. Larry has escaped from Michigan to become chairman of the B.U. physics department, and he is frantically making it into one of the best physics departments anywhere. Everyone wins but the airline companies.

Larry has guided Boston University into a huge collaborative effort to search for magnetic monopoles. This ambitious monopole hunt, nicknamed MACRO, will be carried out in the brand-new Italian Underground Laboratory thousands of feet below the Gran Sasso mountain, not so far from Rome. The project involves dozens of physicists from Italy and America, and millions of dollars, and billions of lira, to fabricate and deploy an enormous array of sensitive particle detectors in a drive-in underground cavern. I certainly hope that Larry and his buddies will have better luck finding monopoles than they have had so far looking for proton decay. Thank God there are people like that who will spend years of the lives of their graduate students in dank discomfort following up the wild speculations of armchair theorists. And He alone knows whether or not the search will be fruitful.

Proton decay may or may not be observed, and magnetic monopoles might or might not be there to be seen and counted. Even if some version of grand unification turns out to be essentially correct, it does not solve all our problems. We are still left with the mysterious and superfluous replication of fermion families. Furthermore, the number of arbitrary parameters goes *up* rather than down with grand unification. We would be no better off at all in our understanding of why the particle masses are as they are. Also, there is a new and immediate Large Number Problem.

We saw that the weakness of the weak force comes about because the W- and Z-bosons are heavy—around 100 GeV. The proton-decay force is far weaker yet, so that the X-boson must weigh at least 10^{14} GeV. That is to say, the X-boson is at least a *million million* times heavier than the W and Z. Is there a natural origin to such a very large number? Grand unification necessarily leads us to consider energies that are almost as large as the Planck mass, the mass at which quantum gravity rears its unfathomable head. Perhaps, to make further progress, we simply have got to include gravity as well. This was always Einstein's belief, and for the last thirty years of his life it led him down the Princeton garden path.

Can mere mortals succeed where the great Einstein failed? They are trying hard, and they've come up with something new and promising and devilishly hard to understand: the Superstring.

Ouroboros

SCIENCE, IN ITS VARIED DISCIPLINES, is called upon to explain all of the phenomena of nature. The grandest spectacles in the heavens and the most minute twinges of the ultimate particles are equally the subjects of its analysis.

Cosmologists deal with the biggest questions of all: the birth of the universe and the origin and development of its billions of galaxies. Astronomers work their way down to things as small as our solar system. Geologists are concerned with all the nooks and crannies of our planet. Biologists study the things on earth that creep and crawl and swim and fly and infect one another, from wee viruses to the great whales they attack. Next come the chemists and most of the physicists whose job is to explain the bulk properties of matter in terms of the minute atoms of which everything on earth is made. They will tell you why copper is red, why the sky is blue, how a candle burns, and what makes dew.

The first generation of subatomic specialists is the nuclear physicists. Their domain is the central core of atoms, the key to nuclear weapons, to nuclear power, and the marvelous nuclear furnace we call the sun. Elementary-particle physicists study the constituents of atomic nuclei, the neutrons and protons, and all of the other particles that were once thought to be elementary but are now known to be made up of quarks.

The microscopes of the particle physicist are high-energy accelerators. The higher the energy, the smaller the structures they can discern. With the accelerators available today, things as small as 10^{-16} centimeter can be observed, things like the W and Z particles, things that are a hundred million times smaller than the atom!

Theoretical physicists speculate on the existence of even smaller entities, things so small that no conceivable accelerator could spot them. In their grand unification theories (GUTs), all of the elementary-particle forces are seen to result from a simple underlying struc-

THE SNAKE OF SIZES

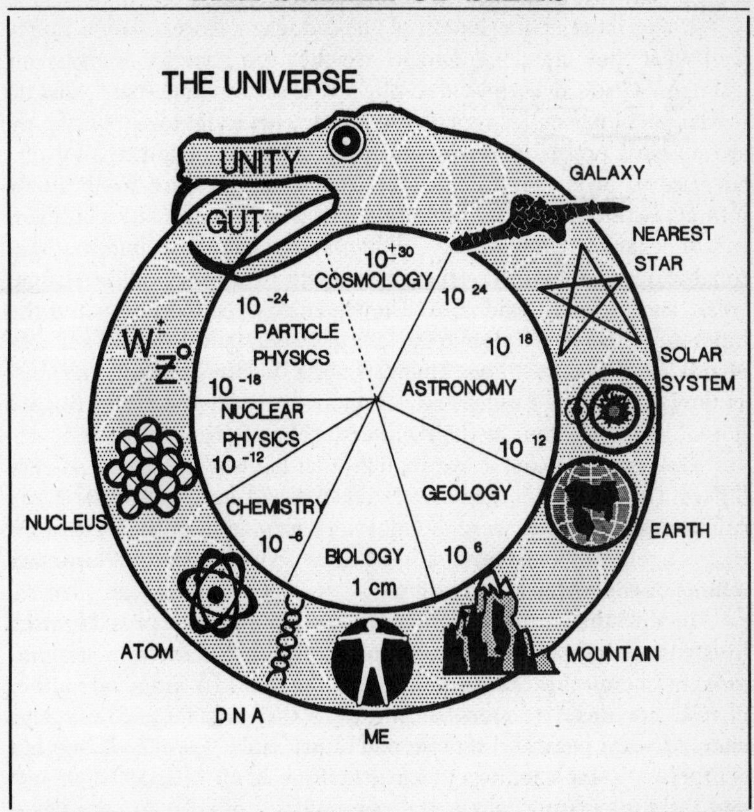

ture. The symmetry among the forces is apparent only at very short distances, and these theories require the existence of new particles that are very heavy and incredibly tiny. The length scale associated with grand unified theories is only 10^{-28} centimeter (a trillion times smaller than W and Z particles). At smaller distances yet, distances of the order of 10^{-33} centimeter, we finally reach a natural endpoint. Quantum gravity becomes the dominant force, and we have no agreed-upon theory that can deal with it. This "Planck length" defines, for now, the very bottom of our cosmic ladder. Suppose that each rung corresponds to an increase in size by a factor of ten. There are fifty-two rungs from the Planck mass, up through the world of inches and feet, to the size of the entire visible universe.

Along most of the length of the ladder, scientists know pretty well what they are doing and where they are going. Chemists, for example, agree that their discipline is closed in the sense that no fundamental principles remain unknown. This is not to say that there are no hard problems remaining to be solved or unanticipated discoveries to be made. It's just that the basic rules are fixed: atoms interact with one another according to the well-known rules of quantum mechanics, rules whose validity is established beyond any reasonable doubt. Chemists can entirely neglect the effects of the strong, weak, and gravitational forces. They have no need to know about the inner structure of the atomic nucleus or understand Einstein's theory of gravity. (Of course, *good* chemists often do know these things, just as they may speak French, read Milton, and play the oboe. But it's not all that important to their professional activities.)

Once upon a time it was thought that life was something so very different from the inorganic world that it had its own special set of rules. Organic bodies were thought to be formed from their elements by the agency of a mysterious "vital force" lying beyond the ordinary realms of chemistry and physics.

In 1828 the German chemist Friedrich Wöhler surprised himself by synthesizing the organic compound urea by means of a straightforward chemical procedure, "without the need for kidneys, neither of man nor dog." It soon became clear that organic and inorganic chemistry are peas in the same pod of principle. Yet it took another century or so for scientists to agree that life in all its aspects, though endlessly fascinating, obeys the same old set of rules as snowflakes or stars do.

The last decisive step was taken in the 1950s with Watson and Crick's momentous discovery of the structure of DNA and the method by which genetic information is encoded. A rabbit, we now know, is just an extraordinary complex concatenation of atoms and molecules doing the same things that atoms and molecules always do. Even the mystery of human cognition and the vagaries of memory are sure to have an explanation in terms of physical laws that are already common knowledge. Finding that explanation, though, will be both hard work and good fun.

The wild things are at the extreme ends of the cosmic ladder. My fascination with cosmology and particle physics lies in the fact that they are open-ended disciplines. We really don't know all the rules of those games, how and why the universe evolved as it did, and how it will behave in the distant future. We do know that it all began about fifteen billion years ago in the big bang. Since then, the universe has been expanding and cooling. Will it continue to expand and cool indefinitely (the Big Chill) or will it one day turn around, shrink, and eventually implode (the Big Squeeze)? The answer to this question depends on just how much junk is in the universe.

Decades ago, astronomers believed that most of the matter in the universe had formed into luminous stars. (After all, 99.99 percent of the mass of our solar system is clumped in the sun.) By counting up the stars, they could estimate the total amount of matter in the universe. Their conclusion *was* that it is to be the Big Chill for sure. More recently, however, we have learned that most of the matter of the universe is not in the form of luminous stars. Somewhere between 90 and 99 percent of the matter is elsewhere. The stars and galaxies that astronomers look at are only the trimmings. Where's the beef?

What is the nature of the mysterious dark matter that constitutes the vast bulk of matter in the universe? All manner of things has been suggested: black holes, small dark stars, neutrinos, photinos, axions, cosmions, sphalerons, and so on. That there are hundreds of scholarly papers addressing the question is a sure sign that nobody knows the answer.

In the visible universe there are about ten billion galaxies like our Milky Way, each containing about a hundred billion stars. How are they distributed in space? They are not at all scattered about at random, but are gathered together into small groups and giant clusters. Very recently, Margaret Geller and her collaborators at the Harvard-Smithsonian Center for Astrophysics discovered that the pattern of galaxies in a slice of the sky resembles a cross section through the suds in a kitchen sink! That is, the galaxies seem to lie upon the surfaces and within the interstices of a network of giant spherical voids.

How did this come to be? Are the empty voids relics of ancient and titanic explosions? What exploded? No familiar forms of matter or energy could have produced this remarkable pattern. Again, something important is missing from our picture of the universe.

The story is the same at either end of the cosmic ladder. What are the ultimate forms of matter? This question applies both to the top and to the bottom. Why are the fermion families as they are, and

A SLICE THROUGH THE UNIVERSE

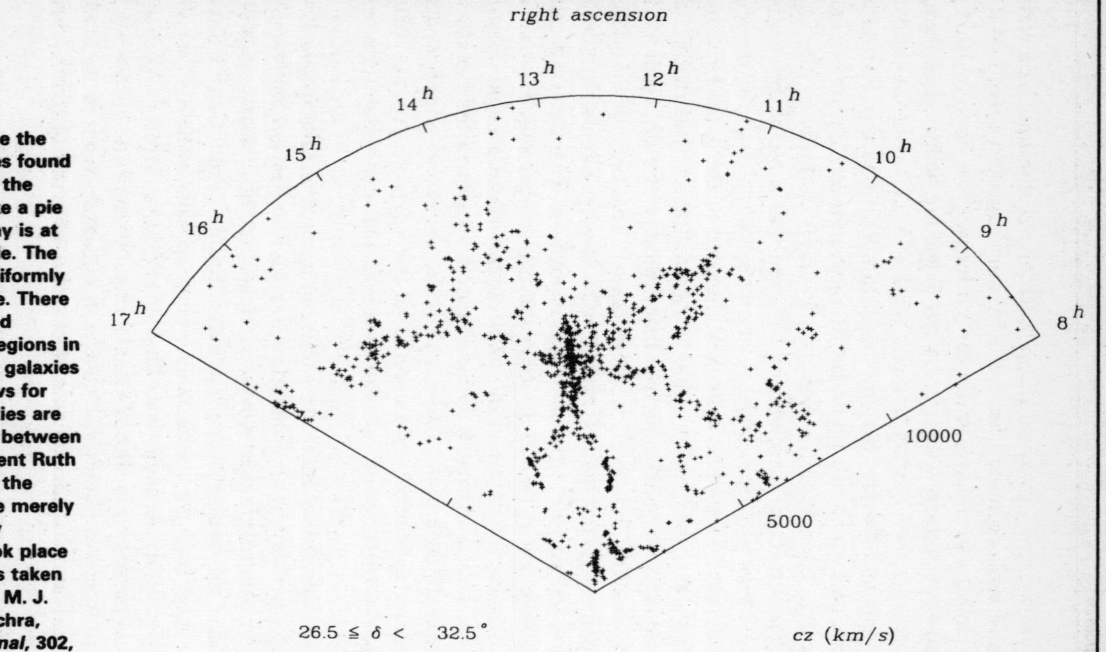

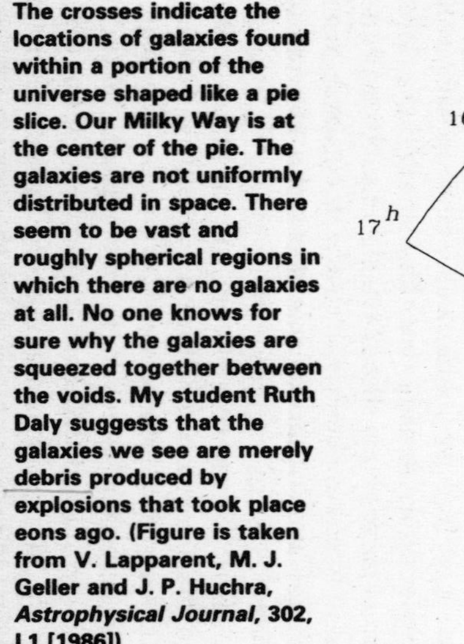

The crosses indicate the locations of galaxies found within a portion of the universe shaped like a pie slice. Our Milky Way is at the center of the pie. The galaxies are not uniformly distributed in space. There seem to be vast and roughly spherical regions in which there are no galaxies at all. No one knows for sure why the galaxies are squeezed together between the voids. My student Ruth Daly suggests that the galaxies we see are merely debris produced by explosions that took place eons ago. (Figure is taken from V. Lapparent, M. J. Geller and J. P. Huchra, *Astrophysical Journal*, 302, L1 [1986])

why are the galaxies as they are? Neither question can be answered in terms of known constructs. At the very largest scale of distances, as at the very smallest, is where our understanding, even in principle, is dramatically incomplete. Crazy new ideas are welcome, the crazier the better.

Gravity is the odd man out of our unified scheme of elementary-particle physics. Its effects are entirely irrelevant for particle physics at accessible energies. Yet gravity is the dominant force at the higher rungs of the cosmic ladder. Gravity determines whether or not the universe will expand endlessly. It binds the stars into galaxies and controls their motions within. On a smaller scale, gravity holds the Sun together and keeps its family of planets firmly in orbit. It glues the Moon to Earth and holds our oceans and atmosphere in their proper places.

But when it comes to things a few kilometers in size, the force of gravity must yield to electromagnetism. Gravity tries, but it cannot stop a tree from growing or a mountain from forming. Here at last, gravity is successfully defied. Our everyday world is essentially electromagnetic. Our men and women in space, quite free of Earth's loving gravitational embrace, can eat and drink and even grow their food and beget their children in the complete absence of gravitational forces. Gravity provides us with a comfortable home, but is otherwise pretty much superfluous. Its manifest destiny lies in the heavens.

Each time we arise in the morning, each time a spring flower sprouts, the supremacy of electromagnetism is displayed. Its majesty ranges from things the size of mountains down the ladder to elephants, microbes, molecules and to their component atoms. After about seventeen rungs of the cosmic ladder, electromagnetism is superseded in power by the stronger but shorter-range color force. In its limited domain, this force is all-powerful. As the basis for the strong nuclear force, it holds protons and neutrons together to form the atomic nucleus, much against the inclination of electromagnetism, which would as soon blow the nucleus apart. What we call protons and neutrons are composites made up of quarks, again bound by the color force. But after another four rungs, the color force gives way to the weak force, which had been quietly awaiting its role. The physics at 100 GeV is dominated by the effects of the W and Z particles, the carriers of the weak interactions.

We have reached the point on our ladder corresponding to the highest energy accelerators at our disposal. Our knowledge about the physics of the lower seventeen rungs is scanty and speculative. Because of the spectacular successes of the standard theory, and because of the lack of surprising new discoveries in recent years, some theorists suggest that nothing at all of great interest takes place at distance scales ranging down seventeen powers of ten, from 10^{-16} centimeter to the Planck length at 10^{-33} centimeter. The strong, weak and electromagnetic forces are all equally effective, but there are no new particles, no new forces and no interesting phenomena to observe.

According to this philosophy of despair, no surprises await us at energies much higher than those that have been explored already. We may have entered upon a great desert that extends seventeen rungs of the cosmic ladder, or equivalently, from energies of 100 GeV all the way past the Planck mass. According to this viewpoint, the most fundamental laws of physics are apparent only at energies approaching 10^{19} GeV, where structures the size of the Planck mass can be observed and the effects of quantum gravity are all-important. At this forever inaccessible energy, the workings of our "low energy" world are determined.

I doubt that there is such a great desert, and I am confident that many new and unexpected phenomena will be discovered at accessible energies, no more than one or two rungs down the ladder from where we are today. Of course, if we don't look for them (thereby saving a great deal of money!), they won't be found.

A few grains of sand does not make a great desert, but a pessimistic philosophy can make the desert a self-fulfilling prophecy. Fortunately for science, several brand-new accelerators will soon be completed, and we will be able to see for ourselves if we are facing a Sahara or the shores of a new, uncharted, exciting ocean.

Three great electron-positron colliders are being deployed: one in Japan, one at Stanford and one at CERN. All of them are more powerful than today's retired champion, the German collider called PETRA. Something exciting is sure to be found.

Until very recently, the CERN proton-antiproton collider provided the highest collision energies that could be explored. This device led to the discovery of the W and Z particles. Our country has entered the high-energy sweepstakes with a similar, but more powerful accelerator: the Tevatron collider at Fermilab. Operating at a collision energy of 2000 GeV, three times higher than CERN, it is the world's best hope for new discoveries in particles physics in the near future.

Where do we go from here? has been the subject of dozens of conferences, workshops and seminars here and abroad. It is clear that an even larger accelerator is required if we are to make substantial progress in understanding the microworld. The community of American high-energy physicists has focused unanimously on its need for a machine twenty times more powerful than the Tevatron. This proposed accelerator is called the Superconducting SuperCollider, or SSC.

Design studies have been completed, and the cost of the device has been reliably estimated to be about $4 billion. Admittedly, this is a large amount of money. In this time of private wealth but public poverty, there are many pressing needs for scarce government money. Nonetheless, it seems to be the price we must pay for progress in particle physics. I have no doubt that this machine, or one like it, will be built—if not now, soon; if not in the United States, then in another country. CERN's dream machine, slightly smaller than the SSC, is called the Large Hadron Collider (LHC). Italy pursues its ELOISATRON, and the Soviet entry is nicknamed UNK. Possibly some or all of the countries of the world will band together and build a machine cooperatively.

It is impossible for me to believe that humankind will suddenly abandon its quest to understand the ultimate nature of matter. We have come a very long way from the birth of the atomic hypothesis in ancient Greece. The last few decades have seen an explosion of discovery resulting in the enormously successful, but clearly incomplete, standard theory of particle physics. I am convinced that the best is yet to come.

Ultimately, we must confront the physics at the bottom of the ladder, where quantum gravity becomes important. And we must admit that accelerators can never be built that can explore energies great enough to reveal this ultimate domain. We must be more clever, and we can! While we have written of the scale of sizes in terms of a cosmic ladder, we have drawn a snake swallowing its own tail, in which the largest and the smallest sizes come together. Both for the physics of the very small and for the physics of the very large, gravity is the dominant and controlling force. This is one reason the snake swallows its own tail. There is another.

Today the universe is ancient and cold. At an age of about fifteen billion years, its temperature is now a scant three degrees above absolute zero, the point at which molecular motions cease. In the

THE UNIVERSE ON LOG TIME

Life passes more quickly as we grow older. A year, to a five-year-old, is like a decade to a fifty-year-old. In the figure below we show the history of human development as an ordinary linear plot, beginning, at the top, with the moment of conception.

Figure 1 shows that most of our span on Earth is spent as an adult. Yet the most important developments (forming a placenta, growing fingers, being born, etc.) are squinched into the top of the figure. For this reason, scientists often use a different kind of plot where equal distances correspond to multiples, and the distance between five and six years equals that between fifty and sixty. Shown in **Figure 2** is life on log time.

The moment of conception is pushed up to heaven, and the stage of human adulthood is recognized to be merely the last of many developmental phases.

The history of the universe is much like the history of a human being. It all began in the big bang, just as life began at conception, and most of the interesting things happened at the beginning. **Figure 3** is the universe on log time.

Once again, the big bang at time zero is pushed off into infinity, and the development of the universe separates into well-defined stages. Between the age of one and two minutes, many of the atomic nuclei we see about us were assembled. Subsequently, nothing much happened until the universe was a few hundred thousand years old. At that time, electrons and nuclei combined together into atoms, and all of a sudden the universe became transparent, as it is today.

The earth was formed when the universe was about five billion years old, or about a third of its present age. Logarithmically, this important event took place very recently, as the figure shows. The age of dinosaurs, a paltry few hundred million years ago, is too current an event to display on this figure.

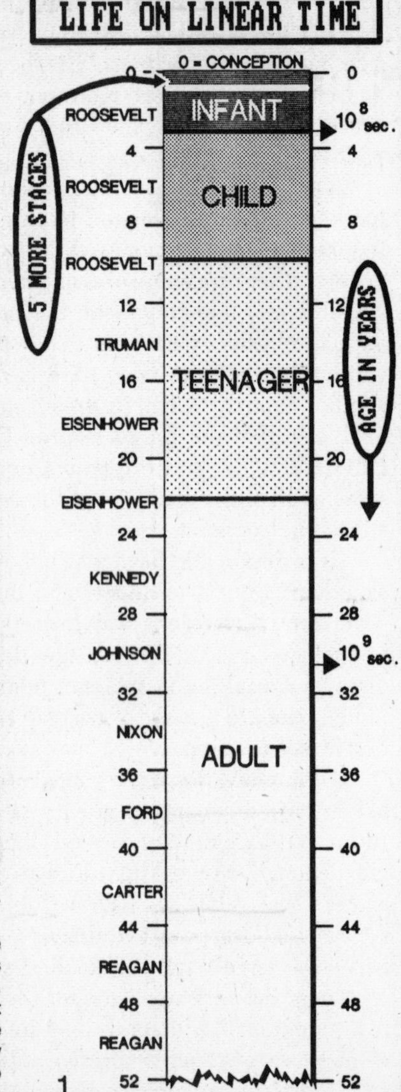

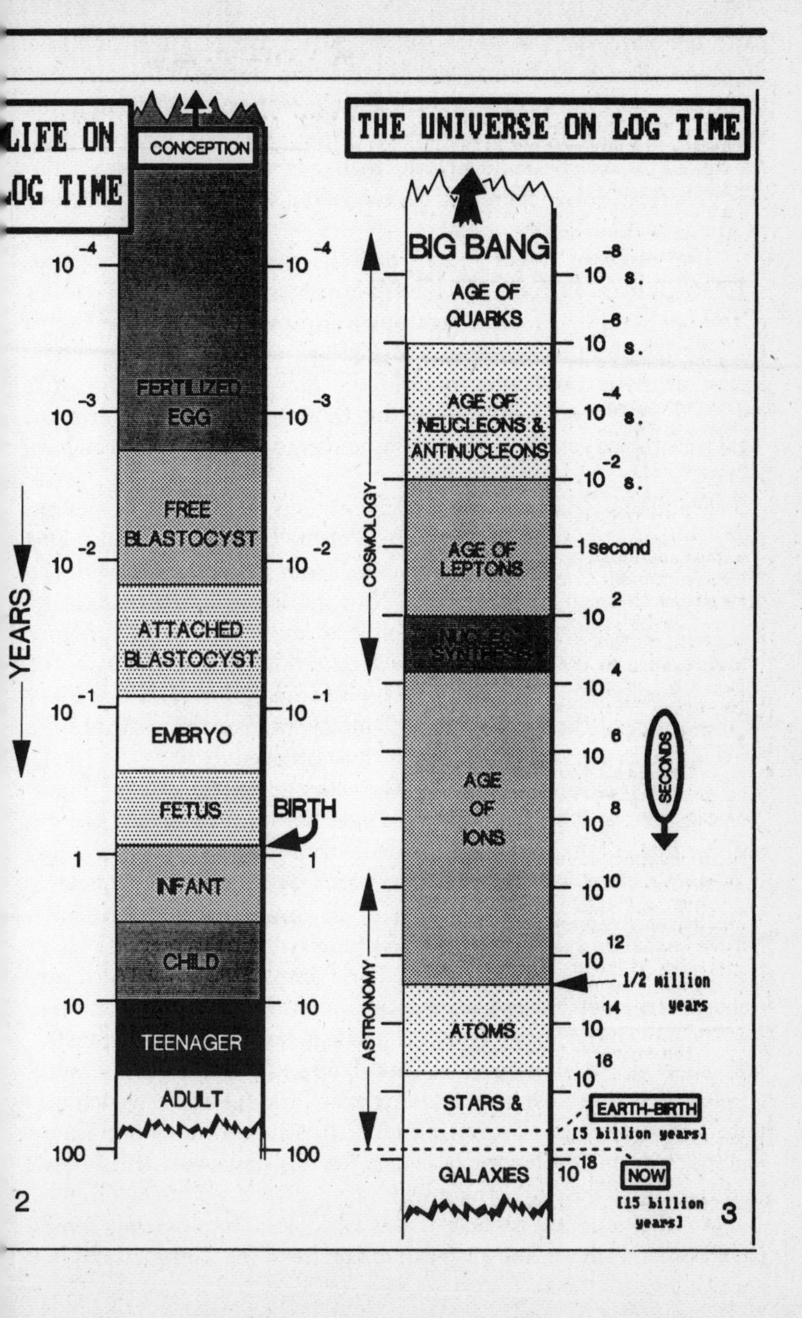
LIFE ON
LOG TIME
CONCEPTION
FERTILIZED EGG
FREE BLASTOCYST
ATTACHED BLASTOCYST
EMBRYO
FETUS
BIRTH
INFANT
CHILD
TEENAGER
ADULT
YEARS
10^{-4}
10^{-3}
10^{-2}
10^{-1}
1
10
100
2
THE UNIVERSE ON LOG TIME
BIG BANG
AGE OF QUARKS
AGE OF NEUCLEONS & ANTINUCLEONS
AGE OF LEPTONS
AGE OF IONS
ATOMS
STARS &
GALAXIES
COSMOLOGY
ASTRONOMY
10^{-8} s.
10^{-6} s.
10^{-4} s.
10^{-2} s.
1 second
10^{2}
10^{4}
10^{6}
10^{8}
10^{10}
10^{12}
10^{14}
10^{16}
10^{18}
SECONDS
1/2 million years
EARTH-BIRTH
[5 billion years]
NOW
[15 billion years]
3

past, the universe was much hotter. At an age of a few hundred thousand years the universe was as hot as the surface of the sun, and the mean energy of a particle was a few electron volts. A few minutes after the big bang, the typical energies were millions of electron volts. As we approach closer and closer to the moment of Creation, the energies relentlessly increase, far exceeding what mere mortals can produce in their laboratories.

The universe is—or more precisely, *was*—the most powerful accelerator ever. Perhaps, if scientists are ingenious enough, the relics of the big bang that we see about us will provide us with the key to the physics of ultra-high energies. Perhaps we shall observe superheavy magnetic monopoles, or cosmic strings, or other wonders that could be produced only during the birth throes of the universe. Again, cosmology and particle physics converge, and the snake swallows its tail.

Elementary-particle physics has reached either a curious crossroads or an insurmountable impasse. For the first time ever, it seems as if all the known particle phenomena can be described within a consistent theoretical framework. No experimental data contradicts the standard theory, nor does it indicate the existence of structure lying outside its domain. Yet, the standard theory leaves many of the juiciest questions unanswered. Foremost among them is the mystery of the origin of the masses of our fundamental particles. There are several promising directions in which to proceed:

1. *Mopping-up Operations*. The most important missing ingredient of the standard theory is the Higgs boson. It is the last observable remnant of the conventional symmetry-breaking mechanism, and it is the weakest link in our theory. Moreover, it is the agency responsible for the origin of masses of all known particles. Possibly, nature has taken a different road in symmetry breaking, and the Higgs boson is not quite what we expect. For example, there may be a very rich spectrum of Higgs-like particles awaiting discovery. Or there may be no Higgs boson at all! The Higgs boson, or its surrogates, may show up in experimental data done at the next generation of accelerators. The Tevatron and LEP may luck out and see such particles. However, our dream machine, the Superconducting Super-Collider, is the accelerator of choice for this endeavor. If it doesn't find the Higgs boson, nothing will.

Another missing particle is the top quark. Its discovery would round out the three known fermion families. The Germans, by *not* discovering this particle at PETRA, have proven that the top quark is heavier than about 20 GeV. Otherwise they would have seen its effects. If top is lighter than 30 GeV, it will certainly show up at the new Japanese machine named Tristan. If it lies between 30 and 50 GeV, the Stanford Linear Collider should find it. If it is heavier than that, then the Tevatron (near Chicago) or LEP (near Geneva, Switzerland) will have a shot at it. As you can see, the search for the top quark is quite an international horse race.

Theorists have a lot of mopping up to do as well. We cannot yet do precise calculations with QCD, our theory of the strong force. We say that it explains the properties of nuclei and hadrons, but we cannot put our money where our mouth is. We ought to be able to compute the masses and properties of the hadrons from first principles, and in terms of the few parameters of our theory, but we have been unable to perform the necessary calculations. Here we are on common ground with chemists and condensed-matter physicists. Given the rules, how do we explain the quantitative data? The answer may lie in the exciting new interface between physics and computer science. What appears at first as a technically intractable problem can often be solved with today's supercomputers, or if not, with tomorrow's.

Our faith in the validity of the standard theory is not justified by its limited empirical successes. We believe in QCD because of its simplicity, its elegance, its uniqueness, and the success of a few qualitative calculations. Clearly, we shall have to do better. Many physicists have accepted this challenge. Within a few years they should develop computer-assisted techniques to obtain precise quantitative predictions from the theory. Then, and only then, can we be absolutely certain that we've got the right theory of the strong force.

2. *The High-Precision Frontier*. The remaining mysteries of elementary-particle physics live at energies higher than those that can be studied with existing facilities. Yet it may be possible to discern hints of the new physics by doing very precise experiments at accessible energies. The muon, for example, acts like a microscopic magnet. Its strength has been measured to ten decimal places and the result agrees perfectly with its calculated value. The next decimal place is crucial, for at that point the effects of the weak force come into play. If the result to a proposed experiment to measure this next decimal place disagrees with the standard theory, we shall have a clue to the nature of its successor.

According to the standard theory, a muon should never decay

into three electrons, and a kaon should never decay into an electron and a muon. Indeed, these and other "forbidden" decay modes have never been observed to take place. Much more ambitious searches for forbidden processes may eventually find them, and such searches are being carried out today. A positive result would prove that a theory more elaborate than today's standard is required. The accuracy that is required in these experiments is truly mind-boggling. We already know that "forbidden" muon decays, if they take place at all, are rarer than one in 100 billion! How much more precise an experiment can be done?

The decisive prediction of grand unified theories is the ultimate instability of all matter. The proton, a basic building block of all nuclear matter, is not eternal, but must be subject to radioactive decay. Experimenters have already shown that its lifetime is greater than 10^{30} years (a hundred quintillion times longer than the age of the universe!). This means that less than one proton per *ton* of matter decays in a year. Experiments to find proton decay can and will be pursued to greater sensitivity by a factor of a hundred or so. The observation of proton decay would be a giant step in our understanding of nature's most fundamental laws. Physicists, eternally optimistic, anxiously await the results to the next generation of proton-decay experiments.

3. *The High-Energy Frontier.* Since the days when Rutherford was bouncing alpha particles off atoms of gold, every time that physicists have looked at higher energies they have discovered something new and wonderful. Everyday phenomena have to do with particle energies of an electron volt (eV) or less. One eV is the typical amount of energy involved when a single chemical process, like the oxidation of a single atom of carbon, takes place. It is the typical energy of a single visible light quantum, or photon. It was from the study of light emitted by excited atoms, from what is known as atomic spectroscopy, that we learned about quantum mechanics and the structure of atoms.

Thousands of electron volts (KeV) is the domain of X-rays. These high-energy photons are used routinely for purposes of medical diagnosis, since they may pass through the body and reveal its internal structures. In the same way, X-rays pass through the outer shells of atoms to reveal their inner mysteries. It was by means of X-ray spectroscopy that the most closely guarded secret of the atom was revealed, the central electric charge on the nucleus that determines the chemical properties of an atom.

With a leap by another factor of a thousand to the MeV domain, the world of the interior of the atomic nucleus is revealed. It is at these energies that nuclear physicists learn the nature of the nuclear world. When a radioactive nucleus explodes, or when a uranium nucleus fissions in a reactor, or when two protons fuse in the sun, millions of electron volts are released. Great as these energies are, they are not sufficient to show the inner structure of the nucleons themselves.

The next step, once again a thousand times higher in energy, is not an easy one. Energies of billions of electron volts, or GeV, are not easy to come by. There are occasional cosmic rays that reach us from outer space, but they are few and far between. Progress at these energies depended upon the invention and construction of large particle accelerators. It was with these instruments that today's candidates for the title of ultimate elementary particles were discovered: the leptons and the quarks. It is the physics of these energies that has led us to the standard theory.

It seems that each giant step in physics requires a jump by a factor of 1000 in energy. The next frontier lies at trillions of electron volts, or TeV. It is a frontier we are just beginning to approach. The great CERN collider achieves about half a TeV, and the American Tevatron almost two TeV. So far, physicists have seen no indications of new and surprising phenomena lying beyond the standard theory. Yet, we have just begun to look. The next accelerator, the proposed Superconducting SuperCollider, will get to 40 TeV. Surely that is where the wild beasties are!

4. *The Cosmological Frontier*. Cosmology and elementary-particle physics deal with phenomena at the opposite extremes of the cosmic ladder. Yet they have converged upon one another. The early universe was supernally hot. Particle energies were well beyond the reach of any man-made accelerator. The mysterious force of gravity is another common link between the universe as a whole and the very smallest elementary particle. It is the reason that the snake swallows its tail. We have got to understand gravity if we are to have a complete theory of the microworld and the macroworld and all that lies between.

Particle physicists have already explored the high-energy world up to energies of hundreds of GeV. Aside from tiny corrections or rare phenomena, there are probably no great surprises lurking at lower energies beyond those encompassed in our standard theory. In

THE CHANGING HIGH-ENERGY FRONTIER

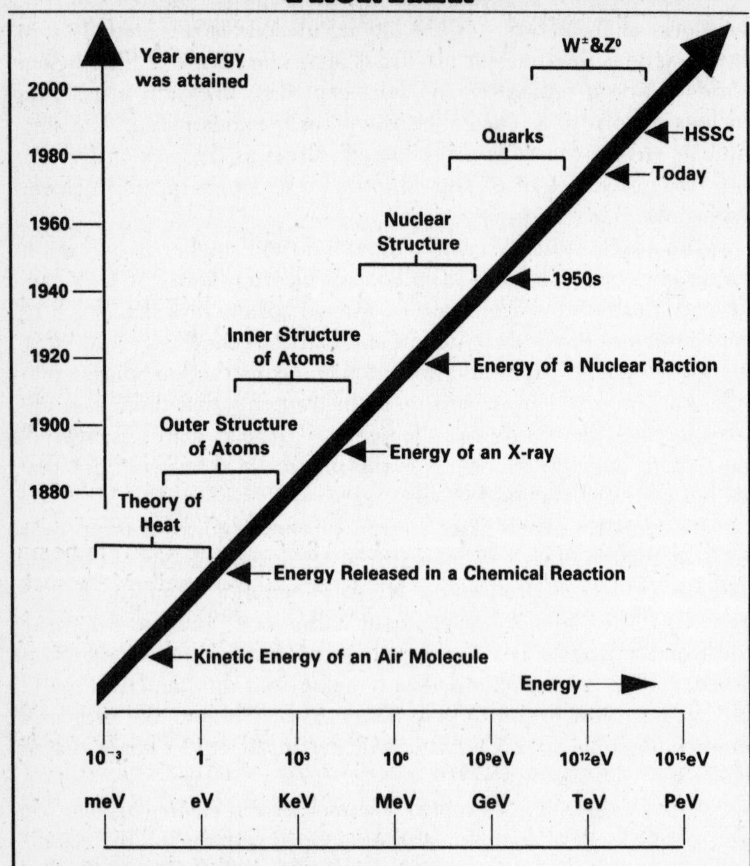

m for milli- from Latin meaning thousand (th)
K for kilo- from Greek meaning thousand
M for mega- from Greek meaning big
G for giga- from Greek meaning giant
T for tera- from Latin meaning monster
P for peta- from Latin meaning fart

This figure summarizes the history of fundamental-particle physics in terms of a race to higher energies. The higher the energy, the smaller the structures that can be studied. In the past century we have progressed from optical spectroscopy at energies of a few electron volts to today's great accelerators at energies of nearly a trillion eV. Each increase in energy by a factor of a thousand seems to reveal a profound new secret about the structure of matter.

much the same way, cosmologists have a more or less sound understanding of the birth of the universe in the primordial big bang. At least they understand things from the present time back to the time when the universe was about a thousandth of a second old. Of course, the cosmologists are not satisfied. They would like to push their frontier back to the first microsecond, or the first trillionth of a second, and ultimately to the moment of Creation itself. Earlier times mean higher energies. The more we know of high-energy physics, the more we know of the early universe, and conversely.

Microphysics and macrophysics must be consistent with one another. The mysterious nature of the dark matter in the universe must be reconciled with our understanding of particle physics: the celestial and the terrestrial are one. The origin of the galaxies must be explicable in terms of microscopic laws. Many links have already been established between the two disciplines: cosmologists declare that there cannot exist more than four different kinds of neutrinos. So far, particle physicists have found only three. Cosmologists conclude that neutrinos cannot be too massive, and particle physicists concur. (Indeed! Neutrinos may even be massless.) So far, there seems to be no conflict between our theories of the large and the small. More such connections are sure to be found, and each is another sign that we are on the right track toward a more profound understanding of nature.

Strange beasts live in the sky, far stranger than the bears and swans and crabs that ancient men imagined. Neutron stars are made of matter so compressed that a teaspoonful weighs a trillion tons! Spinning neutron stars produce cosmic beacons, called pulsars, that can be seen throughout the Milky Way. Their signals, repeated every fraction of a second, come to us with a regularity that rivals our finest clocks. One of the known pulsars spins on its axis almost a thousand times a second! Another is wedded to a second compact star, the two stars revolving about each other three times a day. The analysis of this binary pulsar shows that the two stars are gradually approaching one another, thereby presenting us with incontrovertible evidence for the existence of gravitational waves, a consequence of Einstein's theory that has defied verification by other means.

Occasionally a star in our galaxy suffers a titanic explosion and becomes a supernova. It becomes brighter than the full Moon and visible in daylight. The most recent supernova in the Milky Way was seen on Earth and studied by Kepler in 1604. The relics of another are visible today as the giant Crab Nebula. The light of its explosion first reached Earth in the year 1054. The event is well documented in ancient Chinese literature, and American Indians living in the southwest desert of the United States recorded it in drawings on rocks that survive today. But there is no European chronicle of what must have been a stupendous show. The Dark Ages were dark indeed.

On February 23, 1987, a bright new star suddenly appeared in the sky: a supernova in our neighboring galaxy, the Greater Magellanic Cloud. It is the first supernova visible to the naked eye for almost three centuries. Too bad it can be seen only in the Southern Hemisphere. The event took place 160,000 years ago, and it took all that time for the light produced by the explosion to reach our eyes. At last astronomers have an opportunity to study a newly born and "nearby" supernova.

Our Milky Way is an average-sized galaxy among about ten billion. Extragalactic supernovae are often seen and studied. Galaxies contain billions of stars, but when one star goes supernova, it outshines its entire galaxy. In the long history of our own galaxy there may have been hundreds of millions of supernovae. The heavy elements that we find on Earth, elements like bismuth, thorium and uranium, were created in these ancient explosions. Our solar system and we ourselves are made up of recycled matter that was once part of other stars with violent histories.

The science of radio astronomy was born in 1931 when Karl Jansky first detected radio emissions from the stars. With the impetus of radar research during World War II, radio astronomy emerged as a serious discipline. Astronomers discovered radio galaxies that emit ten million times as much radio power as our galaxy, and galaxies that are spewing forth enormous jets of high-energy particles, as if they were gigantic rocket ships on some incredible voyage. There even seem to be galaxies in the throes of a horrible disease, in which they are being eaten up, from the inside out, by an insatiable giant black hole.

And then there are quasars, the inhabitants of the universe that lived long ago when it was only a mere billion years old. Today they appear as faint but peculiar stars. They are faint only because they are so far away—billions of light-years. Intrinsically, they are hundreds of times brighter than a galaxy. Yet they are tiny objects, only a bit larger than a solar system. We haven't got the foggiest idea of what

SUPERNOVA 1987

Although the stupendous stellar happening caused the parent star to become ten thousand times brighter than it had been, the vast majority of the energy of the blast, about 99.9 percent, was released in the form of neutrinos. This time humans were prepared, even if somewhat serendipitously. Large neutrino detectors in America, Japan, Italy and Russia, originally deployed to search for proton decay, responded to these ghostly particles from the supernova. Eleven neutrino collisions were spotted by a Japanese-American team working in the Japan Alps. At the same time, eight events were observed at the large IMB proton-decay detector in an Ohio salt mine. Here are just a few of the wonderful things we have learned:

1. Astrophysicists know what they are doing. The new supernova produced a tremendous burst of neutrinos, as they had predicted.

2. Some of the supernova neutrinos succeeded in traversing intergalactic space to light up detectors like video games. This proves that electron neutrinos are stable particles, or at least *very* long-lived.

3. The neutrinos got here promptly. Indeed, they were detected hours before the supernova was visible on Earth. This is because the light from the explosion had to work its way out of the giant star, whereas the neutrinos raced through the stellar material much more easily. An analysis of the arrival times of the neutrinos proves that they cannot be very heavy. The upper limit we may establish on the mass of the electron neutrinos from the supernova is stronger than the results of three decades of earthbound laboratory experiments.

The behavior of the new supernova was different in detail from what the astrophysicists anticipated. For one thing, the parent star, a blue giant, is simply not the kind of star that was expected to explode. Moreover, the supernova seems to have a mysterious and unexplained companion (a sun, not a daughter) that is a light-month away, a tenth as bright, and *was certainly not there before the explosion.*

they are, or what incredible machine could have produced their enormous luminance.

First there was optical astronomy, then radio astronomy. Many more windows to the universe have recently opened in the space age.

Many forms of electromagnetic radiation cannot easily pass through our atmosphere: infrared radiation, ultraviolet, X-rays and gamma rays. These can be studied with rockets and satellites. And the new science of neutrino astronomy has just emerged with a bang. Every new window has revealed new marvels. Astronomy today is in a process of vigorous growth and startling discovery. We particle physicists have finally realized that we have got to understand what is going on Out There before we can pretend to understand our own little microcosm.

5. *The High Road*. Thus far, in my survey of the Future of Particle Physics, I have been describing what I like to call "the low road." It is the path from the laboratory to the blackboard, from experiment to theory, from hard-won empirical observations to the mathematical framework in terms of which they are described, explained and ultimately understood. This is the traditional path that science has so successfully followed since the Renaissance. Newton's theory of universal gravitation was an attempt to explain the observed motions of bodies in the heavens and on Earth. Lavoisier's formulation of the conservation of matter followed from his precise measurements of masses before and after chemical reactions. Maxwell's theory of electromagnetism was designed to encompass the careful experiments of his predecessors, Oersted, Faraday, Ampère and so on. Einstein's revolutionary new view of space and time, the special theory of relativity, did not and could not emerge from mere philosophical musings on how things "had" to be. Einstein was simply trying to reconcile Galileo's principle of relativity with Maxwell's electromagnetism. He wanted to (and did!) formulate a theory of electromagnetism that could be applied to moving bodies.

So it goes today. Gell-Mann's quarks were strictly an invention "to save the phenomenon." In searching for the ultimate structure of matter, we came up with more "elementary" particles than there are animals in the Bronx Zoo. Gell-Mann, in searching for an underlying simplicity in the data, showed how all of these particles could be built from just three species of quarks, if only we could figure out what force could bind the quarks together. The resolution to this problem arose from the analysis of electron-scattering experiments. Quantum chromodynamics, our successful theory of the strong force, resulted from the collaboration between experimental physicists and theorists. The electroweak theory, in whose construction I played a role, was an attempt to find a suitable famework to interpret the variety of weak-interaction phenomena. In each of these cases, scientists built their theories upon a scaffold of experimental data. The standard theory could not have been invented by theorists, however brilliant, just sitting around and thinking.

Sometimes scientists have followed a different road. The high road tries to avoid the morass of mundane experimental data. The ancient Greeks tried it once. They invented the notion of the atom and the element, but they never got past the point of arguing whether there were four elements (fire, water, earth and air), or just two, or maybe only one. Although Thales noticed that a piece of amber became mysteriously electrified when rubbed with fur, neither he nor his successors bothered to follow up this profound clue to the structure of matter. They preferred to sit in their hot tubs and think, rather than testing their hypotheses by doing careful experiments. The Greek "wisdom" led alchemists astray for centuries.

It happened again in the eighteenth century with Ruggero Boscovitch's theory of immutable and pointlike atoms. I bet you never heard of him!

The young Einstein was a devout follower of the low road. He even performed experiments! The papers he wrote in the miraculous year of 1905 at the age of twenty-six are enough to establish him as the greatest theoretical physicist ever. Then he did something that no physicist before or since has succeeded at. He took the high road and got somewhere.

Einstein's conversion began in November 1907. He later wrote:

> I was sitting in a chair in the patent office at Bern when all of a sudden a thought occurred to me: "If a person falls freely he will not feel his own weight." I was startled. This simple thought made a deep impression on me. It impelled me toward a theory of gravitation.

Out of the brilliant clarity of his incomparable mind, and over the next eight years, Einstein labored to construct his general theory of relativity. The mathematics he needed he invented as he went along. Through the power of positive thinking alone and without any hints from experiment, he invented a generalized version of his theory of relativity in which the force of gravity appears naturally, as a consequence of the geometry of space and time. By November 1915 Einstein's new theory was completed. While it was conceptually revolutionary, the general theory of relativity differs only very slightly from Newtonian theory in its consequences. There seemed to be only three experiments that could distinguish Einstein's theory from Newton's: the behavior of Mercury in its orbit, the bending of light by the Sun, and the effect of gravity upon the rate of a clock.

Einstein's theory cleared up a long-standing discrepancy in Mercury's behavior, a problem that seemed to have no other reasonable explanation. Overjoyed at this empirical triumph, Einstein became certain of the truth of his new theory. The rest of the world was convinced when his second prediction, the bending of light by the Sun, was quantitatively confirmed by a British eclipse expedition in 1919. "NEW THEORY OF THE UNIVERSE," proclaimed the headline of the London *Times*; "NEWTONIAN IDEAS OVERTHROWN."

Later, Einstein was to recant his earlier faith in the low road. "I do not consider the main significance of the general theory of relativity to be the prediction of some tiny observable effects, but rather the simplicity of its foundations and its consistency."

Einstein was never able to regain the low road. After his success with general relativity, he devoted the remaining decades of his life to an entirely unsuccessful quest for the unified field theory, by which he meant a single theory that could explain all of the forces of nature, a theory whose truth would be evident from its own simplicity and elegance. To Einstein, the forces of nature were gravity and electromagnetism. He never bothered to follow the new discoveries concerning the structure of the nucleus. He was never convinced of the existence of fundamental strong and weak nuclear forces.

That even the great Einstein should fail in his quest is no surprise. His chosen road had departed from the world of perceptions and observations. Einstein, in his later years, simply didn't know enough about physics. There is simply no way he could have succeeded.

In this time of challenge and frustration, many of the best and brightest young theoretical physicists have once again chosen Einstein's high road. Stubbornly clinging to the low road, I find myself a dinosaur in a world of upstart mammals. Superstring is the name of the game, an unlikely synthesis of some of the most bizarre theoretical notions ever put forward.

According to this new religion, space has nine dimensions, not just the three we see. Six of them are curled up into a tiny ball whose radius is the Planck length. The extra dimensions are far too small ever to be noticed by our big and clumsy species. The basic language of particle physics for the past half century has been quantum field

FIELD THEORY VERSUS STRING THEORY

According to quantum field theory, a fundamental particle is like a mathematical point, and its trajectory through space is a line. Field theory deals with diagrams such as Figure 1, below, where two initial point particles, A and B, exchange two other particles to become particles C and D.

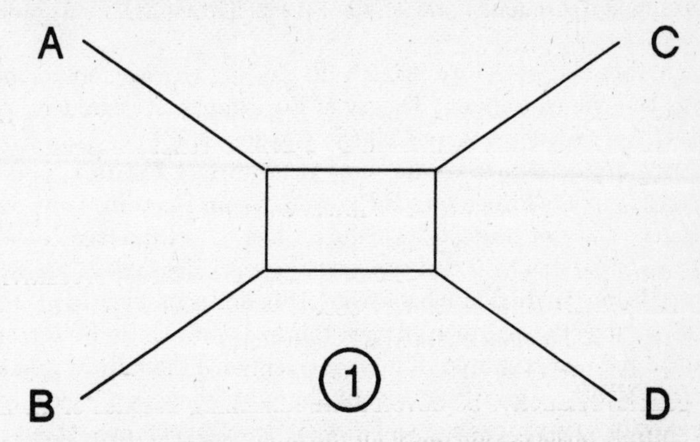

In string theory, an elementary particle can be likened to a tiny closed loop of string. Its motion sweeps out a tube in space. In Figure 2, the tubes corresponding to two string loops, A and B, engage one another, forming a structure with a hole in the middle, and emerge as string loops C and D.

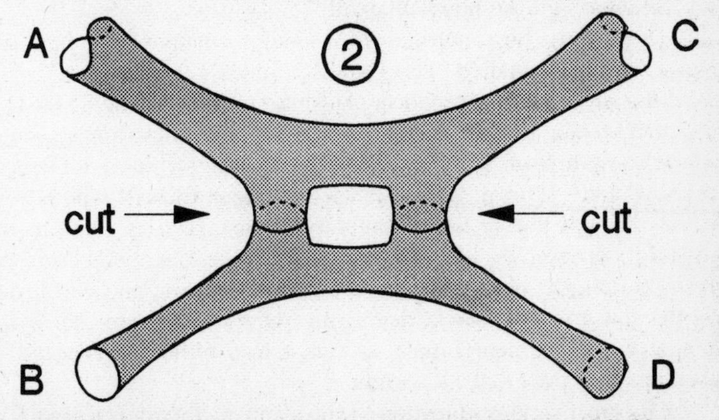

Imagine cutting the figure as shown. You can see that the two original loops have exchanged two loops. Thus Figure 2 is the string theory equivalent of the field theory diagram, Figure 1. Of course, there's much more to string theory, but this is about as much of it as I understand.

theory, in which a particle is regarded to be a pointlike structure. In superstring theory, the pointlike structure is abandoned and a fundamental particle is identified as a tiny loop of string. Of course, we can't see the string: its typical dimension is, again, the Planck length. There is only one kind of string, and the wiggles of the loop in nine-dimensional space determine whether it acts like a quark, or a photon, or whatever.

The popular literature brims with glowing articles about superstrings and the consequent Theory of Everything. Its founders, Michael Green of Queen Mary College, London, and John Schwartz of CalTech, are acclaimed as the new inheritors of Einstein, and its leading disciple is Princeton's Ed Witten (whom Harvard, to its eternal regret, did not promote to tenure when it had the chance). But the truth of the matter is that superstring theory does not as yet make any verifiable prediction whatsoever. It is not even clear that it includes a correct description of such things as protons and electrons, let alone the past triumphs of more conventional theoretical physics. Worst of all, the superstring theory does not follow as a logical consequence of some appealing and elegant set of hypotheses about nature. Unlike the general theory of relativity, superstring neither "predicts tiny observable effects" nor can it be admired for "the simplicity of its foundations."

So what's all the hooplah about?

There must be something to it since so many of the brainiest physicists are enthralled. The Princeton physics department has become the world's largest producer of young, strung and gung-ho Ph.D.s. Harvard's recent theoretical appointment got himself strung up somewhere en route from Stanford. While my old mentor Julian Schwinger remains aloof, Murray Gell-Mann is a string enthusiast who is convinced he will live to see elementary-particle physics come to the end of its string. My high school buddy Steve Weinberg leads the Texas contingent of the string wagon while keeping one foot firmly on the low road. He feels that string theory will be tested in the laboratory in the near future so that it may either be adopted or abandoned. I pray that he is right.

The most serious objection to the standard theory is its inability to describe gravity quantum-mechanically. Even though there are absolutely no verifiable consequences of a consistent quantum theory of gravity, it's just something we have got to have to be satisfied. "We couldn't make progress without a successful quantum theory of

Let us honor the forces of unification
Believing in things such as nu oscillation
And mourn not the monojet that couldn't be deader
For Carlo is coming with something much better.
The Seventh WOGU we have held in Toyama
While waiting for word of the death of all matta.
We must pity the student in his deep dark hole
Whose thesis depends on that one monopole,
Or on solar neutrinos that wriggle about
Unless they are saying our sun has gone out.
Some of us wonder how all things came to be
Leaving nary a clue but for old gravity,
And just seventeen particles, some of them quarks,
Maybe seventeen more and some of them squarks.
Something happened, they say, out in Cygnus the Swan,
Don't bother to look 'cause now it's all gone.
The Universe from Harvard looks like suds in the sink,
So crash your computers and take time to think.
The Theory of Everything, if you dare to be bold,
Might be something more than a string orbifold.
While some of your leaders have got old and sclerotic,
Not to be trusted alone with things heterotic,
Please heed our advice that you too are not smitten—
The Book is not finished, the last word is not Witten.

—From the Proceedings of the Seventh Workshop on Grand Unification (WOGU) held at Toyama, Japan, in April 1986. I presented this bit of doggerel in lieu of a conference summary.

gravity," says Steve Weinberg, "and string theories gave what seemed to be the only hope." Internal consistency is an obligatory criterion for a theory of nature. Superstrings, for the first time ever, appear to present us with a theoretically acceptable quantum theory of gravity. Maybe there is another way to fix gravity up, but nobody has found it yet. I've got to give superstring theory a brownie point for this.

Furthermore, superstring theory seems to be unique or almost unique. In contrast to the standard theory, or any of the grand unified theories, it may not have any adjustable parameters at all. In principle, the electron mass, the proton mass, the strength of the electromagnetic coupling, and all other measurable parameters are calculable from the theory without resort to experiment. In practice, however, no such calculation can be imagined, let alone performed.

The three gauge theories, and the three families of fermions that constitute the standard theory, are expected to emerge from superstring theory as unique and necessary logical consequences. Followers of the high road are so optimistic as to believe that this is not merely the best of all possible worlds, but the only possible one. String physicists are overjoyed at their good fortune to be pursuing physics at this very moment. They are convinced that they are on the verge of the ultimate breakthrough, a complete understanding of the nature of the physical universe.

The trouble is that the going is rough. Previous endeavors in theoretical physics usually involved mathematical analysis that most mathematicians found quaint and old-fashioned. This is certainly not the case for superstring theory, which involves several disciplines of mathematics so new that they don't yet exist. For the first time in decades, mathematicians and physicists are working hand in hand. They are proving a lot of new theorems, and having lots of fun. Mathematicians are delighted to have been given a new toy.

But physicists have not yet shown that superstring theory really works. They cannot even show that the standard theory, our successful description of the "low energy" world, is a necessary and logical consequence of string theory. They can't even be sure that their formalism includes a description of such things as protons and electrons. There is not yet even one teeny-tiny experimental prediction. Why, you may ask, do the stringers insist that space is nine-dimensional? It is not a consequence of elegant arguments nor compelling philosophy—it is simply that string theory doesn't make sense in any other kind of space.

Will today's superstring hordes succeed where the great Einstein failed? Is there gold at the end of the high road, or simply a morass of ever more abstruse mathematics? Even the most ambitious of the string advocates believe that it will take decades of study before we have learned enough to tell for sure, or even to make any experimental predictions at all.

Meanwhile, the historical connection between experimental physics and theory has been lost, as far as superstring theory is concerned. Until the string people can explain and interpret perceived properties of the real world, they are simply not doing physics. Should they be paid by physics departments and be permitted to pervert impressionable students? Will young Ph.D.s whose expertise is limited to superstring theory be employable if and when the string snaps? String thoughts may be more appropriate to departments of mathematics or even to schools of divinity. How many angels can dance on the head of a pin? How many dimensions are there in a compactified manifold thirty powers of ten smaller than a pinhead? Has medieval theology been resurrected?

Here's a riddle: Name two grand designs that are incredibly complex, require decades of research to develop, and may never work in the real world? Star Wars and string theory. Both are ambitious schemes designed to extricate us from long-standing problems. Neither ambition can be accomplished with existing technology, and neither may achieve its stated objectives. Both adventures are costly in terms of scarce human resources. And, in both cases, the Russians are trying desperately to catch up.

The last great experimental surprise in particle physics, the discovery of the third family of quarks and leptons, took place during the celebration of our nation's Bicentennial. The tau lepton was found in California, and the bottom, or "beauty," quark showed up in Illinois.

More than a decade has passed, and there has been no world-shaking discovery in particle physics. A conference was held in early 1987 in Santa Monica entitled "The Fourth Family of Quarks and Leptons," even though there is not a bit of experimental evidence or theoretical motivation for such particles. Physicists are so desperate to have something to talk about that they are playing "Let's Pretend."

History is on our side. Every few years there has been a world-shaking new discovery in fundamental physics or cosmology. There have been well over a hundred such discoveries since Newton's day. Can anyone really believe that nature's bag of tricks has run out? Have we finally reached the point where there is no longer a new particle, a "fifth" force, or a bewildering new phenomenon to observe? Of course not.

Let the show go on!

Appendix I

ETYMOLOGY OF PARTICLE NAMES

ALPHA PARTICLE: The particle emitted by a nucleus in the radioactive process called alpha decay. It is an energetic helium nucleus made up of two protons and two neutrons.

ANTIPARTICLE: To each type of particle there corresponds an antiparticle with identical mass but opposite electric and magnetic properties. The antiparticle of the electron (the positron) was discovered in 1932, the antiproton in 1955. Antimatter, composed of positrons and antinucleons, would annihilate ordinary matter upon contact, releasing scads of energy. There doesn't seem to be much antimatter in the universe.

BARYON: From Greek *barys* meaning "heavy" + *-on*. Coined in 1953 as a collective word including nucleons, hyperons and yet undiscovered heavy particles of the same ilk. Today the word signifies any particle made up of three quarks.

BETA PARTICLE: An energetic electron emitted by a nucleus in the radioactive process called beta decay.

BOSON: From the Indian physicist S. N. *Bose* + *-on*, as coined by P. Dirac in 1947. Any particle whose wave function must be symmetric. This means that it must have whole-number spin and does not satisfy the exclusion principle. Mesons and photons are bosons.

CHARMONIUM: A collective word coined in 1974 by A. De Rujula in analogy with positronium to describe any meson consisting of a charmed quark bound to its antiquark. Eleven states of charmonium, including the J/psi particle, have been found to date.

D PARTICLE: A charmed meson, which is to say, any meson containing one charmed quark. First observed at the Stanford Linear Accelerator Center in 1976 by G. Goldhaber and collaborators.

DEUTERIUM: From Greek *deuteros* meaning "second" + *-um*. The isotope of hydrogen with atomic weight 2, also known as heavy hydrogen. Discovered and named by H. C. Urey in 1933.

DEUTERON: The deuterium nucleus contains one proton and one neutron. Originally called the deuton, it was renamed the deuteron by H. C. Urey in 1933.

ELECTRON: The name suggested by G. Johnstone Stoney in 1891 for the natural unit of electricity. The particle that plays this role was discovered by J. J. Thomson in 1897. Electron, electricity et cetera derive from *electrica*, coined by W. Gilbert in 1600 to denote substances like amber (*electrum* in Greek) that attract light objects when rubbed.

FERMION: From the Italian physicist E. *Fermi* + *-on*, as coined by P. Dirac in 1947. Any particle whose wave function must be antisymmetric. This means that it must have a spin that is half a whole number and must satisfy the exclusion principle. Electrons and protons are fermions.

GAMMA RAY: An energetic photon emitted by a nucleus in the radioactive process called gamma decay.

GAUGE BOSON: A particle whose existence is demanded by a principle of gauge invariance. This category includes photons, gluons, W-bosons, Z-bosons and gravitons.

GLUON: The fundamental boson mediating the force that "glues" quarks together to form protons and other hadrons, and that produces the strong forces among them. Neither quarks nor gluons exist as isolated particles, but only as parts of hadrons.

GRAVITON: According to quantum theory, gravity must display both wave- and particle-like behavior. Neither aspect has been observed in the laboratory as yet. The particle avatar of gravity, whose exchange can be thought to generate the gravitational force, is known as the graviton.

HADRON: From Greek *adros* meaning "thick and bulky" + *-on*, as coined by the Russian physicist L. Okun' in 1962. Originally, any strongly interacting elementary particle. Today, any elementary particle made up of quarks. Baryons, mesons and antibaryons are the three known types of hadron.

HIGGS BOSON: The hypothetical remnant particle associated with spontaneous symmetry breaking, the search for which is among the greatest open challenges of high-energy physics. The theory underlying this particle was developed by P. Higgs and others in 1964.

HYPERON: From Greek *iper* meaning "above" or "beyond" + *-on*. Word originated ca. 1953 to describe heavy unstable baryons. Today, any baryon containing at least one strange quark.

J/PSI PARTICLE: Discovered simultaneously in November 1974 at

Brookhaven National Laboratory by S. Ting and at the Stanford Linear Accelerator Center by B. Richter, and their collaborators. The one group called it the J particle, the other the psi particle, and the joint name has stuck. It is a vector meson made of a charmed quark and a charmed antiquark, one of many quantum states of charmonium.

KAON: Several varieties of strange particles were discovered in 1947. Some were called K-mesons, a term that was shortened to kaons ca. 1958. The kaon contains one ordinary quark and one strange antiquark, and comes in two varieties: positively charged and neutral. Kaons weigh about half as much as nucleons and decay by means of the weak interactions.

LEPTON: From Greek *leptos* meaning "small" or "slight" + *-on*, as coined by L. Rosenfeld in 1948 to designate any particle of small mass like the electron or neutrino. Today, any fermion lacking strong interactions. There are three known charged leptons: the electron, muon and tau lepton. To each of these there corresponds a light, and perhaps massless, neutral lepton or neutrino.

MAGNETIC MONOPOLE: P. Dirac suggested in the 1930s that there should exist in nature a fundamental unit of magnetism, just as there exists a fundamental unit of electricity, the electron. This hypothetical particle is called the magnetic monopole because, unlike any known particle or magnet, it has only one magnetic pole. Searches for magnetic monopoles, so far, have been fruitless.

MESON: Word coined in 1939 by the Indian physicist H. J. Bhaba, who wrote: "The name 'mesotron' has been suggested by Anderson and Neddermeyer for the new particle found in cosmic radiation with a mass intermediate between that of the electron and proton. It is felt that the 'tr' in the word is redundant since it does not belong to the Greek root 'meso' for middle; the 'tr' in neutron and electron belong, of course, to the roots 'neutr' and 'electra.' . . . It would therefore be more logical and also shorter to call the new particle a meson instead of a mesotron." Today the word *meson* is reserved for any elementary particle containing one quark and one antiquark. Equivalently, a meson is a particle that is both a hadron and a boson. Ironically, the particle discovered by Anderson and Neddermeyer is neither. Today it is known as the muon, and it is certainly not any longer a meson.

MUON: A charged lepton about 200 times heavier than the electron. It was originally called the mu-meson, but is now no longer classed as a meson. Term originated ca. 1953 as a shortened form.

NEUTRINO: From Italian *neutro* meaning neutral + the diminutive *ino*, hence "little neutral one," as coined by E. Fermi in 1933. Any neutral lepton, of which three species are known. *Time* magazine, July 2, 1956, reported: "For 20 years nuclear physicists have used neutrinos in their calculations. . . . But no apparatus has ever detected neutrinos. . . . Last week from the Atomic Energy Commission came big news. Neutrinos do exist."

NEUTRON: This term seems to have been invented by E. Rutherford in 1921 to signify a close association of a proton and an electron which could act as a building block of larger nuclei. A few years after its discovery in 1932, the neutron was recognized to be as elementary a particle as the proton. But no more so: like the proton and other baryons, the neutron is now known to be made up of three quarks.

NUCLEON: Physicist C. Moller wrote in 1941: "Following the original proposal of Belinfante, the writer has used the word 'nuclon' as a common notation for neutrons and protons. In the meantime, however, it has been pointed out to me that, since the root of the word nucleus is 'nucle', the notation nucleon would from a philological point of view be more appropriate."

NUCLEUS: From Latin *nucleus* meaning "inner part." E. Rutherford, upon discovering the atomic nucleus in 1912, wrote: "I have given reasons for believing that the atom consists of a positively charged nucleus of very small dimensions surrounded by a distribution of electrons in rapid motion."

OMEGA-MINUS: Named and predicted to exist by M. Gell-Mann in 1962 as the last particle in his bestiary remaining to be discovered. Perhaps that is why he designated it by the last letter of the Greek alphabet. The *-minus* denotes its electric charge. It was first spied in 1964 by N. Samios and collaborators at Brookhaven National Laboratory. A baryon made up of three strange quarks, it is certainly not the last interesting new particle to be discovered.

PHOTON: From Greek combining form *photo-* meaning "light" + *-on*, as coined by Californian G. N. Lewis in 1926 to describe his kooky construct of light as an indestructible atom of a new kind. His theory is forgotten, but the word survives to describe a quantum unit of electromagnetic radiation, or, less accurately, a particle of light.

PION: By 1948, physicists recognized that secondary cosmic radiation consisted of what were then called pi-mesons, and their decay products, mu-mesons. The shortened form pion appeared ca. 1951. The force between nucleons comes about partly by the exchange of pions as H. Yukawa had predicted in the 1930s. However, the pion is only one of many mesons that are involved. Today we know that the pion is made up of an ordinary quark and an ordinary antiquark.

POSITRON: Shortened form of *posi(tive elec)tron* introduced in 1933 by C. D. Anderson, who was the first scientist to observe the track of this particle in a cloud chamber a year earlier. The positron is the antiparticle of the electron whose existence had been prophesized by Dirac.

POSITRONIUM: A short-lived neutral system consisting of a positron and an electron bound together. Name introduced by A. E. Ruark in 1945.

PREON: Invented by A. Salam to designate the hypothetical constituents of quarks and leptons. Alternatively called straton (by Russians) and maon (by this author). Many words for the concept, but no evidence at all for such substructure.

PROTON: From Greek *protos* meaning "first" + *-on*, as coined by E. Rutherford in 1920. It is the simplest atomic nucleus, that of the common isotope of hydrogen with atomic weight one. The proton is the only hadron that is not evidently unstable. However, grand unified theories demand that it, and consequently all matter, is ultimately radioactive and must eventually decay.

QUARK: Introduced by M. Gell-Mann in 1963 for the fundamental constituent of hadrons. Although quark rhymes with fork, MGM claims inspiration from "Three quarks for muster Mark . . ." from Joyce's *Finnegans Wake*. Quarks come in five known flavors: up, down are the everyday flavors from which ordinary matter is built. Then there are strange, charmed and bottom. A sixth flavor, top, must exist but has not yet been seen. There may be even more flavors to come. Baryons (like the proton) are made of three quarks, mesons of one quark and one antiquark, antibaryons of three antiquarks. Individual quarks cannot be produced.

RHO MESON: One of several unstable mesons with unit spin which play essential roles in the electrical structure of nucleons. Rho decays into two pions by the strong interaction in the absurdly short time of a septillionth of a second. Its existence was predicted by theorists shortly before its discovery in 1961.

SQUARKS AND SLEPTONS: Bosonic counterparts of quarks and leptons, which must exist according to supersymmetry theory. Conversely, the fermionic counterparts of gluons and photons are called gluinos and photinos. None of these peculiar perversions have been seen in the laboratory. Perhaps supersymmetry theory is wrong.

TACHYON: From Greek *tachis* meaning "fast" + *-on*, as coined by G. Feinberg in 1967 to signify a hypothetical particle that travels faster than light. Tachyons do not seem to exist in nature, which is a good thing because they would have led to serious problems if they did.

TAU LEPTON: The third and heaviest known charged lepton, weighing about twice as much as a proton. Discovered and named by M. Perl at the Stanford Linear Accelerator Center in 1975.

TOPONIUM: Hypothetical particle made of a top quark bound to its antiquark. Toponium will get a better name if and when it is discovered.

TRITIUM: From Greek *tritos* meaning "third" + *-on*. This isotope of hydrogen with atomic weight 3 was named by Urey and Murphy in 1933, a year prior to its discovery. Since ca. 1942 the tritium nucleus has been called the triton. It is made of a proton and two neutrons.

UPSILON: The first particle found consisting of a bottom quark bound to its antiquark, hence, the fifth-quark analog to the J/psi particle. Discovered and named by L. Lederman and collaborators at Fermilab in 1977. Eleven additional particles with this quark structure (sometimes called bottomonium) have been identified so far.

VECTOR BOSON: Any particle of unit spin, like the W-boson, the photon or the rho meson.

W-BOSON: The long-sought charged intermediate vector boson of the weak force, and a principal subject of this book. It was discovered by C. Rubbia and collaborators at the European Center for Nuclear Research (CERN) in 1983.

X-RAY: Electromagnetic radiation produced by energetic electrons striking a metal target. Discovered and named by W. C. Roentgen in 1895.

Z-BOSON: The neutral-current weak force was first detected at CERN in 1973. The Z-boson, the particle that mediates this force, was discovered a decade later, again at CERN, by C. Rubbia and his collaborators.

Appendix II

CHRONOLOGY OF MAJOR DISCOVERIES

Astonishing developments in particle physics and cosmology take place almost every year. Here are some of the most important discoveries of the past century, together with the principal scientist(s) involved and the country in which the discovery took place:

Year	Discovery	Scientist(s)	Country
1894	The Inert Gases	William Ramsey & Lord Rayleigh	England
1895	X-rays	Wilhelm K. Roentgen	Germany
1896	Radioactivity	Henri Becquerel	France
1897	The Electron	J. J. Thomson	England
1898	Radium	Marie Curie	France
1900	The Quantum	Max Planck	Germany
1902	The Law of Radioactive Decay	Ernest Rutherford & Frederick Soddy	Canada
1905	Special Theory of Relativity	Albert Einstein	Switzerland
1911	The Atomic Nucleus	Ernest Rutherford	England
1912	Cosmic Rays	Victor F. Hess	Austria
1913	Nuclear Charge	H. G. J. Moseley	England
	Quantum Rules for Atomic Structure	Niels Bohr	Denmark
1915	General Theory of Relativity (Gravity)	Albert Einstein	Germany
1917	Artificial Disintegration of the Nucleus	Ernest Rutherford	England
1923	Particle-Wave Duality	Louis de Broglie	France
1924	Electron Spin	Samuel Goudsmit & George Uhlenbeck	Holland
1925	The Exclusion Principle	Wolfgang Pauli	Switzerland
1926	Matrix Mechanics (One Version of Quantum Theory)	Werner Heisenberg	Germany
	Wave Mechanics (Another Version of Quantum Theory)	Erwin Schroedinger	Germany

1929	The Expanding Universe	Edwin P. Hubble	U.S.A.
1931	Pluto, the Planet	Clyde Tombaugh	U.S.A.
	Birth of Radio Astronomy	Karl Jansky	U.S.A.
	Prediction of the Positron	Paul A. M. Dirac	England
1932	The Positron	Carl D. Anderson	U.S.A.
	The Neutron	Sir James Chadwick	England
1934	Meson Theory of the Nuclear Force	Hideki Yukawa	Japan
1937	First Man-made Chemical Element	Emilio Segré	Italy
	Discovery of the Muon	Anderson, et al.	U.S.A.
1938	Nuclear Fission	O. Hahn & F. Strassman	Germany
1939	Nuclear Origin of Stellar Energy	Hans Bethe	U.S.A.
1942	First Nuclear Reactor	Enrico Fermi	U.S.A.
1947	Observation of Strange Particles	George D. Rochester & Clifford C. Butler	England
	The Lamb Shift Is Measured	Willis E. Lamb & Robert C. Retherford	U.S.A.
1948	The Charged Pion	C. F. Powell, et al.	England
	The Lamb Shift Is Calculated	Julian Schwinger	U.S.A.
1950	The Neutral Pion	Jack Steinberger	U.S.A.
1954	Gauge Theory	Chen Ning Yang & R. L. Mills	U.S.A.
1955	Antiproton	Emilio Segré & Owen Chamberlain	U.S.A.
1956	The Electron Neutrino	Clyde Cowan & Frederick Reines	U.S.A.
	Mirror Symmetry Questioned	Tsung Dao Lee & Chen Ning Yang	U.S.A.
1957	Mirror Symmetry Violated	Mme. Chien-Shiung Wu, et al.	U.S.A.
	V-A Model of Weak Force	R. P. Feynman & M. Gell-Mann, R. E. Marshak & E. C. G. Sudarshan	U.S.A.
1958	Cabibbo Theory	Nicola Cabibbo	Italy
1959	Slowing of Clocks by Gravity observed	Robert J. Pound	U.S.A.
1961	The Muon Neutrino	G. Danby, et al.	U.S.A.
	Electroweak Model with Z-nought	S. Glashow	Denmark
	The Eightfold Way	M. Gell-Mann & Y. Ne'eman	U.S.A. & Israel
1963	Quarks Invented	M. Gell-Mann & G. Zweig	U.S.A.
1964	Cosmic Background Radiation	Arno A. Penzias & Robert W. Wilson	U.S.A.
	Violation of Time Reversal Symmetry	Val Fitch & James Cronin	U.S.A.

	Omega Particle	Nick Samios, et al.	U.S.A.
	Higgs Mechanism	Peter Higgs, et al.	Scotland
1967	Partons Seen in Nucleons	Henry Kendall, Jerry Friedman & Richard Taylor	U.S.A.
	The First Pulsar	Jocelyn Bell Burnell	England
	Electroweak Model with Higgs Mechanism	Steven Weinberg & Abdus Salam	U.S.A. & England
1970	GIM Mechanism	Glashow, John Iliopoulos & Luciano Maiani	U.S.A.
1973	Neutral Currents	F. J. Hasert, et al.	CERN (Switzerland)
1974	J/Psi Particle	Burton Richter & Samuel C. C. Ting	U.S.A.
	Grand Unified Theory	Howard Georgi & Glashow	U.S.A.
1975	Neutrinos from the Sun	Raymond Davis	U.S.A.
	Tau Lepton	Martin J. Perl	U.S.A.
1976	Charmed Mesons	Gerson Goldhaber	U.S.A.
1977	Upsilon Particle	Leon Lederman, et al.	U.S.A.
1978	Atomic Parity Violation	C. Y. Prescott, et al.	U.S.A.
1979	Gluon Jets	Mme. Sau-Lan Wu, et al.	Germany
1980	Beauty Particles	Workers at CESR	U.S.A.
1981	Renormalizability of Electroweak Theory	Gerhard 'tHooft	Holland
1983	W-bosons	Carlo Rubbia, et al.	CERN (Switzerland)
	Z-bosons	Carlo Rubbia, et al.	CERN (Switzerland)
1985	The Superstring	Michael Green & John Schwartz	U.S.A. & England
1986	Intergalactic Voids	Margaret Geller, et al.	U.S.A.
1987	Neutrinos from the New Supernova	Kamiokande Groups & IMB	Japan & U.S.A.

A Selection of Writings by S. L. Glashow

1. "Hunting of the Quark," *New York Times Magazine*, July 18, 1975.
2. "Grand Unification: Tomorrow's Physics," *New Scientist*, September 18, 1980.
3. Introduction to Howard Georgi, *Lie Algebras in Particle Physics* (Reading, Mass.: Benjamin-Cummings, 1982).
4. "Passing the Torch," *Physics Today*, April 1983.
5. Letter to Los Angeles *Times*, March 30, 1986.
6. "Desperately Seeking Superstrings?", *Physics Today*, May 1986.
7. Letter to *The New York Times*, May 14, 1987.

pg. 151 "Gauge Theories of Vector Particles
Glashow, Gell Mann
Anals of Physics, June 1961

pg. 155 156
{ Marius Sophus
Lei — Group
Elie Cartan — re. Lei groups

pg. 201 Bob Hofstadter — Stanford

pg. 229 "Assymptotically Free—"
David Gross. (Charmonium)

pg. 231
{ "Quantum Chromodynamics
QCD — ~~gluons~~
"Standard Theory of
Elementary Particle
Physics"

pg. 232 Experimental Meson Spectroscopy

pg. 234 Resonance vs Lifetime ?

pg. 281
{ Resonant Hadron "States"
"Renormalizable gauge couplings.
Paper — "Hadron Masses in a Gauge
Theory"

pg. 232
{ Charmonium — "massive" rest
energy/Mass of 3.1 GeV.
Charmonium system — gluons

pg: 262 Bottomonium

pg: 265 Manifest asymmetry between matter & antimatter.

pg: 265 & 267 Violation of CP invariance. CP asymmetry

pg: 269 Polarized electrons -

pg: 278 Local symmetry principle

pg: 292 "Fundamental Act of Becoming"

pg: 301 Allowed patterns of symmetry → representations of the group.